PROBLEM SOLVING
IN CHEMICAL ENGINEERING
WITH NUMERICAL METHODS

PROBLEM SOLVING

IN CHEMICAL ENGINEERING

WITH NUMERICAL METHODS

Michael B. Cutlip
University of Connecticut
Mordechai Shacham
Ben-Gurion University of the Negev

Prentice Hall PTR
Upper Saddle River, NJ 07458
http://www.phptr.com

Library of Congress Cataloging-in-Publication Data

Cutlip, Michael B.
    Problem solving in chemical engineering with numerical methods /
    Michael B. Cutlip, Mordechai Shacham
      p. cm. -- (Prentice Hall international series in the physical
    and chemical engineering sciences)
    Includes bibliographical references and index.
    ISBN 0-13-862566-2
    1. Chemical Engineering--Problems, excercises, etc. 2. Chemical
    engineering--Data processing. 3. Problem Solving. 4. Numerical
    analysis I. Shacham, Mordechai. II. Title. III. Series.
TP168.C88 1999   98-26056
660'.01'5194--dc21   CIP

Production Editor: Kerry Reardon
Editor in Chief: Bernard M. Goodwin
Cover Design Director: Jerry Votta
Manufacturing Manager: Alan Fischer

 ©1999 Prentice Hall PTR
Prentice-Hall, Inc.
A Simon & Schuster Company
Upper Saddle River, New Jersey 07458

Prentice Hall books are widely used by corporations and government agencies for training, marketing, and resale. The publisher offers discounts on this book when ordered in bulk quantities.
For more information, contact Corporate Sales Department, Phone: 800-382-3419;
Fax: 201-236-7141; E-mail: corpsales@prenhall.com
Or write: Prentice Hall PTR, Corp. Sales Dept., One Lake Street, Upper Saddle River, NJ 07458.

Product names mentioned herein are the trademarks or registered trademarks of their respective owners.

Printed in the United States of America
10 9 8 7 6 5 4 3 2 1

ISBN 0-13-862566-2

Prentice-Hall International (UK) Limited, *London*
Prentice-Hall of Australia Pty. Limited, *Sydney*
Prentice-Hall Canada Inc., *Toronto*
Prentice-Hall Hispanoamericana, S.A., *Mexico*
Prentice-Hall of India Private Limited, *New Delhi*
Prentice-Hall of Japan, Inc., *Tokyo*
Simon & Schuster Asia Pte. Ltd., *Singapore*
Editora Prentice-Hall do Brasil, Ltda., *Rio de Janeiro*

To Our Parents : Wilma Sampson Cutlip, Sidney B. Cutlip,
Lusztig Erzsèbet, Schwarczkopf Zoltán

Contents

Chapter 2 Regression ad Correlation of Data   41

Chapter 5 Fluid Mechanics   **159**

Chapter 6 Heat Transfer   209

Chapter 7 Mass Transfer   259

Chapter 8 Chemical Reaction Engineering   321

PREFACE

Intended Audience

This book is for the chemical engineering student or the professional engineer who is interested in solving problems that require numerical methods by using mathematical software packages on personal computers. This book provides many typical problems throughout the core subject areas of chemical engineering. Additionally, the "nuts and bolts" or practical applications of numerical methods are presented in a concise form during example problem solving, which gives detailed solutions to selected problems.

Background

The widespread use of personal computers has led to the development of a variety of tools that can be utilized in the solution of engineering problems. These include mathematical software packages like MathCAD,[*] Maple,[†] Mathematica,[‡] Matlab,[•] Polymath,[#] and spreadsheets like Excel.[¶] While there is great potential in the hands of individual PC users, often this potential is not well developed in current engineering problem solving.

In the past the computer was used only for the difficult tasks of rigorous modeling and simulation of unit operations, processes, or control systems, while the routine calculations were carried out using hand-held calculators (or spreadsheet programs more recently), using essentially the same techniques that were used in the slide rule era. Limiting the use of the computer solely to the difficult tasks was justified before the introduction of interactive numerical packages because the use of the computer was very time consuming.

A typical computer assignment in that era would require the student to carry out the following tasks: (1) Derive the model equations for the problem at hand, (2) find an appropriate numerical method to solve the model, (3) write and debug a FORTRAN program to solve the problem using the selected numerical algorithm, and (4) analyze the results for validity and precision.

It was soon recognized that the second and third tasks of the solution were minor contributions to the learning of the subject matter in most chemical engineering courses, but they were actually the most time consuming and frustrating parts of a computer assignment. The computer enables students to solve realistic problems, but the time spent on the technical details that were of minor rele-

---

[*] MathCAD is a trademark of Mathsoft, Inc. (http://www.mathsoft.com)
[†] Maple is a trademark of Waterloo Maple, Inc. (http://maplesoft.com)
[‡] Mathematica is a trademark of Wolfram Research, Inc. (http://www.wolfram.com)
[•] MATLAB is a trademark of The Math Works, Inc. (http://www.mathworks.com)
[#] POLYMATH is copyrighted by M. B. Cutlip and M. Shacham (http://www.polymath-software.com)
[¶] Excel is a trademark of Microsoft Corporation (http://www.microsoft.com)

vancy to the subject matter was much too long.

The introduction of interactive numerical software packages brought about a major change in chemical engineering calculations. This change has been called a "paradigm shift" by Fogler.[2] Using those packages the student's (or the practicing engineer's) main task is to set up the model equations. The interactive program provides accurate solutions to these equations in a short time, displaying the results in graphical and numerical forms. The meaning of the paradigm shift, however, is that using the old calculation techniques with the new computer tools brings very little benefit. This is emphasized in the following observation made by deNevers and Seader[1]: "Since the advent of digital computers, textbooks have slowly migrated toward computer solutions of examples and homework problems, but in many cases the nature of the examples and problems has been retained so that they can be solved with or without a computer."

In most of the examples and problems provided in this book, new solution techniques are presented that require the use of the computer. Thus the full benefits of a computer solution can be gained even for routine, simple problems, not just for complicated ones.

In spite of many available numerical problem-solving packages, advanced problem solving via personal computers continues to be under utilized in chemical engineering education. A recent survey by Jones[5] has indicated that "across the country, computers are usually not used effectively in undergraduate engineering science courses. Often they are not used at all. Problem solving approaches and calculation methods are little influenced by the availability of computers." There are several major reasons for this situation.

First, many of the current engineering textbooks and reference books have been very slow to react to the enhanced problem-solving capabilities that are currently available. Unfortunately, the current textbooks in most engineering subject areas have been slow to react to this emerging capability. The lack of properly framed standard problems in various engineering disciplines is accompanied by a lack of faculty interest in the use of new tools and the creation of appropriate problems that utilize these tools.

Another important reason for the lack of mathematical software usage for advanced problem solving is the actual cost of the software for individual students. While there are many educational benefits to having problem solving close at hand on student-owned personal computers, often the cost to the individual students is prohibitively high. Fortunately, the costs to major colleges and universities for site licenses for the use of software only in computer labs is much more reasonable. However, this pricing structure forces students to use problem-solving software only in computer labs and does not allow interactive use of the software at other locations. Thus advanced problem-solving capabilities are not currently as close at hand as the nearest personal computer.

Finally, there is a significant learning curve to most of the advanced problem-solving software. This requires users to become familiar with a command structure that is often not intuitive and thus difficult to use. This is a significant impediment to student, professional, and faculty use of many packages.

Purposes of This Book

The main purpose of this book is to provide a comprehensive selection of chemical engineering problems that require numerical solutions. Many problems are completely or partially solved for the reader. This text is intended to be supplementary to most of the current chemical engineering textbooks, which do not emphasize numerical solutions to example and posed problems. This book is highly indexed, as indicated in Tables 1-9 at the end of this preface. The reader can only consider a particular subject area of interest or the application of a particular numerical method in actual problem solving. In either area, problems or methods, the book gives concise and easy-to-follow treatments.

The problems are presented in a general way so that various numerical problem-solving computer packages can be utilized. Many of the problems are completely solved so as to demonstrate a particular problem-solving approach. In other cases, problem-solving skills of the reader need to be applied.

This book has been designed for use with any mathematical problem solving package. The reader is encouraged to use the mathematical software package of his or her choice to achieve problem solutions. However, the POLYMATH package has been used as an example package, and a complete version of POLYMATH is included in the CD-ROM that accompanies this text. This allows the convenient use of POLYMATH throughout the book, as many of the problems have some part of the solution in POLYMATH files that are available on the CD-ROM.

General Problem Format

All problems presented in the book have the same general format for the convenience of the reader. The concise problem topic is first followed by a listing of the chemical engineering concepts demonstrated by the problem. Then the numerical methods utilized in the solution are indicated just before the detailed problem statement. Typically a particular problem presents all of the detailed equations that are necessary for solution, including the appropriate units in a variety of systems, with Système International d'Unités (SI) being the most commonly used. Physical properties are either given directly in the problem or in the appendices.

Students

Students will find the chapter organization of the book, by chemical engineering subject areas, to be convenient. The problems are organized in the typical manner in which they are covered in most courses. Complete solutions are given to many of the problems that demonstrate the appropriate numerical methods in problem solving. Practice and application of various numerical methods can be accomplished by working through the problems as listed in Table 9.

Practicing Engineers

Engineers in the workplace face ever-increasing productivity demands. Thus the concise framework of the problems in this book should aid in the proper formulation of a problem solution using numerical methods.

Faculty

This book can assist faculty members in introducing numerical methods into their courses. This book is intended to provide supplementary problems that can be assigned to students. Many of the problems can be easily extended to open-ended problem solving so that critical thinking skills can be developed. The numerical solutions can be used to answer many "what if … " type questions so students can be encouraged to think about the implications of the problem solutions.

Chemical Engineering Departments

Departments are encouraged to consider adopting this book during the first introductory course in chemical engineering and then utilize the book as a supplement for many of the following courses in the curriculum. This allows an integrated approach to the use of numerical methods throughout the curriculum. This approach can be helpful in satisfying the Accreditation Board for Engineering and Technology (ABET) requirements for appropriate computer use in undergraduate studies.

A first course in numerical methods can also utilize many of the problems as relevant examples. In this application, the book will supplement a standard numerical methods textbook. Students will find the problems in this book to be more interesting than the strictly mathematical or simplified problems presented in many standard numerical analysis textbooks.

Educational Resources on CD-ROM

A CD-ROM is provided that contains additional learning resources including a complete operational version of the POLYMATH Numerical Computation Package which can be installed on a reader's personal computer to enable efficient interactive problem solving. All illustrative solved problems are available from the CD-ROM for execution and modification using POLYMATH. Ten representative book problems have also been solved by knowledgeable professionals with Excel, Maple, Mathcad, Mathematica, MATLAB, and POLYMATH. Detailed writeups and the files to solve these problems with these packages are included on the CD-ROM. The icon at the beginning of this paragraph is used to designate a CD-ROM resource throughout the book. For many problems, tabulated data for individual problems are provided as input files to POLYMATH, thereby eliminating time-consuming data entry. The complete details on the CD-ROM are given in Appendix F.

Book Organization

Chapter 1, "Basic Principles and Calculations," serves a dual purpose. The chapter introduces the reader to the subject material that is typically taught in a first chemical engineering course (in most universities called Material and Energy Balance, or Stoichiometry). Additionally, this chapter introduces numerical solutions that are presented using the POLYMATH Numerical Computation Package. This material can also be used in a separate "Introduction to Personal Computers" course that can be given in parallel to the first chemical engineering course. For the past three years at Ben-Gurion University, the material from Chapter 1 of this book has been taught in the second semester of the first year, in parallel with the second part of the material and energy balance course. The students are introduced to the POLYMATH software in two two-hour lectures and two one-hour computer lab sessions. During the lectures and lab sessions, Problems 1.1, 1.3 and 1.13 are presented to introduce students to the different programs of POLYMATH. After this workshop, students are expected to use POLYMATH without additional help.

Chapters 2 and 3 are not associated with any particular required course in the chemical engineering curriculum. Chapter 2, "Regression and Correlation of Data," presents advanced statistical techniques for regression of experimental data. Students can be encouraged to complete this chapter as part of a statistics course or as preparation for the chemical engineering laboratory. Chapter 3, "Advanced Techniques in Problem Solving," provides the background necessary for solution of more complicated problems, such as stiff differential equations, two-point boundary value problems, and systems of differential-algebraic equations using interactive numerical software packages. This chapter can be integrated into the curriculum or covered as part of a separate numerical analysis course. The titles of the remaining chapters clearly indicate in which courses the problems can be used.

The fully or partially solved problems demonstrate solution methods that are not included in regular textbooks. Some of them also show advanced solution techniques that may not be obvious to the casual user. Table 3 lists these special techniques and the problem numbers in which they are demonstrated or required.

Book Notation

Because of the wide variety of problems posed in this book, the notation used has been standardized according to one of the major Prentice Hall textbooks in the various subject areas whenever possible. These books are summarized in Table 10.

The POLYMATH Numerical Computation Package

We have authored the POLYMATH package to provide convenient solutions to many numerical analysis problems, including the chemical engineering problems that are presented in this book.

The first PC version of POLYMATH was published in 1984, and it has been in use since then in over one hundred universities and selected industrial sites mainly in the United States and Israel. The initial version included with this book, POLYMATH 4.02, was released in May of 1998. This version executes in computers with DOS (and Windows) operating systems. The package contains the following programs:

- Ordinary Differential Equations Solver
- Nonlinear Algebraic Equations Solver
- Linear Algebraic Equations Solver
- Polynomial, Multiple Linear, and Nonlinear Regression Program

The programs are extremely easy to use, and all options are menu driven. Equations are entered in standard form with user-defined notation. Results are presented in graphical or tabular form. No computer language is used, and a manual is not required. All problems can be stored on disk for future use. A sophisticated calculator and a general unit conversion utility are available from within POLYMATH upon request.

Current information on the latest POLYMATH software is available from http://www.polymath-software.com/.

Web Site: http://www.polymath-software.com/book

This site on the World Wide Web (WWW) will be maintained by the book authors to provide any corrections or updates to this book. The site will also provide information about where the CD-ROM may be reordered in the event that it has become damaged, outdated, or lost. Details on the latest POLYMATH software will also be available on the WWW, allowing inexpensive software upgrades to be downloaded. Additionally, the site may provide computer files for various solved problems for the convenient of readers to wish to use other mathematical software packages with this book.

Acknowledgments

We would like to express their appreciation to our wives and families who have shared the burden of this effort, which took longer than anticipated to complete. We particularly thank Professor H. Scott Fogler for his encouragement with this book effort and with the continuing development of the POLYMATH. Numerical Analysis Package. We are indebted to our colleagues from the American Society for Engineering Education (ASEE) Chemical Engineering Summer School who permitted reproduction of their problem solutions on the CD-ROM. We thank Nancy Neborsky Pickering for initially learning the FrameMaker desktop publishing package and for entering the initial materials into the book format. Leslie Wang provided considerable valuable feedback on most chapters of the book. Additionally, we appreciate the input and suggestions of our students, who have been subjected to preliminary versions of the problems and have endured the various prerelease versions of the POLYMATH software over the years.

During the 17 years that POLYMATH has been in use, many of our colleagues provided advice and gave us help in revising and improving this software package. In particular, we would like to acknowledge the assistance of Professors N. Brauner, H. S. Fogler, B. Carnahan, D. M. Himmelblau, J. D. Seader, and E. M. Rosen. H. S. Fogler and N. Brauner have also provided some of the problems included in the book and assisted with their solutions.

Development of a package such as POLYMATH and this book is an expensive endeavor in both resources and time. We are indebted to our universities: The Ben-Gurion University of the Negev and the University of Connecticut for the continuous support we have received. M. S. spent several summers and a sabbatical year at the University of Michigan. The first draft of this book was written during the stays at Michigan, and the support of the College of Engineering of the University of Michigan is sincerely appreciated. M. B. C. spent much of a sabbatical from the University of Connecticut and several summers on the preparation of book materials.

The routine maintenance and development of the POLYMATH package has been done by Orit Shacham. For the last 10 years she has been spending most of her vacations fixing bugs and writing new code for still another version of POLYMATH. She continues to amaze us by the speed and precision with which she converts ideas into computer code.

The first draft of this book was typed (and retyped) by Michal Shacham. She took several months of vacation from her job to learn to use various word processors and graphic programs and type the book. The draft she typed became the basis for class testing and refinement of the book.

Michael B. Cutlip
Mordechai Shacham

Table 1 Introductory Problems

NO.	INTRODUCTORY PROBLEMS
1.1	Molar Volume and Compressibility Factor from van der Waals Equation
1.2	Molar Volume and Compressibility Factor from Redlich-Kwong Equation
1.3	Fitting Polynomials and Correlation Equations to Vapor Pressure Data
1.4	Vapor Pressure Correlations for Sulfur Compounds Present in Petroleum
1.5	Steady-State Material Balances on a Separation Train
1.6	Mean Heat Capacity of n-Propane
1.7	Vapor Pressure Correlation by Clapeyron and Antoine Equations
1.8	Gas Volume Calculations Using Various Equations of State
1.9	Bubble Point Calculation for an Ideal Binary Mixture
1.10	Dew Point Calculation for an Ideal Binary Mixture
1.11	Bubble Point and Dew Point for an Ideal Multicomponent Mixture
1.12	Adiabatic Flame Temperature in Combustion
1.13	Unsteady-State Mixing in a Tank
1.14	Unsteady-State Mixing in a Series of Tanks
1.15	Heat Exchange in a Series of Tanks

Table 2 Problem in Regression and Correlation of Data

NO.	PROBLEMS IN REGRESSION AND CORRELATION OF DATA
2.1	Estimation of Antoine Equation Parameters Using Nonlinear Regression
2.2	Antoine Equation Parameters for Various Hydrocarbons
2.3	Correlation of Thermodynamic and Physical Properties of n-Propane
2.4	Temperature Dependency of Selected Properties
2.5	Heat Transfer Correlations from Dimensional Analysis
2.6	Heat Transfer Correlation of Liquids in Tubes
2.7	Heat Transfer Correlation in Fluidized Bed Reactor
2.8	Correlation of Binary Activity Coefficients Using Margules Equations
2.9	Margules Equations for Binary Systems Containing Trichloroethane
2.10	Rate Data Analysis for a Catalytic Reforming Reaction
2.11	Regression of Rate Data-Checking Dependency among Variables
2.12	Regression of Heterogeneous Catalytic Rate Data
2.13	Variation of Reaction Rate Constant with Temperature
2.14	Calculation of Antoine Equation Parameters Using Linear Regression
4.10	Correlation of Activity Coefficients with the van Laar Equations
8.7	Differential Method of Rate Data Analysis in a Batch Reactor
8.8	Integral Method of Rate Data Analysis in a Batch Reactor
8.9	Integral Method of Rate Data Analysis—Bimolecular Reaction
8.10	Initial Rate Method of Data Analysis
8.11	Half-Life Method for Rate Data Analysis
8.12	Method of Excess for Rate Data Analysis in a Batch Reactor
8.13	Rate Data Analysis for a CSTR
8.14	Differential Rate Data Analysis for a Plug-Flow Reactor
8.15	Integral Rate Data Analysis for a Plug-Flow Reactor
8.16	Rate Data Analysis for a Catalytic Reforming Reaction
8.17	Determination of Rate Expressions for a Catalytic Reaction

Table 3 Problem Solving Techniques

NO.	ADVANCED TECHNIQUES IN PROBLEM SOLVING
3.1	Demonstration of Iterative Methods for Solving a Nonlinear Equation (Terminal Velocity of Falling Particles)
3.2	Solution of Stiff Ordinary Differential Equations (A Biochemical Batch Reactor)
3.3	Stiff Ordinary Differential Equations in Chemical Kinetics (Gear's Stiff Problem in Chemical Kinetics)
3.4	Multiple Steady States in a System of ODEs (Transient Behavior of a Catalytic Fluidized Bed Reactor)
3.5	Single-Variable Optimization (Heat Transfer with Conduction and Radiation)
3.6	Shooting Method for Solving Two-Point Boundary Value Problems (Diffusion with First-Order Reaction in a Layer)
3.7	Expediting the Solution of Systems of Nonlinear Algebraic Equations (Complex Chemical Equilibrium)
3.8	Solving Differential Algebraic Equations (Binary Batch Distillation)
3.9	Method of Lines for Partial Differential Equations (Transient Heat Conduction in a Slab)

Table 4 Problems in Thermodynamics

NO.	PROBLEMS IN THERMODYNAMICS
1.1	Molar Volume and Compressibility Factor from van der Waals Equation
1.2	Molar Volume and Compressibility Factor from Redlich-Kwong Equation
1.3	Fitting Polynomials and Correlation Equations to Vapor Pressure Data
1.4	Vapor Pressure Correlations for Sulfur Compounds Present in Petroleum
1.6	Mean Heat Capacity of n-Propane
1.7	Vapor Pressure Correlation by Clapeyron and Antoine Equations
1.8	Gas Volume Calculations Using Various Equations of State
1.9	Bubble Point Calculation for an Ideal Binary Mixture
1.10	Dew Point Calculation for an Ideal Binary Mixture
1.11	Bubble Point and Dew Point for an Ideal Multicomponent Mixture
1.12	Adiabatic Flame Temperature in Combustion
2.1	Estimation of Antoine Equation Parameters Using Nonlinear Regression
2.2	Antoine Equation Parameters for Various Hydrocarbons
2.3	Correlation of Thermodynamic and Physical Properties of n-Propane
2.4	Temperature Dependency of Selected Properties
2.8	Correlation of Binary Activity Coefficients Using Margules Equations
2.14	Calculation of Antoine Equation Parameters Using Linear Regression
3.9	Expediting the Solution of Systems of Nonlinear Algebraic Equations (Equilibrium Problem)
4.1	Compressibility Factor Variation from van der Waals Equation
4.2	Compressibility Factor Variation from Various Equations of State
4.3	Isothermal Compression of Gas Using Redlich-Kwong Equation of State
4.4	Thermodynamic Properties of Steam from Redlich-Kwong Equation
4.5	Enthalpy and Entropy Departure Using the Redlich-Kwong Equation
4.6	Fugacity Coefficients of Pure Fluids from Various Equations of State
4.7	Fugacity Coefficients for Ammonia—Experimental and Predicted
4.8	Flash Evaporation of an Ideal Multicomponent Mixture
4.9	Flash Evaporation of Various Hydrocarbon Mixtures
4.10	Correlation of Activity Coefficients with the van Laar Equations
4.11	Vapor Liquid Equilibrium Data from Total Pressure Measurements I
4.12	Vapor Liquid Equilibrium Data from Total Pressure Measurements II

Table 6 Problems in Heat Transfer

NO.	PROBLEMS IN HEAT TRANSFER
1.15	Heat Exchange in a Series of Tanks
3.5	Single Variable Optimization for a Heat Transfer Problem
3.9	Method of Lines for Unsteady State Heat Conduction in a Slab
6.1	One-Dimensional Heat Transfer through a Multilayered Wall
6.2	Heat Conduction in a Wire with Electrical Heat Source and Insulation
6.3	Radial Heat Transfer by Conduction with Convection at Boundaries
6.4	Energy Loss from an Insulated Pipe
6.5	Heat Loss through Pipe Flanges
6.6	Heat Transfer from a Horizontal Cylindrical Attached to a Heated Wall
6.7	Heat Transfer from a Triangular Fin
6.8	Single-Pass Heat Exchanger with Convective Heat Transfer on Tube Side
6.9	Double-Pipe Heat Exchanger
6.10	Heat Loss from an Uninsulated Tank Due to Convection
6.11	Unsteady-State Radiation to a Thin Plate
6.12	Unsteady-State Heat Conduction within a Semi-Infinite Slab
6.13	Cooling of a Solid Sphere in a Finite Water Bath
6.14	Unsteady-State Conduction in Two Dimensions
8.23	Material and Energy Balances on a Batch Reactor
8.24	Operation of a Cooled Exothermic CSTR
8.25	Exothermic Reversible Gas-Phase Reaction in a Packed Bed Reactor
8.26	Temperature Effects with Exothermic Reactions

Table 7 Problems in Mass Transfer

NO.	PROBLEMS IN MASS TRANSFER
3.6	Shooting Method for Solving Two-Point Boundary Problems
7.1	One-Dimensional Binary Mass Transfer in a Stefan Tube
7.2	Mass Transfer in a Packed Bed with Known Mass Transfer Coefficient
7.3	Slow Sublimation of a Solid Sphere
7.4	Controlled Drug Delivery by Dissolution of Pill Coating
7.5	Diffusion with Simultaneous Reaction in Isothermal Catalyst Particles
7.6	General Effectiveness Factor Calculations for First-Order Reactions
7.7	Simultaneous Diffusion and Reversible Reaction in a Catalytic Layer
7.8	Simultaneous Multicomponent Diffusion of Gases
7.9	Multicomponent Diffusion of Acetone and Methanol in Air
7.10	Multicomponent Diffusion in a Porous Layer Covering a Catalyst
7.11	Second-Order Reaction with Diffusion in Liquid Film
7.12	Simultaneous Heat and Mass Transfer in Catalyst Particles
7.13	Unsteady-State Mass Transfer in a Slab
7.14	Unsteady-State Diffusion and Reaction in a Semi-Infinite Slab
7.15	Diffusion into a Falling Laminar Liquid Film of Finite Thickness
8.4	Catalytic Reactor with Membrane Separation
8.27	Diffusion with Multiple Reactions in Porous Catalyst Particles

Table 8 Problems in Chemical Reaction Engineering

NO.	PROBLEMS IN CHEMICAL REACTION ENGINEERING
2.10	Rate Data Analysis for a Catalytic Reforming Reaction
2.11	Regression of Rate Data—Checking Dependency among Variables
2.12	Regression of Heterogeneous Catalytic Rate Data
2.13	Variation of Reaction Rate Constant with Temperature
3.6	Shooting Method for Solving Two-Point Boundary Value Problems (Diffusion with First-Order Reaction in a Layer)
4.13	Complex Chemical Equilibrium
4.14	Reaction Equilibrium at Constant Pressure or Constant Volume
8.1	Plug-Flow Reactor with Volume Change during Reaction
8.2	Variation of Conversion with Reaction Order in a Plug-Flow Reactor
8.3	Gas-Phase Reaction in a Packed Bed Reactor with Pressure Drop
8.4	Catalytic Reactor with Membrane Separation
8.5	Semibatch Reactor with Reversible Liquid-Phase Reaction
8.6	Operation of Three Continuous Stirred Tank Reactors in Series
8.7	Differential Method of Rate Data Analysis in a Batch Reactor
8.8	Integral Method of Rate Data Analysis in a Batch Reactor
8.9	Integral Method of Rate Data Analysis—Bimolecular Reaction
8.10	Initial Rate Method of Data Analysis
8.11	Half-Life Method for Rate Data Analysis
8.12	Method of Excess for Rate Data Analysis in a Batch Reactor
8.13	Rate Data Analysis for a CSTR
8.14	Differential Rate Data Analysis for a Plug-Flow Reactor
8.15	Integral Rate Data Analysis for a Plug-Flow Reactor
8.16	Rate Data Analysis for a Catalytic Reforming Reaction
8.17	Determination of Rate Expressions for a Catalytic Reaction
8.18	Packed Bed Reactor Design for a Gas-Phase Catalytic Reaction
8.19	Catalyst Decay in a Packed Bed Reactor Modeled by a Series of CSTRs
8.20	Design for Catalyst Decay in a Straight-Through Reactor
8.21	Enzymatic Reactions in a Batch Reactor
8.22	Isothermal Reactor Design for Multiple Reactions
8.23	Material and Energy Balances on a Batch Reactor

Table 8 Problems in Chemical Reaction Engineering

NO.	PROBLEMS IN CHEMICAL REACTION ENGINEERING
8.24	Operation of a Cooled Exothermic CSTR
8.25	Exothermic Reversible Gas-Phase Reaction in a Packed Bed Reactor
8.26	Temperature Effects with Exothermic Reactions
8.27	Diffusion with Multiple Reactions in Porous Catalyst Particles

Table 9 List of Advanced Solution Techniques Demonstrated in Various Problems

ADVANCED SOLUTION TECHNIQUES	PROBLEM NO.
Ordinary Differential Equations	
Plotting Solution Trajectory for an Algebraic Equation Using the ODE Solver	3.1, 4.1, 4.5
Solution of Stiff Differential Equations	3.2, 3.3, 3.4
Solution of Two-Point Boundary Value Problems by Shooting Methods	3.5, 3.6, 5.1, 5.2, 5.3, 5.4, 5.18, 6.2, 6.5, 6.6, 6.7, 7.1, 7.3, 7.5, 7.6, 7.7, 7.8, 7.9, 7.10, 7.12
Conversion of Higher-Order Differential Equations to System of First-Order ODEs	3.6, 5.16, 5.18, 6.2, 6.5
Solution of Differential Algebraic System of Equations	3.8
Using the l'Hôpital's Rule for Undefined Functions at the Beginning or End Point of Integration Interval	4.11, 4.12
Using "If" Statement to Avoid Division by Zero	5.1
Switching Variables On and Off during Integration	5.16
Retaining a Value when a Condition Is Satisfied	5.16
Generation of Error Function	5.17
Functions Undefined at the Initial Point	6.2, 6.5
Using "If" Statement to Match Different Equations to the Same Variable	6.2, 7.4
Implicit Finite Difference Techniques	7.7, 7.11
Partial Differential Equations	
Numerical Method of Lines	3.9, 5.17, 6.12, 6.13, 6.14, 7.13, 7.14, 7.15
Transformation to an ODE	5.18

Table 9 List of Advanced Solution Techniques Demonstrated in Various Problems

ADVANCED SOLUTION TECHNIQUES	PROBLEM NO.
Algebraic Equations	
Using "If" Statement to Match Different Equations to One Variable	3.1, 5.6, 5.10
Ill-Conditioned Systems	3.4
Conversion of a System of Nonlinear Equations into a Single Equation	3.4
Selection of Initial Estimates	3.6, 4.12, 7.2
Modification of Strongly Nonlinear Equations for Easier Solution	3.7, 4.13
Conversion of a Nonlinear Algebraic Equation to a Differential Equation	4.1, 4.2, 4.5, 4.6
Data Modeling and Analysis	
Using Residual Plot for Data Analysis	1.3, 1.4, 2.1, 2.3, 2.5, 2.8, 2.14, 4.10
Using Confidence Intervals for Checking Significance of Parameters	1.3, 1.4, 2.1, 2.3, 2.8, 4.10
Transformation of Nonlinear Models for Linear Representations	1.3, 2.3, 2.5, 2.8
Differentiation and Integration of Tabular Data	1.6, 8.7, 8.12, 8.14, 8.15
Checking Linear Dependency among Independent Variables	2.11
Comparison of Linear and Nonlinear Regression	2.12, 2.13, 8.7

Table 10 Prentice Hall Textbooks for Notation

Prentice Hall Textbooks	
Author	**Title**
Himmelblau[4]	*Basic Principles and Calculations in Chemical Engineering*
Kyle[6]	*Chemical and Process Thermodynamics*
Geankoplis[3]	*Transport Processes and Unit Operations*
Fogler[2]	*Elements of Chemical Reaction Engineering*

REFERENCES

1. deNevers, N.,and Seader, J.D. "Helping Students to Develop a Critical Attitude Towards Chemical Process Calculations", *Chem. Engr. Ed.*, 26 (2), pp. 88-93 (1992).
2. Fogler, H.S. Elements of Chemical Reaction Engineering, 2nd ed., Prentice Hall, Englewood Cliffs: NJ, 1992.
3. Geankoplis, C. J. *Transport Processes and Unit Operations*, 3rd ed, Englewood Cliffs, NJ: Prentice-Hall, 1993.
4. Himmelblau, D. M., *Basic Principles and Calculations in Chemical Engineering*, 6th ed, Englewood Cliffs, NJ: Prentice-Hall, 1996.
5. Jones, J. B. "The Non-Use of Computers in Undergraduate Engineering Science Courses", *J. Engr. Ed.*, 87(1), 11 (1998).
6. Kyle, B. G. *Chemical and Process Thermodynamics*, 2nd ed, Englewood Cliffs, NJ: Prentice-Hall, 1992.

PROBLEM SOLVING IN CHEMICAL ENGINEERING WITH NUMERICAL METHODS

Basic Principles and Calculations

1.1 MOLAR VOLUME AND COMPRESSIBILITY FACTOR FROM VAN DER WAALS EQUATION

1.1.1 Concepts Demonstrated

Use of the van der Waals equation of state to calculate molar volume and compressibility factor for a gas.

1.1.2 Numerical Methods Utilized

Solution of a single nonlinear algebraic equation.

1.1.3 Problem Statement

The ideal gas law can represent the pressure-volume-temperature (PVT) relationship of gases only at low (near atmospheric) pressures. For higher pressures more complex equations of state should be used. The calculation of the molar volume and the compressibility factor using complex equations of state typically requires a numerical solution when the pressure and temperature are specified.

The van der Waals equation of state is given by

$$\left(P + \frac{a}{V^2}\right)(V - b) = RT \tag{1-1}$$

where

$$a = \frac{27}{64}\left(\frac{R^2 T_c^2}{P_c}\right) \tag{1-2}$$

and

$$b = \frac{RT_c}{8P_c} \tag{1-3}$$

The variables are defined by

P = pressure in atm

V = molar volume in L/g-mol

T = temperature in K

R = gas constant (R = 0.08206 atm·L/g-mol·K)

T_c = critical temperature (405.5 K for ammonia)

P_c = critical pressure (111.3 atm for ammonia)

Reduced pressure is defined as

$$P_r = \frac{P}{P_c}$$ (1-4)

and the compressibility factor is given by

$$Z = \frac{PV}{RT}$$ (1-5)

> **(a)** Calculate the molar volume and compressibility factor for gaseous ammonia at a pressure P = 56 atm and a temperature T = 450 K using the van der Waals equation of state.
> **(b)** Repeat the calculations for the following reduced pressures: P_r = 1, 2, 4, 10, and 20.
> **(c)** How does the compressibility factor vary as a function of P_r?

1.1.4   Solution

Equation (1-1) cannot be rearranged into a form where V can be explicitly expressed as a function of T and P. However, it can easily be solved numerically using techniques for nonlinear equations. In order to solve Equation (1-1) using the POLYMATH *Simultaneous Algebraic Equation Solver*, it must be rewritten in the form

$$f(V) = \left(P + \frac{a}{V^2}\right)(V - b) - RT$$ (1-6)

where the solution is obtained when the function is close to zero, $f(V) \approx 0$. Additional explicit equations and data can be entered into the POLYMATH program in direct algebraic form. The POLYMATH program will reorder these equations as necessary in order to allow sequential calculation.

The POLYMATH equation set for this problem is given by

```
Equations:
 f(V)=(P+a/(V^2))*(V-b)-R*T
 P=56
 R=0.08206
 T=450
 Tc=405.5
```

```
Pc=111.3
Pr=P/Pc
a=27*(R^2*Tc^2/Pc)/64
b=R*Tc/(8*Pc)
Z=P*V/(R*T)
Search Range:
V(min)=0.4, V(max)=1
```

In order to solve a single nonlinear equation with POLYMATH, an interval for the expected solution variable, V in this case, must be entered into the program. This interval can usually be found by consideration of the physical nature of the problem.

(a)   For part (a) of this problem, the volume calculated from the ideal gas law can be a basis for specifying the required solution interval. The POLYMATH *Calculator* is convenient to calculate the molar volume from $V = RT/P$ at the specified temperature and pressure follows:

0.08206*450/56
CALCULATOR: Enter an expression and press <ENTER> to evaluate it.
= 0.659410714

Thus the estimated molar volume using the ideal gas law is $V = 0.66$ L/g-mol. An interval for the expected solution for V can be entered as between 0.4 as the lower limit and 1.0 as the higher limit. The POLYMATH solution, which is given in Figure 1–1 for $T = 450$ K and $P = 56$ atm, yields $V = 0.5749$ L/g-mol, where the compressibility factor is $Z = 0.8718$.

The POLYMATH problem solution file for part (a) is found in the *Simultaneous Algebraic Equation Solver Library* located in directory CHAP1 with file named P1-01A.POL. This problem is also solved with Excel, Maple, MathCAD, MATLAB, and POLYMATH as problem 1 in the Set of Ten Problems discussed in Appendix F.

(b)   Solution for the additional pressure values can be accomplished by changing the equations in the POLYMATH program for P and P_r to

```
Pr=1
P=Pr*Pc
```

Additionally, the bounds on the molar volume V may need to be altered to obtain an interval where there is a solution. Subsequent program execution for the various P_r's is required.

The POLYMATH problem solution file for part (b) is found in the *Simultaneous Algebraic Equation Solver Library* located in directory CHAP1 with file named P1-01B.POL. This problem is also solved with Excel, Maple, MathCAD, MATLAB, Mathematica, and POLYMATH as problem 1 in the Set of Ten Problems discussed in Appendix F.

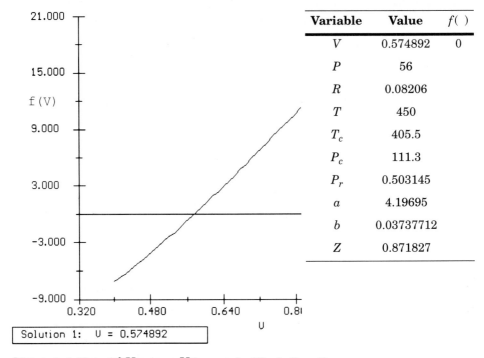

Variable	Value	$f(\ )$
V	0.574892	0
P	56	
R	0.08206	
T	450	
T_c	405.5	
P_c	111.3	
P_r	0.503145	
a	4.19695	
b	0.03737712	
Z	0.871827	

Solution 1: $V = 0.574892$

Figure 1–1 Plot of $f(V)$ versus V for van der Waals Equation

(c) The calculated molar volumes and compressibility factors are summarized in Table 1-1. These calculated results indicate that there is a minimum in the compressibility factor Z at approximately $P_r = 2$. The compressibility factor then starts to increase and reaches $Z = 2.783$ for $P_r = 20$.

Table 1-1 Compressibility Factor for Gaseous Ammonia at 450 K

P(atm)	P_r	V	Z
56	0.503	.574892	0.871827
111.3	1.0	.233509	0.703808
222.6	2.0	.0772676	0.465777
445.2	4.0	.0606543	0.731261
1113.0	10.0	.0508753	1.53341
2226.0	20.0	.046175	2.78348

1.2 MOLAR VOLUME AND COMPRESSIBILITY FACTOR FROM REDLICH-KWONG EQUATION

1.2.1   Concepts Demonstrated

Use of the Redlich-Kwong equation of state to calculate molar volume and compressibility factor for a gas.

1.2.2   Numerical Methods Utilized

Solution of a single nonlinear algebraic equation.

1.2.3   Problem Statement

The Redlich-Kwong equation of state is given by

$$P = \frac{RT}{(V-b)} - \frac{a}{V(V+b)\sqrt{T}} \tag{1-7}$$

where

$$a = 0.42747\left(\frac{R^2 T_c^{5/2}}{P_c}\right) \tag{1-8}$$

$$b = 0.08664\left(\frac{RT_c}{P_c}\right) \tag{1-9}$$

The variables are defined by

P = pressure in atm

V = molar volume in L/g-mol

T = temperature in K

R = gas constant (R = 0.08206 atm·L/g-mol·K)

T_c = the critical temperature (405.5 K for ammonia)

P_c = the critical pressure (111.3 atm for ammonia)

Repeat Problem 1.1 using the Redlich-Kwong equation of state.

1.3 FITTING POLYNOMIALS AND CORRELATION EQUATIONS TO VAPOR PRESSURE DATA

1.3.1   Concepts Demonstrated

Use of polynomials, the Clapeyron equation, and the Riedel equation to correlate vapor pressure versus temperature data.

1.3.2   Numerical Methods Utilized

Regression of polynomials of various degrees and linear regression of correlation equations with variable transformations.

1.3.3   Problem Statement

Table 1-2 presents data of vapor pressure versus temperature for benzene. Some

Table 1-2 Vapor Pressure of Benzene
(Perry[5] with permission)

Temperature, T (°C)	Pressure, P (mm Hg)
−36.7	1
−19.6	5
−11.5	10
−2.6	20
+7.6	40
15.4	60
26.1	100
42.2	200
60.6	400
80.1	760

design calculations require these data to be correlated accurately by algebraic equations.

Polynomial Regression Expression

A simple polynomial is often used as an empirical correlation equation. This can be written in general form as

$$P(x) = a_0 + a_1x + a_2x^2 + a_3x^3 + \dots + a_nx^n \tag{1-10}$$

where $a_0 \dots a_n$ are parameters, also called coefficients, to be determined by regression and n is the degree of the polynomial. Typically the degree of the polynomial is selected that gives the best data correlation when using a least-squares objective function.

The Clapeyron equation is given by

$$\log(P) = A + \frac{B}{T}$$

(1-11)

where T is the absolute temperature in K and both A and B are the parameters of the equation that are typically determined by regression.

The Riedel equation (Perry et al.[5]) has the form

$$\log(P) = A + \frac{B}{T} + C\log(T) + DT^{\beta}$$

(1-12)

where A, B, C, and D are parameters determined by regression and β is an integer exponent. According to the recommendation in Perry et al.[5], $\beta = 2$ can be used for this exponent.

(a) Correlate the data with polynomials of different degrees by assuming that the absolute temperature is the independent variable and P is the dependent variable. Determine what degree of polynomial fits the data best.

(b) Correlate the data using the Clapeyron equation.

(c) Correlate the data using the Riedel equation.

(d) Discuss which of the preceding correlations best represents the given data set.

1.3.4   Solution

(a) Data Correlation by a Polynomial   The POLYMATH *Polynomial, Multiple Linear and Nonlinear Regression Program* can be used to solve this problem. First, the data must be entered and a name should be assigned to each variable (column). Let us denote the column of temperature in °C as TC and the column of pressure as P. A new column can be created that calculates the absolute temperature with the function given by $TK = TC + 273.15$. These columns can be used to obtain the polynomials that represent the data of P (dependent variable) versus data of TK (independent variable). The POLYMATH program simultaneously regresses the dependent variable using first- to fifth-degree polynomials of the form

$$P_{(calc)} = a_0 + a_1 TK + a_2 TK^2 + a_3 TK^3 + a_4 TK^4 + a_5 TK^5$$

(1-13)

to the dependent variable data and presents the parameter values. The least-squares objective function that is minimized is given by

$$\sum_{i=1}^{N} (P_{(obs)} - P_{(calc)})^2 \tag{1-14}$$

where N is the number of data points and (obs) and (calc) refer to observed and calculated values of the dependent variable P in this case.

Table 1-3 summarizes the results for this problem when the independent

Table 1-3 Coefficients and Variance of Different Degree Polynomials for Vapor Pressure Data

Degree	1	2	3	4	5
a_0	−1544.64	5862.7	−12540.8	15917.8	21157.1
a_1	5.89072	−44.998	146.383	−248.69	−339.588
a_2		0.0861526	−0.570969	1.47168	2.09924
a_3			0.000744911	−0.00391702	−0.00607228
a_4				3.96315e−06	7.64532e−6
a_5					−2.50368e−09
Var.	14823.8	1216.79	34.1222	0.39792	0.485903

variable TK is fitted to dependent variable P. In addition to the coefficients, Table 1-3 presents the value of the variance (σ^2) for each polynomial. The variance is one of the indicators that can help to indicate what degree of the polynomial best represents the data. The variance is mathematically defined as

$$\sigma^2 = \sum_{i=1}^{N} \frac{(P_{(obs)} - P_{(calc)})^2}{\nu} \tag{1-15}$$

where ν is the degrees of freedom determined by taking the number of data points and subtracting the number of model parameters. For a polynomial, $\nu = N - (n+1)$, where N is the number of data points and n is the degree of the polynomial.

In this case the fourth-degree polynomial has the smallest variance, and this is one indication that this polynomial fits the data best. The POLYMATH program automatically highlights this polynomial by putting a frame around it. An additional indication for the goodness of the fit is the plot of the calculated curve (using the fourth-degree polynomial) together with the experimental data points, as shown in Figure 1–2, which indicates that there is close agreement between the experimental values (circles) and the calculated values (curve).

When dealing with models (equations) with many parameters (e.g., five parameters in the fourth degree polynomial), it is important to consider the con-

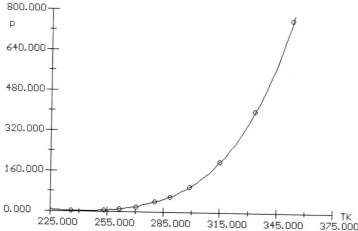

Figure 1–2 Observed Data Points and Calculated Curve for the Fourth-Degree
Polynomial

fidence intervals of the parameter values. While the statistical definition of a
confidence interval is outside the scope of this introduction, it can be generally
said that it represents the uncertainty associated with a particular parameter.
(Problem 2.14 provides a detailed explanation of confidence intervals and their
calculation.) The confidence intervals for the parameters of the fourth-degree
polynomial can be obtained by requesting the *statistical analysis* option from the
POLYMATH program. The parameter values together with the 95% confidence
intervals are shown in Table 1-4.

Table 1-4 Parameter and Confidence Interval Values for the Fourth-Degree
Polynomial

Parameter	Value	0.95 Confidence Interval
a_0	15917.8	3258.15
a_1	−248.69	45.1284
a_2	1.47168	0.232953
a_3	−0.00391701	0.000531162
a_4	3.96315e−06	4.51405e−07

The meaning of the confidence interval is that there is uncertainty with
regard to parameter a_0, and that this value should actually be presented as
$a_0 = 15917.8 \pm 3258.15$. The confidence interval is a function of the precision of
the data, the number of the data points, and the agreement between the model
(equation) and the data. A poor fit between the model and the data is often indicated by confidence intervals that include the value '0' (zero) for one or more

parameters inside the interval. It can be seen in Table 1-4 that none of the parameter confidence intervals include zero for the fourth-degree polynomial. The *statistical analysis* option from the POLYMATH program for the fifth-degree polynomial demonstrates that all parameter confidence intervals include zero, and this indicates a less satisfactory representation of the data.

An additional important indicator for the fit between the experimental data and the model is the 'residual plot' In such a plot the error in the dependent variable, defined as error = $P_{(obs)} - P_{(calc)}$, is plotted versus $P_{(obs)}$. For a good fit, the error must be randomly distributed with zero mean. The residual plot for the fourth-degree polynomial is shown in Figure 1–3. This plot shows that the error

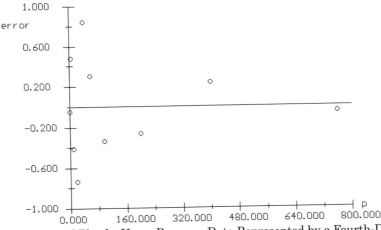

Figure 1–3 Residual Plot for Vapor Pressure Data Represented by a Fourth-Degree Polynomial

is not distributed randomly; rather it is very large for low values of P and very small for large P values. This indicates that even the best polynomial does not represent the vapor pressure data well throughout the entire pressure range.

(b) Clapeyron Equation Data Correlation   Data correlation with the Clapeyron equation can utilize two additional transformed variables (columns) in POLYMATH, which are defined by the relationships $\log P = \log(P)$ and $Trec = 1/TK$. A request for linear regression when the first (and only) independent variable column is $Trec$ and the dependent variable column is $\log P$ yields the following plot and numerical results from POLYMATH, as shown in Figure 1–4. The confidence interval on the parameters A and B is small and the variance is small. However, a detailed examination of Figure 1–4 indicates that the data points should not be represented by a linear relationship. This observation is reinforced by the residual plot shown in Figure 1–5, where the experimental data set exhibits a curvature that is not predicted by the Clapeyron equation.

It should be noted that the variance calculated using dependent variable $\log(P)$ values (as in this case) cannot be compared with the variance calculated using dependent variable P (as calculated for the polynomials). Comparison of variances requires the same form of a variable to be utilized.

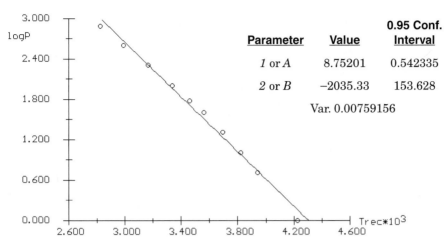

Figure 1–4 Observed Vapor Pressure Data and the Clapeyron Equation Representation

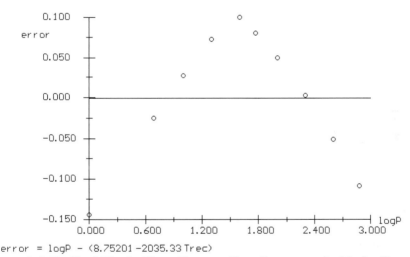

error = logP − (8.75201 −2035.33 Trec)

Figure 1–5 Residual Plot for Vapor Pressure Data Represented with the Clapeyron Equation

(c) **Riedel Equation Data Correlation** The Riedel equation correlation requires two additional columns for transformed variables, $\log T = \log(TK)$ and $T2 = TK \times TK$. Linear regression with $Trec$, $\log T$, and $T2$ as the independent variables and $\log P$ as the dependent variable yields the plot and numerical results presented in Figure 1–6. When there are two or more independent variables, as in this case, POLYMATH places the individual data points on the x axis. As Figure 1–6 shows, there is fairly good agreement between the experimental and calculated values of $\log(P)$. The confidence intervals are much wider than for the polynomials or the Clapeyron equation. The error distribution of the residual plot given in Figure 1–7 is more random than for either the polynomial or the Clapeyron equation.

Parameter	Value	0.95 Conf. Interval
1 or A	216.685	156.815
2 or B	−9317.2	4869.25
3 or C	−75.7355	58.5782
4 or D	4.44453e−05	5.01551e−05

Variance = 0.000296831

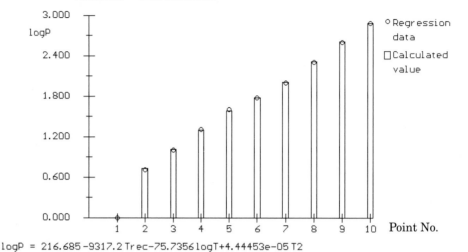

$logP = 216.685 - 9317.2\,Trec - 75.7356\,logT + 4.44453e{-}05\,T2$

Figure 1–6 Observed Vapor Pressure Data and Values Calculated Using the Riedel Equation

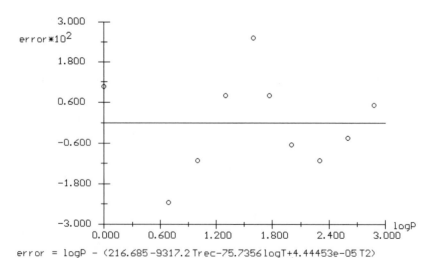

$error = logP - (216.685 - 9317.2\,Trec - 75.7356\,logT + 4.44453e{-}05\,T2)$

Figure 1–7 Residual Plot for Vapor Pressure Data Represented by Riedel Equation

(d) Comparison of Data Correlations An overall comparison of the different models can be made with a variance based on P (instead of log P). This is accomplished by calculating Equation (1-15) for each model. These variances are shown in Table 1-5.

Table 1-5 Variance Based on P for the Different Correlations

Equation	σ^2
fourth-degree polynomial	0.702
Clapeyron	6288.754
Riedel	24.448

It may be concluded that the Clapeyron equation is clearly inappropriate for representing this data set because of the high variance of Table 1-5 and because of the curvature of the residual plot of Figure 1–5. The fourth-degree polynomial has the lowest variance with the largest error at lower pressures, as seen in the residual plot of Figure 1–3. The Riedel equation has a residual plot that is more normally distributed, but it is presented in Figure 1–7 in a $\log(P)$ scale and thus larger errors will exist at larger pressures.

While the polynomial is most useful in representing this data set, it must be used with considerable care as it is a purely ***empirical model that should never be used outside of the region of the input data***. The Riedel model may be more useful in situations where extrapolation must be made.

There is an additional concern about the data set used here. The integer values of the dependent variable values in Table 1-2 indicate that these data have been smoothed and possibly extrapolated before our correlation attempts. Therefore, additional statistical considerations are not justified. The example data set in Problem 1.4 considers similar calculations and appropriate comparisons on more precise experimental data.

The POLYMATH problem solution file for this complete problem is found in the *Polynomial, Multiple Linear and Nonlinear Regression Library* located in directory CHAP1 with file named P1-03.POL. This problem is also solved with Excel, Maple, MathCAD, MATLAB, Mathematica, and POLYMATH as problem 3 in the Set of Ten Problems discussed in Appendix F.

1.4 VAPOR PRESSURE CORRELATIONS FOR SULFUR COMPOUNDS PRESENT IN PETROLEUM

1.4.1   Concepts Demonstrated

Use of polynomials, the Clapeyron equation, and the Riedel equation to correlate vapor pressure versus temperature data.

1.4.2   Numerical Methods Utilized

Regression of polynomials of various degrees and linear regression of correlation equations with variable transformations.

1.4.3   Problem Statement

Tables B–1 to B–4 (in Appendix B) provide data of vapor pressure (P in mm Hg) versus temperature (T in °C) for various sulfur compounds present in petroleum. Descriptions of the Clapeyron and Riedel equations are found in Problem 1.3.

> **(a)** Use polynomials of different degrees to represent the vapor pressure data for one of the compounds in Tables B–1 to B–4. Consider T (°K) as the independent variable and P as the dependent variable. Determine the degree and the parameters of the best-fitting polynomial for your selected compound.
>
> **(b)** Correlate the data with the Clapeyron equation.
>
> **(c)** Correlate the data with the Riedel equation.

 The POLYMATH data file for Tables B–1 to B–4 are found in the *Polynomial, Multiple Linear and Nonlinear Regression Library* located in directory TABLES with files named B-01.POL to B-04.POL.

1.5 STEADY-STATE MATERIAL BALANCES ON A SEPARATION TRAIN

1.5.1 Concepts Demonstrated

Material balances on a steady-state process with no recycle.

1.5.2 Numerical Methods Utilized

Solution of simultaneous linear equations.

1.5.3 Problem Statement

Paraxylene, styrene, toluene, and benzene are to be separated with the array of distillation columns shown in Figure 1–8.

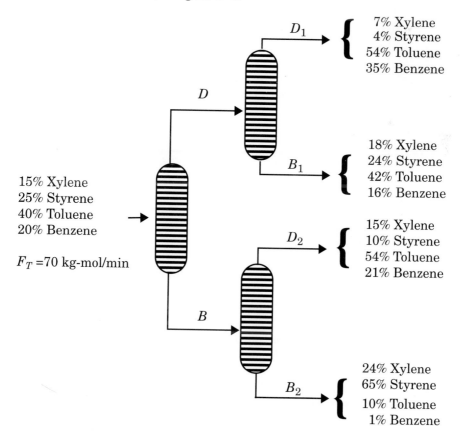

Figure 1–8 Separation Train

(a) Calculate the molar flow rates of D_1, D_2, B_1, and B_2.
(b) Reduce the original feed flow rate to the first column in turn for each one of the components by first 1% then 2% and calculate the corresponding flow rates of D_1, B_1, D_2, and B_2. Explain your results.
(c) Determine the molar flow rates and compositions of streams B and D for part (a).

1.5.4   Solution (Partial)

(a) Material balances on individual components yield

$$\text{Xylene: } 0.07D_1 + 0.18B_1 + 0.15D_2 + 0.24B_2 = 0.15 \times 70$$

$$\text{Styrene: } 0.04D_1 + 0.24B_1 + 0.10D_2 + 0.65B_2 = 0.25 \times 70$$

$$\text{Toluene: } 0.54D_1 + 0.42B_1 + 0.54D_2 + 0.10B_2 = 0.40 \times 70$$

$$\text{Benzene: } 0.35D_1 + 0.16B_1 + 0.21D_2 + 0.01B_2 = 0.20 \times 70$$

The coefficients and the constants in these equations can be directly introduced into the POLYMATH *Linear Equation Solver* in matrix form as follows:

Name	x1	x2	x3	x4	b
1	0.07	0.18	0.15	0.24	10.5
2	0.04	0.24	0.1	0.65	17.5
3	0.54	0.42	0.54	0.1	28
4	0.35	0.16	0.21	0.01	14

The solution is $x1 = 26.25$, $x2 = 17.5$, $x3 = 8.75$, and $x4 = 17.5$, which corresponds to the unknown flow rates of $D_1 = 26.25$ kg-mol/min, $B_1 = 17.5$ kg-mol/min, $D_2 = 8.75$ kg-mol/min, and $B_2 = 17.5$ kg-mol/min.

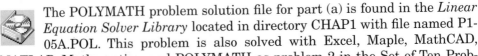

The POLYMATH problem solution file for part (a) is found in the *Linear Equation Solver Library* located in directory CHAP1 with file named P1-05A.POL. This problem is also solved with Excel, Maple, MathCAD, MATLAB, Mathematica, and POLYMATH as problem 2 in the Set of Ten Problems discussed in Appendix F.

(b) The solution can be obtained by changing the vector of constants in the POLYMATH input as required in this problem.

1.6 MEAN HEAT CAPACITY OF *n*-PROPANE

1.6.1   Concepts Demonstrated

Calculation of mean heat capacity from heat capacity versus temperature data.

1.6.2   Numerical Methods Utilized

Regression of polynomials of various degrees to data and integration of fitted polynomials between definite limits.

1.6.3   Problem Statement

The mean heat capacity $(\overline{C}_p)$ between two temperatures T_{ref} and T can be calculated from

$$\overline{C}_p = \frac{\int_{T_{ref}}^{T} C_p \, dT}{T - T_{ref}} \tag{1-16}$$

Use the data in Table 1-6 to complete the empty boxes in the column for the mean heat capacity of *n*-propane. Use 25 °C (298.15 K) as T_{ref}.

Table 1-6 Heat Capacity of Gaseous Propane (Thermodynamics Research Center[7] with permission)

No.	Temperature K	Heat Capacity kJ/kg-mol·K	Mean Heat Capacity kJ/kg-mol·K
1	50	34.06	
2	100	41.30	
3	150	48.79	
4	200	56.07	
5	273.15	68.74	
7	300	73.93	
8	400	94.01	
9	500	112.59	
10	600	128.70	
11	700	142.67	

Table 1-6 Heat Capacity of Gaseous Propane (Thermodynamics Research Center[7] with permission)

No.	Tempera-ture K	Heat Capacity kJ/kg-mol·K	Mean Heat Capacity kJ/kg-mol·K
12	800	154.77	
13	900	163.35	
14	1000	174.60	
15	1100	182.67	
16	1200	189.74	
17	1300	195.85	
18	1400	201.21	
19	1500	205.89	

1.6.4   Solution (Suggestions)

Approach (1)   The preferred approach for solving this problem is to fit polynomials of various degrees to the C_p versus T data. The best-fitting polynomial is then selected as outlined in Problem 1.3. Once the parameters of the best polynomial have been obtained, the analytical expression for the integral of Equation (1-16) can easily be derived. The expression for the integral can be evaluated by breaking it into several terms so that each term can be entered into a separate column in the data table of the polynomial curve-fitting program. The sum of these separate columns divided by $(T - T_{ref})$ yields the mean heat capacity values. This approach uses the data table of the curve fitting program much like a spreadsheet.

Approach (2)   A cubic spline or polynomial can be employed to evaluate Equation (1-16) by using the option *integrate the cubic spline* or *integrate the polynomial* from the *display options* menu of the POLYMATH program. This approach requires the calculation of each data point separately, so it is less convenient than the previous approach.

The POLYMATH problem data file is found in the *Polynomial, Multiple Linear and Nonlinear Regression Library* located in directory CHAP1 with file named P1-06.POL.

1.7 VAPOR PRESSURE CORRELATION BY CLAPEYRON AND ANTOINE EQUATIONS

1.7.1   Concepts Demonstrated

Use of the Clapeyron and Antoine equations for vapor pressure correlation and estimation of latent heat of vaporization from the Clapeyron equation.

1.7.2   Numerical Methods Utilized

Linear regressions after proper transformations to a linear expression.

1.7.3   Problem Statement

The Clapeyron equation is commonly used to correlate vapor pressure (P_v) with absolute temperature (T), where ΔH_v is the latent heat of vaporization.

$$\log P_v = -\frac{\Delta H_v}{RT} + B \tag{1-17}$$

Another common vapor pressure correlation is the Antoine equation, which utilizes three parameters A, B, and C, with P_v typically in mm Hg and T in °C.

$$\log P_v = A + \frac{B}{T + C} \tag{1-18}$$

A particular chemical is to be liquefied and stored in gas cylinders in an outside storage shed. The following data were obtained in the laboratory bomb calorimeter measurements. In this calorimeter, the liquid was slowly heated in a sealed container while the temperature and pressure of Table 1-7 were recorded.

(a)   Determine the heat vaporization and the constant B using the Clapeyron equation.
(b)   Assuming the yearly low and hot temperatures in the storage shed are 10 °F and 120 °F, calculate the expected vapor pressures at these temperature extremes.
(c)   How do your answers to (b) change when you use the Antoine correlation given by Equation (1-18)?
(d)   What do you think about storing the cylinders outside?

Table 1-7 Vapor Pressure Data

T (°C)	17	18	19	21	25	27	28
P (mm Hg)	13.6	22.21	35.54	85.98	413.23	832.62	1160.23

1.7.4   Solution

Finding the heat of vaporization and B in Equation (1-17) requires fitting a straight line to the experimental data. This is accomplished by the regression of $\log(P_v)$ versus $1/T$, where T is the absolute temperature. This is explained in more detail in Problem 1.3. After the parameters of the regression have been determined, the values of ΔH_v and B can be calculated.

The Antoine equation must first be linearized. This is accomplished by multiplication of both sides of Equation (1-18) by $T + C$, yielding

$$(T + C)\log P_v = A(T + C) + B \tag{1-19}$$

Equation (1-19) can be rearranged:

$$\log P_v = A + (AC + B)/T - C \log P_v/T \tag{1-20}$$

Evaluation of the parameters of Equation (1-19) can be accomplished by defining one new dependent and two new independent variables (columns) given by

$$\log P = \log(P_v), \text{Trec} = 1/T \text{ and } \log\text{PonT} = \log(P_v)/T$$

Linear regression with Trec and logPonT as independent and logP as dependent variables will yield the desired parameters.

The linearization of the Antoine equation, in the form of Equation (1-20), is somewhat problematic, in a statistical sense, since the original dependent variable P_v appears in both sides of the equation. However, this linearization usually yields acceptable results. Nonlinear regression will be used in Problem 2.1 to calculate the Antoine equation parameters, and this is the preferred approach in a statistical sense.

Once the constants of the Clapeyron and Antoine equations have been found, the equations can be used to calculate the vapor pressure at different temperatures.

 The POLYMATH problem data file is found in the *Polynomial, Multiple Linear and Nonlinear Regression Library* located in directory CHAP1 with file named P1-07.POL.

1.8 GAS VOLUME CALCULATIONS USING VARIOUS EQUATIONS OF STATE

1.8.1 Concepts Demonstrated

Gas volume calculations using the ideal gas, van der Waals, Soave-Redlich-Kwong, Peng-Robinson, and Beattie-Bridgeman equations of state.

1.8.2 Numerical Methods Utilized

Solution of a single nonlinear algebraic equation.

1.8.3 Problem Statement

It is proposed to use a steel tank to store carbon dioxide at 300 K. The tank is 2.5 m^3 in volume, and the maximum pressure it can safely withstand is 100 atm.

(a) Determine the maximum number of moles of CO_2 that can be stored in the tank using the equations of state which are discussed in the text that follows.

(b) Assuming that the Beattie-Bridgeman equation is the most accurate, what is the percent error in the calculated number of moles using the other correlations?

(c) Repeat (b) for different values of T_r, (T/T_C), and P_r, (P/P_C). How do the accuracies of the different correlations change with T_r and P_r?

Ideal Gas

$$PV = RT \qquad \text{(1-21)}$$

where

P = pressure in atm

V = molar volume in L/g-mol

T = temperature in K

R = gas constant ($R = 0.08206$ L·atm/g-mol·K)

van der Waals equation

See Equations (1-1) through (1-3).

Soave-Redlich-Kwong equation   (see Himmelblau[3] or Felder[1])

$$P = \frac{RT}{V-b} - \left[\frac{\alpha a}{V(V+b)}\right] \qquad \text{(1-22)}$$

where

$$a = 0.42747\left(\frac{R^2 T_C^2}{P_C}\right)$$

$$b = 0.08664\left(\frac{RT_C}{P_C}\right)$$

$$\alpha = [1 + m(1 - \sqrt{T/T_C})]^2$$

$$m = 0.48508 + 1.55171\omega - 0.1561\omega^2$$

T_C = the critical temperature (304.2 K for CO_2)

P_C = the critical pressure (72.9 atm for CO_2)

ω = the acentric factor (0.225 for CO_2)

Peng-Robinson equation[4]

$$P = \frac{RT}{V-b} - \left[\frac{a(T)}{V(V+b)+b(V-b)}\right] \tag{1-23}$$

where

$$b = 0.07780\frac{RT_C}{P_C}$$

$$a(T) = 0.45724\frac{R^2 T_C^2}{P_C}\alpha(T)$$

$$\alpha(T) = [1 + k(1 - (T/T_C)^{0.5})]^2$$

$$k = 0.37464 + 1.54226\omega - 0.26992\omega^2$$

Beattie-Bridgeman equation[2]

$$P = \frac{RT}{V} + \frac{\beta}{V^2} + \frac{\gamma}{V^3} + \frac{\delta}{V^4}$$

where

$$\beta = RTB_0 - A_0 - \frac{Rc}{T^2}$$

$$\gamma = RTB_0 b - A_0 a - \frac{RcBo}{T^2}$$

$$\delta = \frac{RTB_0 bc}{T^2}$$

and $A_0, a, B_0, b,$ and c are constants that depend on the particular gas.

For CO_2, $A_0 = 5.0065$; $a = 0.07132$; $B_0 = 0.10476$; $b = 0.07235$ and $c = 66.0 \times 10^4$.

1.8.4   Solution (Partial)

One solution to this problem is to find the volume of 1 mole of CO_2 at the specified temperature and pressure for each equation of state and then calculate the moles in the 2.5 m^3 volume of the tank.

The first equation of state (ideal gas) can be solved directly. In order to be consistent with the rest of the equations, this one can be rewritten as an implicit expression

$$f(V) = PV - RT \tag{1-24}$$

Equation (1-24) together with the specified numerical values of P, T, and R can be entered into the POLYMATH *Simultaneous Algebraic Equation Solver* as follows:

```
Equations:
f(V)=P*V-R*T
P=100
R=0.08206
T=300
nmoles=2.5*1000/V
Search Range:
V(min)=0.01, V(max)=1
```

This set of equations yields the solution $V = 0.2462$ L/g-mol, and the resultant number of moles in the vessel is 10.155 kg-mol.

The van der Waals equation can be similarly solved as follows:

```
Equations:
f(V)=(P+a/(V*V))*(V-b)-R*T
P=100
R=0.08206
T=300
nmoles=2.5*1000/V
Tc=304.2
Pc=72.9
a=27*R^2*Tc^2/(Pc*64)
b=R*Tc/(8*Pc)
Search Range:
V(min)=0.01, V(max)=1
```

The solution obtained using the van der Waals equation is $V = 0.0796$ L/g-mol and the number of moles = 31.418 kg-mol. Calculations involving the additional equations of state can be carried out in a similar manner.

 The POLYMATH problem solution file for part (a) is found in the *Simultaneous Algebraic Equation Solver Library* located in directory CHAP1 with file named P1-08A.POL.

1.9 BUBBLE POINT CALCULATION FOR AN IDEAL BINARY MIXTURE

1.9.1   Concepts Demonstrated

Calculation of bubble point in an ideal binary mixture.

1.9.2   Numerical Methods Utilized

Solution of a single nonlinear algebraic equation.

1.9.3   Problem Statement

(a) Calculate the bubble point temperature and equilibrium composition associated with a liquid mixture of 10 mol % n-pentane and 90 mol % n-hexane at 1 atm.

(b) Repeat the calculations for liquid mixtures containing 0 mol % up to 100 mol % of n-pentane.

(c) Plot the bubble point temperature and mol % of n-pentane in the vapor phase as a function of the mol % in the liquid phase.

The vapor pressure of n-pentane, P_A^*, in mm Hg can be calculated from the Antoine equation:

$$\log P_A^* = 6.85221 - \frac{1064.63}{T + 232.0} \qquad (1\text{-}25)$$

where T is the temperature in °C.

The vapor pressure of n-hexane, P_B^*, can be calculated from the Antoine equation:

$$\log P_B^* = 6.87776 - \frac{1171.53}{224.366 + T} \qquad (1\text{-}26)$$

1.9.4   Solution

At the bubble point, the sum of the partial vapor pressures of the components must equal the total pressure, which in this case is 1 atm or 760 mm of Hg. Denoting x_A as the mole fraction of n-pentane in the liquid mixture and x_B as the mole fraction of n-hexane, the nonlinear equation to be solved for the bubble point temperature is given by

$$f(T_{bp}) = x_A P_A^* + x_B P_B^* - 760 \qquad (1\text{-}27)$$

At the solution, $f(T_{bp})$ should become very small $[f(T_{bp}) \approx 0]$.

The vapor phase mole fraction of n-pentane, y_A, and the mole fraction of n-hexane, y_B, can be calculated from Raoult's law given by the equations

$$y_A = x_A P_A^* / 760 \qquad \textbf{(1-28)}$$

and

$$y_B = x_B P_B^* / 760 \qquad \textbf{(1-29)}$$

The solution of this problem involves finding the root of a single nonlinear equation given by Equation (1-27), where P_A^* and P_B^* are calculated from rearranged Equations (1-25) and (1-26). The complete set of equations can be entered into the POLYMATH *Simultaneous Algebraic Equation Solver* as follows:

```
Equations:
f(Tbp)=xA*PA+xB*PB-760
PA=10^(6.85221-1064.63/(Tbp+232))
PB=10^(6.87776-1171.53/(224.366+Tbp))
xA=0.1
xB=1-xA
yA=xA*PA/760
yB=xB*PB/760
Search Range:
Tbp(min)=30, Tbp(max)=69
```

A single nonlinear equation requires limits on the unknown between which the solution is to be found. Such limits can usually be found based on the physical nature of the problem. In this case, for example, the normal boiling points of n-pentane (36.07 °C) and n-hexane (68.7 °C) can be the basis for the lower and upper limits. After entering these values, the results are obtained as summarized in Figure 1–9 and Table 1-8.

Table 1-8 Tabulated Results for the Bubble Point Calculation

Solution		
Variable	**Value**	***f(v)***
T_{bp}	63.6645	3.411e−13
P_A	1784.05	
P_B	646.217	
x_A	0.1	
x_B	0.9	
y_A	0.234743	
y_B	0.765257	

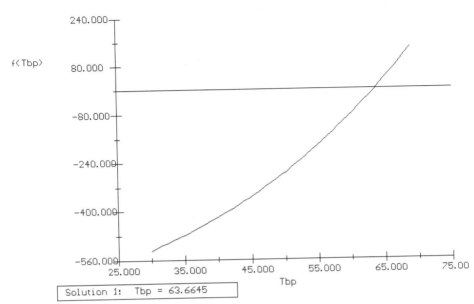

Figure 1–9 Graph of Solution for Bubble Point Temperature

The results indicate that the bubble point temperature is 63.66 °C and at this temperature the vapor is composed of 23.48 mol % of *n*-pentane and 76.52 mol % of *n*-hexane.

The calculations can be repeated for different mol %'s of *n*-pentane by changing x_A in the set of equations and resolving the problem with POLYMATH.

It is important to note that the bubble point can also be considered as the temperature at which the individual mole fractions in the gas phase sum to 1.0 for the given liquid phase composition. Thus this problem can alternately be solved by solving the nonlinear equation given next as an alternate for Equation (1-27).

$$f(T_{bp}) = y_A + y_B - 1 \tag{1-30}$$

 The POLYMATH problem solution file for part (a) is found in the *Simultaneous Algebraic Equation Solver Library* located in directory CHAP1 with file named P1-09.POL.

1.10 DEW POINT CALCULATION FOR AN IDEAL BINARY MIXTURE

1.10.1 Concepts Demonstrated

Calculation of dew point for an ideal binary mixture.

1.10.2 Numerical Methods Utilized

Solution of a single nonlinear algebraic equation.

1.10.3 Problem Statement

(a) Calculate the dew point temperature and the equilibrium liquid composition of a gas mixture containing 10 mol % n-pentane, 10 mol % n-hex-
ane, and the balance nitrogen (noncondensable) at 1 atm.

(b) Repeat the calculation for smaller amounts of nitrogen, as indicated in
Table 1-9.

Table 1-9 Dew Point Calculation for Binary Mixture

y_A	y_B	T_{dp}	x_A	x_B
0.1	0.1			
0.2	0.2			
0.3	0.3			
0.4	0.4			
0.5	0.5			

1.10.4 Solution (Partial)

At the dew point, the mole fractions of the components in the liquid phase, x_A for
n-pentane and x_B for n-hexane, sum to 1.0. This can be expressed by

$$f(T_{dp}) = x_A + x_B - 1 \tag{1-31}$$

The liquid composition at this temperature can be calculated using Raoult's law:

$$x_A = 760(y_A/P_A^*)$$
$$x_B = 760(y_B/P_B^*) \tag{1-32}$$

where y_A and y_B refer to the gas phase mole fractions. The vapor pressures of n-
pentane and n-hexane, P_A^* and P_B^*, can be calculated from Equations (1-25) and
(1-26) as discussed in Problem 1.9.

1.11 BUBBLE POINT AND DEW POINT FOR AN IDEAL MULTICOMPONENT MIXTURE

1.11.1 Concepts Demonstrated

Vapor liquid equilibrium calculations for a multicomponent mixture.

1.11.2 Numerical Methods Utilized

Solution of a single nonlinear algebraic equation.

1.11.3 Problem Statement

A multicomponent mixture, with composition shown in Table 1-10, is being transported in a closed container under high pressure. It has been suggested that transporting the mixture in refrigerated tanks under atmospheric pressure would be less expensive.

(a) Find out whether this suggestion is practical by calculating the bubble and dew point temperatures of the solution at atmospheric pressure.

(b) Repeat these calculations for different pressures.

(c) What is the effect of the pressure on the difference between the bubble and dew point temperatures?

The vapor pressure of the different components can be calculated from the Antoine equation given by Equation (1-18). The Antoine equation constants are given in Table 1-10.

Table 1-10 Liquid Composition and Antoine Constants for the Multicomponent Mixture

	Mole Fraction	A	B	C
Methane	0.1	6.61184	−389.93	266.0
Ethane	0.2	6.80266	−656.4	256.0
Propane	0.3	6.82973	−813.2	248.0
n-Butane	0.2	6.83029	−945.9	240.0
n-Pentane	0.2	6.85221	−1064.63	232.0

1.11.4 Solution

See Problems 1.9 and 1.10 for the method of solution.

1.12 ADIABATIC FLAME TEMPERATURE IN COMBUSTION

1.12.1 Concepts Demonstrated

Material and energy balances on an adiabatic system and calculation of adiabatic flame temperature.

1.12.2 Numerical Methods Utilized

Solution of a single nonlinear algebraic equation and use of logical variable during solution.

1.12.3 Problem Statement

When natural gas is burned with air, the maximum temperature that can be reached (theoretically) is the adiabatic flame temperature (AFT). This temperature depends mainly on the composition of the natural gas and the amount of air used in the burner. Natural gas consists mainly of methane, ethane, and nitrogen. The composition is different for natural gas found in various locations.

Determine the AFT for the following conditions and plot the AFT as a function of mol % CH_4 and the stoichiometric molar air to fuel ratio. The composition of natural gas is given in Table 1-11. The air-to-fuel ratios vary between 0.5 to 2.0. It can be assumed that the air and natural gas enter the burner at room temperature and atmospheric pressure. What composition and air-to-fuel ratio leads to the highest AFT?

Table 1-11 Composition of Natural Gas

Compound	mol %
CH_4	65 – 95
C_2H_6	3 – 33
N_2	2

The molar heat capacity of the reactants and the combustion products can be calculated from the equation

$$C_p^* = \alpha + \beta T + \gamma T^2 \tag{1-33}$$

where T is in K and C_p^* is in cal/g-mol·K. The constants of this equation for the different components are shown in Table 1-12 as given by Smith and Van Ness.[6] The heat of combustion is -212798 cal/g-mol for CH_4 and -372820 cal/g-mol for C_2H_6, as reported by Henley.[2] Assume that both the air and the natural gas enter at the temperature of 298 K and that the N_2 content of the natural gas is always 2.0 mol %. Air is 21 mol % O_2.

Table 1-12 Molar Heat Capacity of Gases (Smith and Van Ness[6] with permission.)

	α	$\beta \times 10^3$	$\gamma \times 10^6$
CH_4	3.381	18.044	-4.30
C_2H_6	2.247	38.201	-11.049
CO_2	6.214	10.396	-3.545
H_2O	7.256	2.298	0.283
O_2	6.148	3.102	-0.923
N_2	6.524	1.25	-0.001

1.12.4   Solution

The stoichiometric equations are

$$CH_4 + 2O_2 \rightarrow CO_2 + 2H_2O$$

$$C_2H_6 + 7/2O_2 \rightarrow 2CO_2 + 3H_2O$$

The actual to theoretical molar air-to-fuel ratio can be denoted by x with the inlet mole fractions of CH_4 and C_2H_6 denoted by y and z, respectively. For 1 mol of natural gas, there would be 0.02 mol N_2, y mol CH_4 and z mol C_2H_6. Therefore, the total moles of air required to react completely with the 1 mol of natural gas would be given by $(2y + [7/2]z)/0.21$.

Material balances for the different compounds using a 1 mol natural gas basis are shown in Table 1-13 for both fuel-rich ($x < 1$) and fuel-lean ($x > 1$) situations.

Table 1-13 Material Balance on the Reacting Species

	Moles in the product (x < 1)		Moles in the product (x > 1)	
	Expression	For y = 0.75	Expression	For y = 0.75
CH_4	$y(1-x)$	$0.75(1-x)$	0	0
C_2H_6	$z(1-x)$	$0.23(1-x)$	0	0
CO_2	$(y+2z)x$	$1.21x$	$y+2z$	1.21
H_2O	$(2y+3z)x$	$2.19x$	$2y+3z$	2.19
O_2	0	0	$\left(2y+\dfrac{7}{2}z\right)(x-1)$	$2.305(x-1)$
N_2	$0.02 + 3.76x\left(2y+\dfrac{7}{2}z\right)$	$0.02 + 8.67x$	$0.02 + 3.76x\left(2y+\dfrac{7}{2}z\right)$	$0.02 + 8.67x$

Since both the gas and the air enter at 298 K, this temperature can be used as a reference for enthalpy calculations. The enthalpy change for the product gases from $T = 298$ K up to the adiabatic flame temperature T_f can be calculated from the following expression:

$$\Delta H_P = \sum_{i=1}^{6} \alpha_i n_i (T_f - 298) + \frac{1}{2} \sum_{i=1}^{6} \beta_i n_i (T_f^2 - 298^2) + \frac{1}{3} \sum_{i=1}^{6} \gamma_i n_i (T_f^3 - 298^3) \quad \textbf{(1-34)}$$

where ΔH_P is the enthalpy change per mole of natural gas fed, and n_i is the number of moles of the different compounds, as shown in Table 1-13.

For $x < 1$ the general energy balance can be written as

$$f(T_f) = -212798xy - 372820xz + \Delta H_P = 0 \quad \textbf{(1-35)}$$

For $x > 1$ the same equation can be used with the value $x = 1$.

Equations (1-34) and (1-35), together with the data from Tables 1-12 and 1-13, can be entered into the POLYMATH *Simultaneous Algebraic Equation Solver.*

The coding for the example case where $y = 0.75$ and $x = 0.5$ is shown below. Note that POLYMATH uses "if ... then ... else ... " statements to provide the logic for the correct value of the molar air-to-fuel ratio.

```
Equations:
f(T)=212798*y*x+372820*z*x+H0-Hf
y=0.75
x=0.5
z=1-y-0.02
CH4=if (x<1) then (y*(1-x)) else (0)
C2H6=if (x<1) then (z*(1-x)) else (0)
CO2=if (x<1) then ((y+2*z)*x) else (y+2*z)
H2O=if (x<1) then ((2*y+3*z)*x) else (2*y+3*z)
N2=0.02+3.76*(2*y+7*z/2)*x
alp=3.381*CH4+2.247*C2H6+6.214*CO2+7.256*H2O+6.524*N2
bet=18.044*CH4+38.201*C2H6+10.396*CO2+2.298*H2O+1.25*N2
gam=-4.3*CH4-11.049*C2H6-3.545*CO2+0.283*H2O-0.001*N2
H0=alp*298+bet*0.001*298*298/2+gam*10**(-6)*298**3/3
Hf=alp*T+bet*0.001*T^2/2+gam*1E-6*T^3/3
Search Range:
T(min)=1000, T(max)=3000
```

For this case, the flame temperature is calculated to be $T = 2198.0$ K. Additional adiabatic flame temperature calculations can be made for specified inlet natural gas compositions and air-to-fuel ratios in a similar manner.

 The POLYMATH problem solution file for the example case is found in the *Simultaneous Algebraic Equation Solver Library* located in directory CHAP1 with file named P1-12.POL.

1.13   UNSTEADY-STATE MIXING IN A TANK

1.13.1   Concepts Demonstrated

Unsteady-state material balances.

1.13.2   Numerical Methods Utilized

Solution of simultaneous ordinary differential equations.

1.13.3   Problem Statement

A large tank is used for removing a small amount of settling solid particles (impurities) from brine in a steady-state process. Normally, a single input stream of brine (20% salt by weight) is pumped into the tank at the rate of 10 kg/min and a single output stream is pumped from the tank at the same flow rate. Normal operation keeps the level constant with the total mass in the tank at 1000 kg which is well below the maximum tank capacity.

At a particular time $(t = 0)$ an operator accidentally opens a valve, which causes pure water to flow continuously into the tank at the rate of 10 kg/min (in addition to the brine feed), and the level in the tank begins to rise.

Determine the amount of both water and salt in the tank as a function of time during the first hour after the pure water valve has been opened. Assume that the outlet flow rate from the tank does not change and the contents of the tank are well mixed at all times.

1.13.4   Solution

A mass balance on the total mass in the tank yields

$$\text{Accumulation} = \text{Input} - \text{Output}$$

$$\frac{dM}{dt} = 10 + 10 - 10 = 10 \tag{1-36}$$

where M is the mass in kg.

A mass balance on the salt in the tank yields

$$\frac{dS}{dt} = 10(0.2) - 10\left(\frac{S}{M}\right) = 2 - 10\left(\frac{S}{M}\right) \tag{1-37}$$

where S is the weight of salt in the tank in kg. Note that S/M represents the mass fraction of salt that is leaving the tank at any time t. This is also the mass fraction of salt within the tank since the tank is well mixed. Thus both M and S are functions of time for this problem. At $t = 0$, the initial conditions are that $M = 1000$ kg and $S = 200$ since the brine contains 20% salt by mass.

Entering Equations (1-36) and (1-37) together with the initial values into the POLYMATH *Simultaneous Differential Equation Solver* yields the following equation set:

```
Equations:
d(M)/d(t)=10
d(S)/d(t)=2-10*S/M
Initial Conditions:
t(0)=0
M(0)=1000
S(0)=200
Final Value:
t(f)=60
```

After the problem is solved, the POLYMATH screen display presented in Table 1-14 summarizes the initial, maximal, minimal, and final values of all the

Table 1-14 Intermediate Results Summary from POLYMATH

Variable	Initial value	Maximum value	Minimum value	Final value
t	0	60	0	60
M	1000	1600	1000	1600
S	200	222.5	200	222.5

problem variables. This numerical solution indicates that total amount of brine in the tank has increased from the initial 1000 kg to 1600 kg after one hour (60% increase). The amount of salt increased much more moderately to 222.5 kg from the initial value of 200 kg.

The different variables can be plotted as a function of time or other variables. For example, the amount of salt, S, as a function of time, t, is shown in Figure 1–10. The increase in the weight of salt in the tank is interesting and not expected. This can be explained by considering that the added water serves to dilute the salt solution in the tank, and the constant outflow from the tank carries out a smaller amount of salt. Since the input of salt to the tank is constant, the amount of salt in the tank increases. This is perhaps made clearer by calculating the percentage of salt in the tank during this same time period. Note that this is also the percentage of salt in the outlet stream from the tank. Let's define an algebraic equation for this percentage of salt.

$$\text{SaltPC} = 100\frac{S}{M} \qquad \text{(1-38)}$$

This algebraic equation can be added to the problem, and the resulting graph of the SaltPC variable is given in Figure 1–11. This shows that the mass percentage of salt is reduced from 20% initially to 13.9% after 20 minutes. Thus, the addition of the pure water input dilutes the brine in the tank and reduces the percent of salt in the output stream. Since the output flow rate was unchanged, the mass of salt in the tank increased.

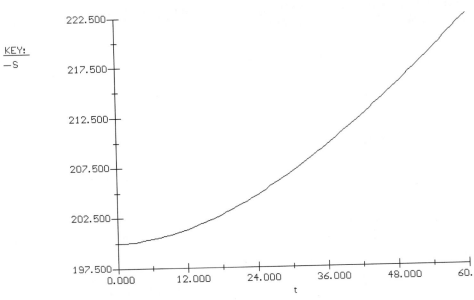

Figure 1–10 Amount of Salt in Tank as a Function of Time

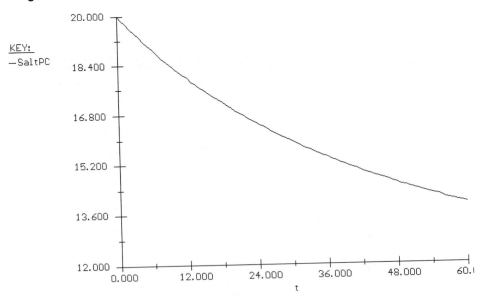

Figure 1–11 Percentage of Salt in the Tank as a Function of Time

 The POLYMATH problem solution file is found in the *Simultaneous Differential Equation Solver Library* located in directory CHAP1 with file named P1-13.POL.

1.14 UNSTEADY-STATE MIXING IN A SERIES OF TANKS

1.14.1 Concepts Demonstrated

Unsteady-state material balances on a series of well-mixed tanks.

1.14.2 Numerical Methods Utilized

Solution of simultaneous first-order ordinary differential equations.

1.14.3 Problem Statement

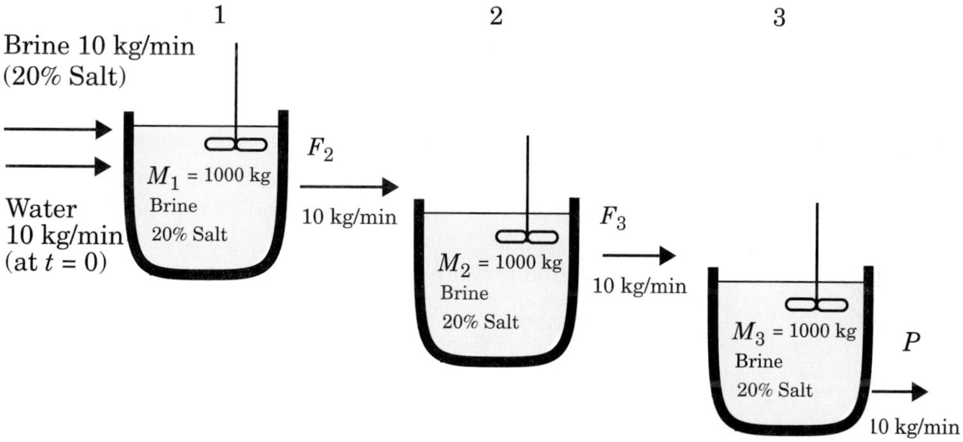

Figure 1–12 A Series of Settling Tanks under Normal Steady-State Operation

    A series of three well-mixed settling tanks (as shown in Figure 1–12) is used for settling solid particles (impurities) from brine that is being fed into a process. Under normal steady-state operation, brine (containing 20% salt by mass) is entering and exiting each tank at the flow rate of 10 kg/min. The three tanks contain 1000 kg brine each, with 20% salt by mass. At a particular time (t = 0), an additional valve is opened through which pure water flows into the first tank at the rate of 10 kg/min.

(a) Assuming that the rest of the flow rates remain the same and that the contents of the tanks are well mixed, determine and plot the amount and mass percent of the salt in the three tanks during the first hour after the opening of the pure water valve.

(b) What will be the mass percent of the salt in the outlet streams of the three tanks after the one-hour period?

1.15   Heat Exchange in a Series of Tanks

1.15.1   Concepts Demonstrated

Unsteady-state energy balances and dynamic response of well-mixed heated tanks in series.

1.15.2   Numerical Methods Utilized

Solution of simultaneous first order ordinary differential equations.

1.15.3   Problem Statement

Three tanks in sequence are used to preheat a multicomponent oil solution before it is fed to a distillation column for separation. Each tank is initially filled with 1000 kg of oil at 20 °C. Saturated steam at a temperature of 250 °C condenses within coils immersed in each tank. The oil is fed into the first tank at the rate of 100 kg/min and overflows into the second and the third tanks at the same flow rate. The temperature of the oil fed to the first tank is 20 °C. The tanks are well mixed so that the temperature inside the tanks is uniform, and the outlet stream temperature is the temperature within the tank. The heat capacity, C_p, of the oil is 2.0 kJ/kg·°C. For a particular tank, the rate at which heat is transferred to the oil from the steam coil is given by the expression

$$Q = UA(T_{steam} - T) \tag{1-39}$$

where UA is the product of the heat transfer coefficient and the area of the coil.

   UA = 10 kJ/min·°C for each tank

   T = temperature of the oil in the tank in °C

   Q = rate of heat transferred in kJ/min

   (a)   Determine the steady-state temperatures in all three tanks. What time interval will be required for T_3 to reach 99% of this steady-state value during startup?
   (b)   After operation at steady state, the oil feed is stopped for three hours. What are the highest temperatures that the oil in each tank will reach during this period?
   (c)   After three hours the oil feed is restored. How long will it take to achieve 99% steady state again for T_3? Will all steady-state temperatures be the same as before in part (a)?

1.15.4   Solution

The sequence of heating tanks is depicted in Figure 1–13.

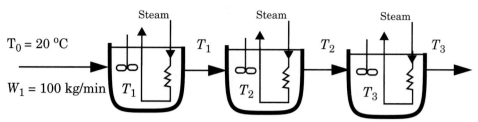

Figure 1–13 Series of Tanks for Oil Heating

Energy balances should be made on each of the individual tanks. In these balances, the mass flow rate to each tank will remain at the same fixed value. Thus $W = W_1 = W_2 = W_3$. The mass in each tank will be assumed constant as the tank volume and oil density are assumed to be constant. Thus $M = M_1 = M_2 = M_3$. For the first tank, the energy balance can be expressed by

$$\text{Accumulation} = \text{Input} - \text{Output}$$

$$MC_p\frac{dT_1}{dt} = WC_pT_0 + UA(T_{\text{steam}} - T_1) - WC_pT_1 \qquad \textbf{(1-40)}$$

Note that the unsteady-state mass balance is not needed for tank 1 or any other tanks since the mass in each tank does not change with time. The preceding differential equation can be rearranged and explicitly solved for the derivative, which is the usual format for numerical solution.

$$\frac{dT_1}{dt} = [WC_p(T_0 - T_1) + UA(T_{\text{steam}} - T_1)]/(MC_p) \qquad \textbf{(1-41)}$$

Similarly, for the second tank,

$$\frac{dT_2}{dt} = [WC_p(T_1 - T_2) + UA(T_{\text{steam}} - T_2)]/(MC_p) \qquad \textbf{(1-42)}$$

For the third tank,

$$\frac{dT_3}{dt} = [WC_p(T_2 - T_3) + UA(T_{\text{steam}} - T_3)]/(MC_p) \qquad \textbf{(1-43)}$$

Equations (1-41) to (1-43), together with the numerical data and initial values given in the problem statement, can be entered into the POLYMATH *Simultaneous Differential Equation Solver.*

   (a) The initial startup will be from a temperature of 20 °C in all three tanks; thus this is the appropriate initial condition for each tank temperature. The final value or steady-state value can be determined by solving the equations to steady state by giving a large time interval for the numerical solution. Alternately, one could set the time derivatives to zero and solve the resulting algebraic equations. In this case, it is easiest just to solve numerically the differential equations to a large value of t, where steady state is achieved. The POLYMATH coding for this aspect of this problem is as follows:

```
Equations:
d(T1)/d(t)=(W*Cp*(T0-T1)+UA*(Tsteam-T1))/(M*Cp)
d(T2)/d(t)=(W*Cp*(T1-T2)+UA*(Tsteam-T2))/(M*Cp)
d(T3)/d(t)=(W*Cp*(T2-T3)+UA*(Tsteam-T3))/(M*Cp)
W=100
Cp=2.0
T0=20
UA=10.
Tsteam=250
M=1000
Initial Conditions:
t(0)=0
T1(0)=20
T2(0)=20
T3(0)=20
Final Value:
t(f)=200
```

The time to reach steady state is usually considered to be the time to reach 99% of the final steady-state value for the variable that is increasing and responds most slowly. For this problem, T_3 increases the most slowly, and the steady-state value is found to be 51.317 °C. In POLYMATH, this can be easily done by displaying the output in tabular form for T_1, T_2, and T_3 so that the approach to steady state can accurately be observed. Thus part (a) can be completed by determining the time when T_3 reaches 0.99(51.317) or 50.804. Again, the tabular form of the output is useful in determining this time.

(b) The steady-state temperatures from part (a) can be used as initial conditions to this problem. The flow rate, W, should be set to zero and the problem solved numerically to a time of three hours entered as 180 minutes. Please note that POLYMATH has a *Problem Option* to restart from current conditions, which can be convenient in this situation.

(c) The return to steady state can be simulated by changing the flow rate, W, to its original value and continuing the numerical integration to large values of time, t, to achieve steady state again. The time for T_3 to be reduced to 101% of the steady-state value can be considered as the time to achieve steady state. This can be determined from the numerical solution to steady state.

The POLYMATH problem solution file for part (a) is found in the *Simultaneous Differential Equation Solver Library* located in directory CHAP1 with file named P1-15A.POL. This problem is also solved with Excel, Maple, MathCAD, MATLAB, Mathematica, and POLYMATH as problem 6 in the Set of Ten Problems discussed in Appendix F.

REFERENCES

1. Felder, R. M., and Rousseau, R. W. *Elementary Principles of Chemical Processes*. 2nd ed New York: Wiley, 1986.
2. Henley, E. J., and Rosen, E. M. *Material and Energy Balance Computation*. New York: Wiley, 1969.
3. Himmelblau, D. M. *Basic Principles and Calculations in Chemical Engineering*, 6th ed Englewood Cliffs, NJ: Prentice Hall, 1996.
4. Peng, D. Y., and Robinson, D. B. *Ind. Eng. Chem. Fundam.*, *15*, 59 (1976).
5. Perry, R. H., Green, D. W., and Malorey, J. D., eds. *Perry's Chemical Engineers Handbook*, 6th ed, New York: McGraw-Hill, 1984.
6. Smith, J. M., and Van Ness, H. C. *Introduction to Chemical Engineering Thermodynamics*, 2nd ed, New York: McGraw-Hill, 1959.
7. Thermodynamics Research Center API44 Hydrocarbon Project, *Selected Values of Properties of Hydrocarbon and Related Compounds*, Texas A&M University, College Station, TX, 1978.

CHAPTER 2

Regression and Correlation of Data

2.1 ESTIMATION OF ANTOINE EQUATION PARAMETERS USING NONLINEAR REGRESSION

2.1.1 Concepts Demonstrated

Direct use of the Antoine equation to correlate vapor pressure versus temperature data.

2.1.2 Numerical Methods Utilized

Nonlinear regression of a general algebraic expression with determination of the overall variance and confidence intervals of individual parameters.

2.1.3 Problem Statement

The Antoine equation is a widely used vapor pressure correlation that utilizes three parameters, A, B, and C. It can be expressed by

$$P_v = 10^{\left(A + \frac{B}{T + C}\right)} \tag{2-1}$$

where P_v is the vapor pressure in mm Hg and T is the temperature in °C.

Vapor pressure data for propane (psia) versus temperature (°F) is found in Table B–5 in Appendix B. Convert this data set to vapor pressure in mm Hg and temperature in °C. Then

(a) Determine the parameters of the Antoine equation and the corresponding 95% confidence intervals for the parameters from the given data set by using nonlinear regression on Equation (2-1).
(b) Calculate the overall variance for the Antoine equation.
(c) Prepare a residual plot for the Antoine equation.
(d) Assess the precision of the data and the appropriateness of the Antoine equation for correlation of the data set.

2.1.4   Solution

The form of the Antoine equation to be considered in this problem is to be regressed in its nonlinear form. An alternative treatment is to use multiple linear regression on the logarithmic form of this equation, but this transformation is not fully suitable, as discussed in Problem 1.7.

Nonlinear Regression

General nonlinear regression can be used to determine the parameters of an explicit algebraic equation, or model equation, as defined by Equation (2-2):

$$y = f(x_1, x_2, ..., x_n; a_1, ..., a_m) \tag{2-2}$$

where the single dependent variable is y, the n independent variables are $x_1, x_2, ..., x_n$, and the m parameters are $a_1, a_2, ..., a_m$. It is usually assumed that the experimental errors in the preceding equation are normally distributed with constant variance. Nonlinear regression algorithms determine the parameters for a particular model by minimizing the least-squares objective function (LS) given by

$$LS = \sum_{i=1}^{N} (y_{i(\text{obs})} - y_{i(\text{calc})})^2 \tag{2-3}$$

where N is the number of data points and (obs) and (calc) refer to observed and calculated values of the dependent variable.

The overall estimate of the model variance, σ^2, is calculated from

$$\sigma^2 = \frac{\sum_{i=1}^{N} (y_{i(\text{obs})} - y_{i(\text{calc})})^2}{\nu} = \frac{LS}{\nu} \tag{2-4}$$

where ν is the degrees of freedom, which is equal to the number of data points less the number of model parameters, $(N - m)$. Thus an algorithm that minimizes the least-squares objective function also minimizes the variance. Often the variance is used to compare the goodness of fit for various models.

A widely used graphical presentation that indicates any systematic difficulties with a particular model is called the residuals plot. This is simply the error, ε_i, in the model plotted versus the observed value of the independent variable, $y_{i(\text{obs})}$, with the error given by

$$\varepsilon_i = (y_{i(\text{obs})} - y_{i(\text{calc})}) \tag{2-5}$$

The POLYMATH *Polynomial, Multiple Linear and Nonlinear Regression Program* can be used to solve this problem directly, as the needed options are readily available.

(a) Nonlinear Regression of the Antoine Equation The first step is to enter the data by assigning a name to each variable (column). The column for the

vapor pressure in psia will be designated as P and the column for the temperature in °F will be denoted as TF. Since the Antoine equation is to correlate the
vapor pressure in mm Hg and the temperature in °C, then a new column denoted
as Pv and another new column denoted as TC can be created that calculate the
pressure and temperature in units of mm Hg and °C by

$$Pv = 51.715*P \tag{2-6}$$

$$TC = (TF - 32)/1.8 \tag{2-7}$$

Once the needed data columns are available within the POLYMATH program,
then the solution option for nonlinear regression can be selected.

The general form of the nonlinear equation using the available variables
should be entered as

$$Pv = 10^{\wedge}(A + B/(TC + C)) \tag{2-8}$$

The initial estimates for the parameters of this equation should be based on good
approximations. In this case, the data of Table 1-10 suggest that the values of A =
6, B = -1000, and C = 200 are appropriate. Figure 2–1 shows the results. The 95%

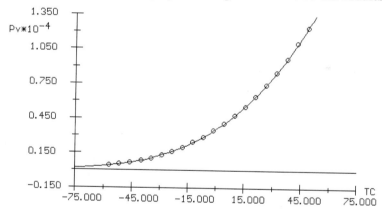

Model: Pv=10^(A+B/(TC+C))
A = 6.99872   C = 260.144
B = -897.603
8 positive residuals, 12 negative residuals.   Sum of squares = 1493.52

Figure 2–1 Nonlinear Regression Results for Antoine Equation

confidence intervals as calculated by the statistical analysis option of POLY-
MATH are given in Table 2–1. (See Problem 2.14 for more on confidence intervals.)

(b) **Overall Variance for the Antoine Equation**   The variance can be easily
calculated from Equation (2-4), where the degrees of freedom are given by (N–m).
For this problem the number of data points, N, is 20 and the number of parame-

Table 2–1 95% Confidence Intervals for Parameters of the Antoine Equation for Propane

Parameter	Converged Value	0.95 Conf. Interval	Lower Limit	Upper Limit
A	6.99872	0.0538121	6.94491	7.05253
B	−897.603	29.7552	−927.358	−867.848
C	260.144	4.56259	255.582	264.707

ters, m, is 3. The LS is the "sum of squares" value of 1493.52 given in Figure 2–1. Thus the overall model variance is 1493.52/(17) or 87.854.

(c) **Residuals Plot for the Antoine Equation**   This plot is defined by Equation (2-5), and creation of this plot is an option available in POLYMATH. The residual plot is given in Figure 2–2.

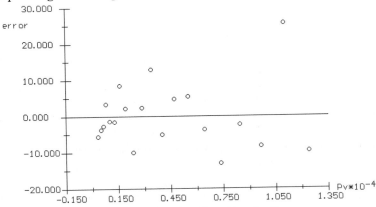

Figure 2–2 Residual Plot of the Antoine Equation for Propane

(d) **Precision of Data and Appropriateness of Antoine Equation**   The residual plot indicates that there is a slight tendency for larger errors for large values of the vapor pressure, P_v. Certainly the small vapor pressure data has smaller errors. This may indicate that the experimental errors are not independent of the measured vapor pressure. A regression of the Antoine equation using $\log P$ as the dependent variable might be useful to consider as an additional nonlinear regression. (This is considered in Problem 2.2 of this chapter.)

The confidence intervals of this nonlinear regression are relatively narrow, which indicates that this form of the Antoine equation provides an adequate correlation for the vapor pressure data of propane over the region of experimental data. Note that POLYMATH allows the calculated model output to be placed in the data table, allowing a direct numerical comparison of P_{obs} and P_{calc}.

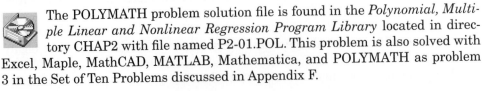

The POLYMATH problem solution file is found in the *Polynomial, Multiple Linear and Nonlinear Regression Program Library* located in directory CHAP2 with file named P2-01.POL. This problem is also solved with Excel, Maple, MathCAD, MATLAB, Mathematica, and POLYMATH as problem 3 in the Set of Ten Problems discussed in Appendix F.

2.2 ANTOINE EQUATION PARAMETERS FOR VARIOUS HYDROCARBONS

2.2.1 Concepts Demonstrated

Direct use of the Antoine equation to correlate vapor pressure versus temperature data.

2.2.2 Numerical Methods Utilized

Nonlinear regression of a general algebraic expression with determination of the overall variance and confidence intervals of individual parameters.

2.2.3 Problem Statement

Table B–7 (Appendix B) presents vapor pressure data for a variety of hydrocarbons as function of temperature. A particular design procedure requires an equation that correlates these data.

(a) Select one of the hydrocarbons in Table B–7. Find the Antoine equation parameters, the corresponding 95% confidence intervals of the parameters, and the overall variance. Prepare the residual plot. The regression should be carried out as discussed in Problem 2.1.

(b) Repeat part (a) with nonlinear regression with the dependent variable as log P.

(c) Repeat part (a) using linear regression as described in Problem 1.7.

(d) Assess the appropriateness of the resulting Antoine equations from (a), (b), and (c) for correlation of the data set. Which equation should be selected?

The POLYMATH data files for Table B–7 is found in the *Polynomial, Multiple Linear and Nonlinear Regression Program Library* located in directory TABLES with files named B-07A.POL through B-07F.POL.

2.3 CORRELATION OF THERMODYNAMIC AND PHYSICAL PROPERTIES OF n-PROPANE

2.3.1   Concepts Demonstrated

Correlations for heat capacity, thermal conductivity, viscosity and latent heat of vaporization.

2.3.2   Numerical Methods Utilized

Linear and nonlinear regression of data with linearization and transformation functions.

2.3.3   Problem Statement

Tables B–8 through B–11 present values for different properties of propane (heat capacity, thermal conductivity for gas, viscosity, and latent heat of vaporization for liquid) as a function of temperature.

> Determine appropriate correlations for the properties of propane listed in Tables B–8 through B–11 using suggested expressions given in the chemical engineering literature.

2.3.4   Solution

A variety of regressions will be used in this problem solution. All of these are available in the POLYMATH *Polynomial, Multiple Linear and Nonlinear Regression Program*. Previous problems that illustrate the regressions that will be used in this solution include Problem 1.3 for polynomial fitting, Problems 1.3 and 1.7 for linear regression, and Problem 2.1 for nonlinear regression. It is important to select among available correlations by using key statistical indicators, such as variance, 95% confidence intervals, and residual plots. This solution indicates how this selection can be accomplished.

Theoretical considerations as well as experience have shown which type of correlations are the best to represent temperature dependence of different physical and thermodynamic properties. This information is available in the literature and will be used to develop the various correlations in this problem solution.

(a) Heat Capacity for a Gas   Heat capacity for propane gas is given in Table B–8 for temperatures between 50 K and 1500 K. According to Perry et al.[4] the heat capacities of gases are most commonly represented as a simple polynomial:

$$C_p = a_0 + a_1 T + a_2 T^2 + a_3 T^3 + \ldots \tag{2-9}$$

Entering the data from Table B–8 into the POLYMATH *Polynomial, Multiple Linear and Nonlinear Regression Program* and selecting the option to "fit a polynomial" yields the results shown in Figure 2–3 and summarized in Table 2–2.

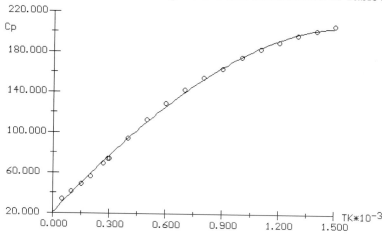

$$P(TK) = 20.5235 + 0.196471 \, TK - 2.6848e{-}05 \, TK^2 - 1.51289e{-}08 \, TK^3$$

Figure 2–3 Third Degree Polynomial Representation for Heat Capacity of *n*-Propane

Table 2–2 Polynomial Coefficients and Variance for n-Propane

Degree	2	3	4	5
a_0	17.7427	20.5235	26.7577	31.0453
a_1	0.217787	0.196471	0.12302	0.0515068
a_2	−6.16428e−05	−2.6848e−05	0.000188971	0.000508754
a_3		−1.51289e−08	−2.35697e−07	−7.90007e−07
a_4			7.22918e−11	4.7874e−10
a_5				−1.0597e−13
variance	6.81537	6.00692	1.86662	0.662227

All of the polynomials with the exception of the first degree visually represent the data in a similar way to Figure 2–3. However, the results in Table 2–2 show that the variance of the fifth-degree polynomial is significantly lower than that of the other polynomials, indicating that this correlation represents the data best. Verification that all the parameters of the fifth-degree polynomial are significantly different from zero can be accomplished with the 95% confidence levels for the parameters available from POLYMATH and summarized in Table 2–3. Since all the confidence intervals are smaller in absolute value than the respective parameter values, this indicates that all of the terms of the polynomial are useful in the data correlation.

Table 2–3 Parameter Values and 95% Confidence Intervals for the Fifth-Degree Polynomial Representing n-Propane Heat Capacity Data

Parameter	Value	0.95 Confidence Interval
a_0	31.0453	2.74852
a_1	0.0515068	0.0353522
a_2	0.000508754	0.000143119
a_3	−7.90007e−07	2.37827e−07
a_4	4.7874e−10	1.71405e−10
a_5	−1.0597e−13	4.44966e−14

In addition to the examination of the confidence intervals, it is very useful to examine the residuals plot which is given in Figure 2–4 for the fifth-degree

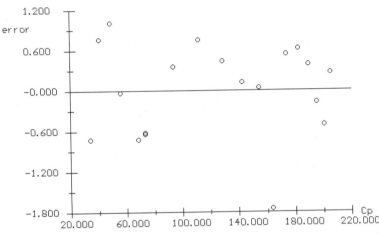

Figure 2–4 Residual Plot for Heat Capacity Represented by Fifth-Degree Polynomial

polynomial. The error, $C_{p(\text{obs})} - C_{p(\text{calc})}$, is found from the plot to be randomly distributed, indicating that there are no definite trends in the error, which further supports the polynomial correlation of the measured heat capacity data.

It is interesting to compare the residual plot of Figure 2–4 with the residual plot of the third-degree polynomial given in Figure 2–5. In this latter correlation there is a clear oscillatory pattern with much larger errors, which indicates that more parameters are probably needed to represent the data satisfactorily.

(b) Thermal Conductivity   Thermal conductivity for gaseous propane is given in Table B–9 for the temperature range –40 °F to 200 °F. Perry et al.[4] note that over small temperature ranges the thermal conductivity of low-pressure gases can be fairly well correlated by a linear equation which is also a first-degree polynomial.

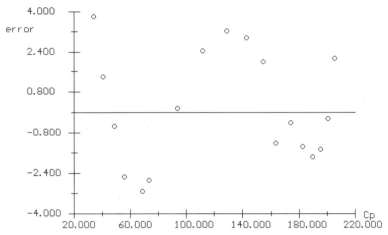

Figure 2–5 Residual Plot for Heat Capacity Represented by Third-Degree
Polynomial

The plots of all the polynomials with second and higher degree are very
similar in their visual representation of the thermal conductivity data. A critical
analysis requires the results tabulated in Table 2–4, which show the coefficient

Table 2–4 Polynomial Representation of Gaseous Thermal Conductivity Data

	First Degree		**Second Degree**		**Third Degree**	
	Value	95% Confidence Interval	Value	95% Confidence Interval	Value	95% Confidence Interval
a_0	32.1462	0.660362	32.2121	0.09897	32.174	0.1212
a_1	0.137157	0.007458	0.119844	0.002402	0.118808	0.003074
a_2			0.115263×10^{-3}	1.417×10^{-5}	1.46389×10^{-4}	6.034×10^{-5}
a_3					-1.29161×10^{-7}	2.436×10^{-7}
σ^2	0.4920		0.01044		0.009509	

values, 95% confidence intervals, and variances for the polynomials of first to
third degree. This table indicates that there is a substantial decrease in the
value of the variances between the first and second-degree polynomials. There is
only a very slight decrease between the second and third-degree polynomials.
The 95% confidence interval of the coefficient a_3 in the third-degree polynomial
is in fact larger in absolute value than the parameter value itself. This indicates
that there is no statistical justification to using the third-degree polynomial
instead of the second-degree polynomial for correlation of these data. Figures 2–
6 and 2–7 show the residual plots for the first-degree and second-degree polynomials, respectively. For the first-degree polynomial there is a clear trend in the

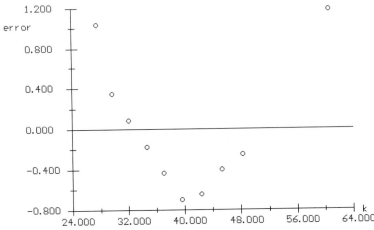

Figure 2–6   Residual Plot for the Thermal Conductivity Represented by a First-degree Polynomial

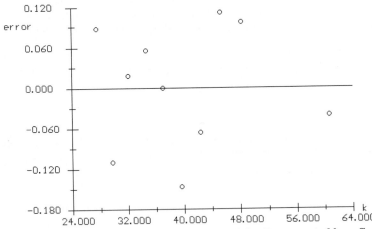

Figure 2–7   Residual Plot for Thermal Conductivity Represented by a Second-Degree Polynomial

residuals, while for the second-degree polynomial the error is randomly distributed. Thus the clear conclusion based upon both the numerical and graphical statistical results is that the second-degree polynomial represents the data adequately.

For a wide range of temperature, Perry et al.[4] recommend the correlation of thermal conductivity, k, with $(T)^n$, where T is the absolute temperature and $n \approx 1.8$. This correlation can be directly evaluated with nonlinear regression using the form of Equation (2-10), where both c and n are parameters.

$$k = cT^n \qquad\qquad (2\text{-}10)$$

The resulting nonlinear correlation result from POLYMATH is given in Figure 2–8.

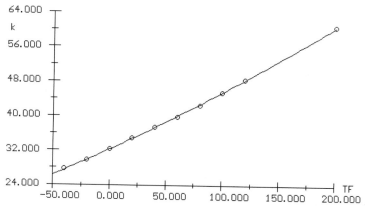

Model: k=c*(TF+460)^n
c = 0.000684662
n = 1.75461
6 positive residuals, 4 negative residuals.   Sum of squares = 0.170874

Figure 2–8 Nonlinear Correlation for Thermal Conductivity of n-Propane

It is interesting that the exponent, n, is indeed approximately 1.8, as suggested by Perry et al.[4]. The entire correlation, including 95% confidence intervals, is found to be

$$k = (6.847{\times}10^{-4} \pm 8.39 \cdot 10^{-5})T^{(1.7546 \pm 0.0195)}$$ **(2-11)**

Figure 2–9 shows the residual plot for the nonlinear equation where the residuals appear to be randomly distributed. Thus the thermal conductivity of n-propane in the region of the data is well correlated by Equation (2-11).

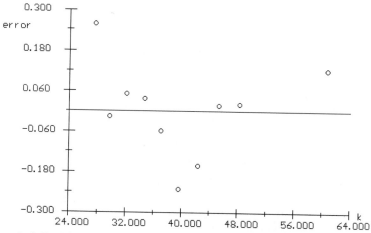

Figure 2–9 Residual Plot for Thermal Conductivity Represented by Nonlinear Equation

The choice between the second-degree polynomial and the nonlinear equation for this thermal conductivity correlation is difficult since both correlations seem quite adequate. Let us now consider the overall variance of each correlation. The variance of the second-degree polynomial is 0.01044 from Table 2–4 as compared to the variance of the nonlinear equation which is 0.02136. The latter variance is calculated from Equation (2-4) using the sum of squares from Figure 2–8 and dividing by 8 for the degrees of freedom (10 data points – 2 parameters). Although this difference in variance is probably not statistically significant, it does indicate that the second-degree polynomial provides a better correlation for this thermal conductivity data set.

(c) Liquid Viscosity   The recommended correlation for viscosity of liquids by Perry et al.[4] is similar to the Antoine equation for vapor pressure:

$$\log(\mu) = A + B/(T + C) \tag{2-12}$$

where μ is the viscosity and A, B and C are parameters. If T is expressed in Kelvin, parameter C can be approximated by $C = 17.71 - 0.19T_b$, where T_b is the normal boiling point in Kelvin. For n-propane the normal boiling point is 231 K; thus the approximate value of C is –26.18.

The nonlinear regression option of POLYMATH can be used directly on Equation (2-12) for calculating the parameters from the data of Table B–10. Alternatively, Equation (2-12) can be linearized as it was done for vapor pressure in Problem 1.7 to determine A, B, and C. Since the nonlinear regression requires good initial estimates of the parameters and estimates are not available for A and B, a multiple linear regression that requires no initial estimates will be used to obtain these estimates. A subsequent nonlinear regression will then be performed with these as initial estimates.

Thus, Equation (2-12) can be linearized by using the value of C to be –26.18. Thus the POLYMATH program needs the variable transformations given by `logmu = log(mu)`, `TK = TC + 273.15`, and `invTKplusC = 1/(TK - 26.18)`. The linearized expression with these new variables becomes

$$\texttt{logmu = A + B*invTKplusC} \tag{2-13}$$

A linear regression with the only independent variable column as `invTKplusC` and the dependent variable column as `logmu` gives A = `-1.3439` and B = `142.07`. These values and C = `-26.18` provide the needed initial estimates for the nonlinear regression of Equation (2-12).

The nonlinear regression results are presented in Figure 2–10, where the apparent representation by the nonlinear model appears adequate. However, the residual plot of Figure 2–11 indicates a clear oscillatory trend in the error distribution. This suggests that a three-parameter equation is probably not adequate for representing viscosity data over such a wide range of temperature.

A four-parameter equation used in Reid et al.[6] provides another possible correlation equation for viscosity of liquids:

$$\log(\mu) = A + B/T + CT + DT^n \tag{2-14}$$

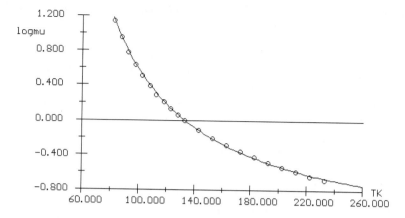

Model: logmu=A+B/(TK+C)
A = -1.40081   C = -22.0591
B = 154.702
9 positive residuals, 12 negative residuals.   Sum of squares = 0.00152025

Figure 2–10 Observed and Calculated Viscosity Data. Antoine Equation Parameters
Calculated by Nonlinear Regression

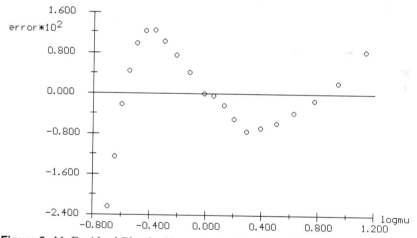

Figure 2–11 Residual Plot for Viscosity for Nonlinear Equation (2-12)

The preceding equation has five parameters that can be fitted with nonlinear
regression.   For such an equation, again it is useful to use linear regression to
provide good initial estimates. Linearization can easily be accomplished by setting the exponent n to some arbitrary value such as 0.5. The final linear regression values from this treatment are $A = -11.1502, B = 467.144, C = -0.028897, D$
$= 0.994892$, with $n = 0.5$. Using these values for a nonlinear regression of Equation (2-14) proves to be difficult with confidence intervals on some parameters,
including zero.

The results for both correlations suggested from the literature are not very
satisfactory, leading to the conclusion that this data set cannot be adequately

represented by the commonly used models for temperature dependence of viscosity. Consideration should be given to a polynomial regression for this data set.

(d) Heat of Vaporization Heat of vaporization data for propane are shown in Table B–11 for the temperature range of –70 °F to 120 °F. Heat of vaporization can be correlated by an equation based on the Watson relation (Perry et al.[4]):

$$\Delta H = A(T_C - T)^n \tag{2-15}$$

Watson's recommended value for n is 0.38, but n can be found by regression of the experimental data. The critical temperature of propane is 666 °R.

The equation can be directly used in nonlinear regression, or it can be linearized by taking the log of each side yielding

$$\log(\Delta H) = \log A + n\log(T_C - T) \tag{2-16}$$

Both regressions will be conducted so the original data from Table B–11 are entered as `TF` and `deltaH`, and the transformations `TR = TF + 460`, `logdeltaH = log(deltaH)`, and `logTCmTR = log(666 - TR)` are used for the linear regression.

The linear regression results are presented in Figure 2-12 and the resulting

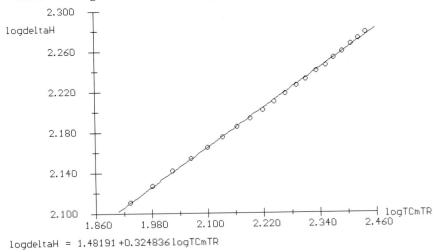

logdeltaH = 1.48191 +0.324836 logTCmTR

Figure 2–12 Heat of Vaporization Data Correlation by Linear Regression

equation for the heat of vaporization is

$$\Delta H = 30.333(666 - T)^{0.3248} \tag{2-17}$$

The nonlinear regression using the linear regression results as initial estimates is as follows:

$$\Delta H = 30.083(666 - T)^{0.3264} \tag{2-18}$$

Note that both equations are very similar. The slight differences in the parameter values arise because the least-squares objective function in the linear regression is calculated from $\log(\Delta H)$, whereas the least-squares objective function in the nonlinear regression is calculated from ΔH.

The residual plot from the nonlinear regression presented in Figure 2–13 shows a cyclic trend, indicating that the data are not represented well enough by this equation. A similar trend is found for the linear regression as well.

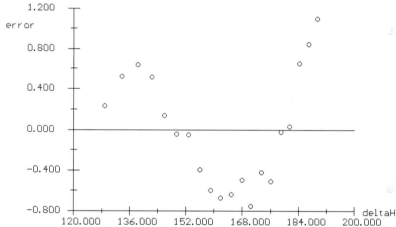

Figure 2–13 Heat of Vaporization Data Correlation by Nonlinear Regression

A fourth-degree polynomial represents the data adequately.

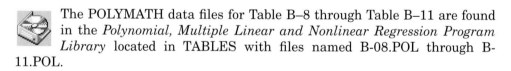

The POLYMATH data files for Table B–8 through Table B–11 are found in the *Polynomial, Multiple Linear and Nonlinear Regression Program Library* located in TABLES with files named B-08.POL through B-11.POL.

2.4 TEMPERATURE DEPENDENCY OF SELECTED PROPERTIES

2.4.1   Concepts Demonstrated

Presentation and correlation of thermodynamic and physical property data.

2.4.2   Numerical Methods Utilized

Fitting different types of curves to experimental data and transformation of variables to obtain linear expressions.

2.4.3   Problem Statement

Tables B–12 through B–15 present values of different properties of various compounds as a function of temperature.

Find the best correlations to represent the temperature dependency of the properties of these compounds.

The POLYMATH data files for Table B–12 through Table B–15 are found in the *Polynomial, Multiple Linear and Nonlinear Regression Program Library* located in directory TABLES with files named B-12A.POL ... B-12D.POL, B-13A.POL ... B-13G.POL, B-14A.POL ... B-14F.POL, and B-15A.POL ... B-15F.POL.

2.5 HEAT TRANSFER CORRELATIONS FROM DIMENSIONAL ANALYSIS

2.5.1 Concepts Demonstrated

Correlation of heat transfer data using dimensionless groups.

2.5.2 Numerical Methods Utilized

Linear and nonlinear regression of data with linearization and use of transformation functions.

2.5.3 Problem Statement

An important tool in the correlation of engineering data is the use of dimensional analysis. This treatment leads to the determination of the independent dimensionless numbers, which may be important for a particular problem. Linear and nonlinear regression can be very useful in determining the correlations of dimensionless numbers with experimental data.

A treatment of heat transfer within a pipe has been considered by Geankoplis[3] using the Buckingham method, and the result is that the Nusselt number is expected to be a function of the Reynolds and Prandtl numbers.

$$Nu = f(Re, Pr) \quad \text{or} \quad \frac{hD}{k} = f\left(\frac{Dv\rho}{\mu}, \frac{C_p\mu}{k}\right) \tag{2-19}$$

A typical correlation function suggested by Equation (2-19) is

$$Nu = aRe^b Pr^c \tag{2-20}$$

where a, b, and c are parameters that can be determined from experimental data.

A widely used correlation for heat transfer during turbulent flow in pipes is the Sieder-Tate equation[7]:

$$Nu = 0.023 Re^{0.8} Pr^{1/3} (\mu/\mu_w)^{0.14} \tag{2-21}$$

in which a dimensionless viscosity ratio has been added. This ratio (μ/μ_w) is the viscosity at the mean fluid temperature to that at the wall temperature.

Table B–16 gives some of the data reported by Katz and Williams[8] for heat transfer external to 3/4-inch outside diameter tubes where the Re, Pr, (μ/μ_w), and Nu dimensionless numbers have been measured.

> (a) Use multiple linear regression to determine the parameter values of
>   the functional forms of Equations (2-20) and (2-21) that represent the
>   data of Table B–16.
> (b) Repeat part (a) using nonlinear regression.
> (c) Which functional form and parameter values should be recommended
>   as a correlation for this data set? Justify your selection.

2.5.4 Solution

For convenience, let the functional forms of Equations (2-20) and (2-21) be written as

$$Nu = a_i Re^{b_i} Pr^{c_i} \tag{2-22}$$

$$Nu = d_i Re^{e_i} Pr^{f_i} Mu^{g_i} \tag{2-23}$$

where $i = 1$ indicates parameter values from linear regression and $i = 2$ indicates
parameter values from nonlinear regression. Mu represents the viscosity ratio.
The POLYMATH *Polynomial, Multiple Linear and Nonlinear Regression Program* can be used to carry out the linear and nonlinear regressions.

(a) A linear regression of either Equation (2-22) or (2-23) requires a transformation into a linear form. This is easily accomplished by taking the ln (natural logarithm) of each side of each equation. The resulting transformed equations
are

$$\ln Nu = \ln a_1 + b_1 \ln Re + c_1 \ln Pr \tag{2-24}$$

$$\ln Nu = \ln d_1 + e_1 \ln Re + f_1 \ln Pr + g_1 \ln Mu \tag{2-25}$$

The data of Table B–16 can be entered directly into POLYMATH under columns defined as Re, Pr, Mu, and Nu. Additional columns can be created that provide the needed ln's for the linear regression by using lnRe = ln(Re), lnPr =
ln(Pr), lnMu = ln(Mu) and lnNu = ln(Nu). The results of the linear regressions are summarized in Table 2–5. Note that the reported values for a_1 and d_1
have been calculated from the regression results that give the ln values of these
parameters. The calculated values from the current linear regression can be
automatically entered into the data sheet by pressing "s" from the Display
Option menu, which gives the regression results.

The linear regression results in Table 2–5 for Equations (2-24) and (2-25)
are consistent when comparing parameters a with d, b with e, and c with f. The
95% confidence intervals are all relatively small except for a, d and g, which have
large intervals that include zero.

Table 2–5 Summary of Parameter Values from Regressions

Parameter	Linear Regression Equations (2-24) and (2-25)		Nonlinear Regression Equations (2-22) and (2-23)	
	Value	95% Confidence Interval	Value	95% Confidence Interval
a	0.662294	3.8784	0.165641	0.126371
b	0.539538	0.116042	0.663548	0.0647257
c	0.245353	0.113922	0.341377	0.0621987
d	0.534712	5.6335	0.149154	0.117509
e	0.558834	0.150682	0.673289	0.0672626
f	0.252375	0.123001	0.328567	0.066708
g	−0.0677231	0.316351	−0.177693	0.332514

(b) Direct nonlinear regressions of both Equations (2-22) and (2-23) can be completed where the converged values from the linear regressions of part (a) can be used as initial parameter estimates. The results are also summarized in Table 2–5. Again the results are consistent when comparing parameters a with d, b with e, and c with f. The 95% confidence intervals are all relatively small except for g which has a large interval that includes zero. Note that the calculated values from the current nonlinear model can be automatically entered into the data sheet by pressing "s" from the Display Option menu, which gives the regression results.

(c) There are a number of considerations when selecting the most appropriate correlation.

Confidence Intervals A major indicator is the 95% confidence interval of each parameter. When the confidence interval is very large relative to the parameter, this suggests that the parameter may not be important in the correlation and perhaps should be set to zero. This seems to be the case in *both* the linear and nonlinear regressions carried out in parts (a) and (b) for the parameter g, which is the exponent of Mu. This is an indication that Mu should not be included in the correlation. An alternate explanation is that Mu may be dependent upon other variables in the regression. An examination of this will be considered later in this section.

Residual Plots It is always very useful to examine residual plots of regressions to determine if there are any obvious trends, as the errors should be random. A typical residual plot is shown in Figure 2–14 for the nonlinear regression of Equation (2-22). This residual plot demonstrates that the error is randomly distributed. Since there are no unusual error patterns in any of the linear and nonlinear regressions considered here, the residual plots are not helpful in selecting a correlation in this case.

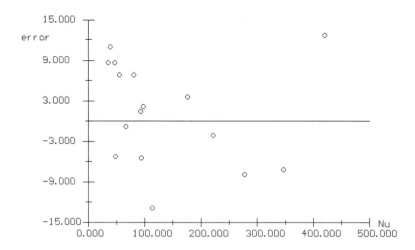

```
error = Nu - (a2*Re^b2*Pr^c2)
a2 = 0.165641                      c2 = 0.341377
b2 = 0.663548
```

Figure 2–14 Residual Plot for Heat Transfer Data Using Nonlinear Regression with Equation (2-22)

Comparison of Variances A comparison of variances for these correlations deserves special attention because the dependent variable in the linear regression is ln(Nu) while in the nonlinear regression it is Nu. It is necessary to use a variance based on the same variable for comparisons. The variance based on the Nusselt number for this problem is defined as

$$\sigma^2 = \frac{1}{\nu} \sum_{i=1}^{N} (Nu_{(\text{obs})} - Nu_{(\text{calc})})^2 \qquad (2\text{-}26)$$

where ν is the degrees of freedom, which is equal to the number of data points less the number of model parameters, $(N - m)$. A relative error variance for this problem can also be defined as

$$\sigma_r^2 = \frac{1}{\nu} \sum_{i=1}^{N} \left(\frac{Nu_{(\text{obs})} - Nu_{(\text{calc})}}{Nu_{(\text{obs})}} \right)^2 \qquad (2\text{-}27)$$

These variances can be calculated within POLYMATH after the various regressions have been completed by defining columns to evaluate the terms in the summation functions. There is a convenient option within POLYMATH that can sum individual columns. The resulting calculations are summarized in Table 2–6.

An examination of the *variance column* in Table 2–6 indicates that both correlations obtained with nonlinear regression are superior to both correlations

Table 2–6 Calculated Variances for Heat Transfer Correlations

Regression Equation	Variance σ^2	Relative Variance σ_r^2
Linear Regression Equation (2-24)	265.60	0.010277
Linear Regression Equation (2-25)	235.83	0.011344
Nonlinear Regression Equation (2-22)	68.275	0.016832
Nonlinear Regression Equation (2-23)	66.685	0.014763

obtained with linear regression. This is because higher Nu values have a greater influence on the regression. Consideration of the *relative variance column* suggests that both linear regressions result in the lowest relative variance. This is because the $\ln(Nu)$ is used as the dependent variable in the regression, which lessens the effect of data points with larger Nu values. This indicates the general conclusion that *logarithmic transformations are useful if relative errors are to be minimized*.

Thus the selection of the regression equation depends on the experimental errors as to whether they are relative or proportional to the measured Nu values. Since this information is not known about this data set, the selection cannot be made on the variance or relative variance calculations.

Possible Interdependency of Variables The linear and nonlinear regressions assume that the independent variables do not depend on each other. Possible dependency between assumed "independent" variables can be examined by plotting one variable with another. In this case, the large confidence interval on Mu, which includes zero, is an indication that Mu may be related to other variables. A regression of Mu versus $\ln(Re)$ shows definite dependence, as shown in Figure 2–15. Since the viscosity ratio apparently was not changed independently of Re during the experiments, its effect on Nu cannot be isolated. The conclusion with regard to the power of the viscosity ratio in Equation (2-23) is that the data of Katz and Williams are insufficient for determining an exponent for this ratio.

Final Correlation Since the nonlinear regression gives the Nu directly and the variance as calculated by Equation (2-26) is usually employed in regression, the recommended correlation for the data set is given by

$$Nu = (0.1656 \pm 0.123)Re^{(0.664 \pm 0.0647)}Pr^{(0.341 \pm 0.0622)} \qquad \textbf{(2-28)}$$

More experiments should be performed to investigate further the Mu ratio and to determine the best variance to use in the regression of this data set.

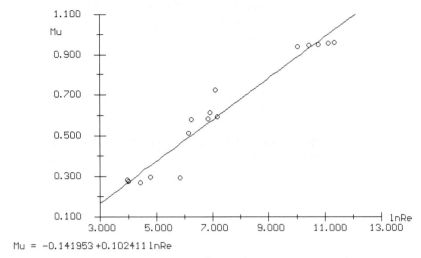

Mu = -0.141953 +0.102411 lnRe

Figure 2–15 Plot of the Viscosity Ratio (μ/μ_w) versus ln of Reynolds Number $\ln(Re)$

 The POLYMATH problem data file is found in the *Polynomial, Multiple Linear and Nonlinear Regression Program Library* located in directory CHAP2 with file named P2-05.POL.

2.6 HEAT TRANSFER CORRELATION OF LIQUIDS IN TUBES

2.6.1 Concepts Demonstrated

Correlation of heat transfer data using dimensionless groups.

2.6.2 Numerical Methods Utilized

Linear and nonlinear regression of data with linearization and use of transformation functions.

2.6.3 Problem Statement

Tables B–17 through B–19 give data reported by Sieder and Tate[7] for heat transfer in a concentric tube heat exchanger with a 0.75-inch No. 16 BWG inside tube and a 1.25-inch outside iron pipe. The measurements were made for three different types of oil. The dimensionless groups identified in the table as Nu, Re, Pr are the Nusselt, Reynolds and Prandtl numbers, respectively. The column labeled (μ/μ_w) is the ratio of viscosity at the mean fluid temperature to that at the wall temperature.

(a) Determine the most appropriate correlation for the heat transfer data in Tables B–17 through B–19. Evaluate expressions that have the general form of the Sieder-Tate equation[7] written in the form

$$Nu = a_0 Re^{a_1} Pr^{a_2} (\mu/\mu_w)^{a_3}$$

where a_0, a_1, a_2, and a_3 are parameters to be determined from the total data set.

(b) Compare your results with the equation suggested by Sieder and Tate, which is given by

$$Nu = 0.023 Re^{0.8} Pr^{1/3} (\mu/\mu_w)^{0.14}$$

The POLYMATH data files for Tables B–17 through B–19 are found in the *Polynomial, Multiple Linear and Nonlinear Regression Program Library* located in directory TABLES with files named B-17.POL through B-19.POL.

2.7 HEAT TRANSFER IN FLUIDIZED BED REACTOR

2.7.1 Concepts Demonstrated

Correlation of heat transfer data using dimensionless groups.

2.7.2 Numerical Methods Utilized

Linear and nonlinear regression of data with linearization.

2.7.3 Problem Statement

Dow and Jacob[2] proposed the following dimensionless equation as a suitable representation for experimental data dealing with heat transfer between a vertical tube and a fluidized air solid mixture:

$$Nu = a_1 \left(\frac{D_t}{L}\right)^{a_2} \left(\frac{D_t}{D_p}\right)^{a_3} \left(\frac{1-\varepsilon}{\varepsilon} \frac{\rho_s C_s}{\rho_g C_g}\right)^{a_4} \left(\frac{D_t G}{\mu_g}\right)^{a_5} \tag{2-29}$$

where

$$Nu = \frac{h_m D_t}{k_g} = \text{Nusselt number}$$

$$h_m = \text{heat transfer coefficient}$$

$$D_t = \text{tube diameter}$$

$$D_p = \text{solid-particle diameter}$$

$$L = \text{heated fluidized bed length}$$

$$\varepsilon = \text{void fraction of fluid bed}$$

$$G = \text{gas mass velocity}$$

$$k_g, \rho_g, C_g, \mu_g = \text{properties of gas phase}$$

$$C_s, \rho_s = \text{properties of solid phase}$$

(a) Use the data in Table B–20 to calculate the parameter values $a_1, a_2,$ ... a_5 for Equation (2-29) using linear regression on $\log(Nu)$ and nonlinear regression on Nu.

(b) Which regression is the most useful in correlation of the data? Justify your selection with confidence intervals, variances, and residual plots.

(c) Explain the differences between the two types of regression utilized in this problem.

The POLYMATH data file for Table B–20 is found in the *Polynomial, Multiple Linear and Nonlinear Regression Program Library* located in directory TABLES with file named B-20.POL.

2.8 CORRELATION OF BINARY ACTIVITY COEFFICIENTS USING MARGULES EQUATIONS

2.8.1 Concepts Demonstrated

Estimation of parameters in the Margules equations for the correlation of binary activity coefficients.

2.8.2 Numerical Methods Utilized

Linear and nonlinear regression, transformation of data for regression; calculation and comparison of confidence intervals, residual plots, and sum of squares.

2.8.3 Problem Statement

The Margules equations for correlation of binary activity coefficients are

$$\gamma_1 = \exp[x_2^2(2B - A) + 2x_2^3(A - B)] \tag{2-30}$$

$$\gamma_2 = \exp[x_1^2(2A - B) + 2x_1^3(B - A)] \tag{2-31}$$

where x_1 and x_2 are mole fractions of components 1 and 2, respectively, and γ_1 and γ_2 are activity coefficients. Parameters A and B are constant for a particular binary mixture.

Equations (2-30) and (2-31) can be combined to give the excess Gibbs energy expression:

$$g = G_E/RT = x_1\ln\gamma_1 + x_2\ln\gamma_2 = x_1 x_2(Ax_2 + Bx_1) \tag{2-32}$$

Activity coefficients at various mole fractions are available for the benzene and n-heptane binary system in Table 2–7, from which g in Equation (2-32) can

Table 2–7 Activity Coefficients for the System Benzene(1) and n-Heptane(2)[a]

No.	x_1	γ_1	γ_2
1	0.0464	1.2968	0.9985
2	0.0861	1.2798	0.9998
3	0.2004	1.2358	1.0068
4	0.2792	1.1988	1.0159
5	0.3842	1.1598	1.0359
6	0.4857	1.1196	1.0676
7	0.5824	1.0838	1.1096
8	0.6904	1.0538	1.1664

Table 2–7 Activity Coefficients for the System Benzene(1) and *n*-Heptane(2)[a]

No.	x_1	γ_1	γ_2
9	0.7842	1.0311	1.2401
10	0.8972	1.0078	1.4038

[a]Brown, *Australian J. Sci. Res., A5*, 530 (1952).

be calculated. A multiple linear regression without the free parameter can be used to estimate the parameter values of A and B. Another method is to sum Equations (2-30) and (2-31) and use nonlinear regression on this sum to determine A and B.

(a)   Use multiple linear regression on Equation (2-32) with the data of Table 2–7 to determine A and B in the Margules equations for the benzene and *n*-heptane binary system.

(b)   Estimate A and B by employing nonlinear regression on a single equation that is the sum of Equations (2-30) and (2-31).

(c)   Compare the results of the regressions in (a) and (b) using parameter confidence intervals, residual plots, and sums of squares of errors (least-squares summations calculated with both activity coefficients).

2.8.4   Solution

(a)   Linear Regression of Excess Gibbs Energy Equation   The data can be entered into the POLYMATH *Polynomial, Multiple Linear and Nonlinear Regression Program* with column headings designated as `x1`, `gamma1` and `gamma2`. A column to calculate the mole fraction `x2` must be defined through the transformation function `x2 = 1 − x1`. Next the Gibbs energy variable can be calculated by a column defined as `g = x1*ln(gamma1) + x2*ln(gamma2)`. Equation (2-32) can be rearranged to a linear form as

$$g = Ax_1x_2^2 + Bx_1^2x_2 = a_1X_1 + a_2X_2 \qquad \text{(2-33)}$$

Thus the final two transformation columns needed for multiple linear regression can be defined as `X1 = x1*x2^2` and `X2 = x1^2*x2`.

All that remains is to request the "Linear regression without the free parameter" solution option in the POLYMATH program. Note that this regression option sets the a_0 parameter in the multiple linear relationship to zero, as desired in Equation (2-33). The multiple linear regression with $X1$ as the first independent column name, $X2$ as the second independent column name, and g as the dependent column name yields the results given in Figure 2–16. The confidence intervals on the parameters A and B are given in Table 2–8, and the residual plot is reproduced in Figure 2–17.

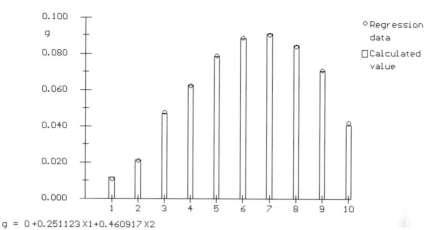

$g = 0 + 0.251123\,X1 + 0.460917\,X2$

Figure 2–16 Multiple Linear Regression for Parameters of Margules Equations

Table 2–8 Multiple Linear Regression Results for Margules Equation Parameters

Parameter	Value	95% Confidence Interval	Lower Limit	Upper Limit
1 or A	0.251123	0.00827717	0.242846	0.2594
2 or B	0.460917	0.00848425	0.452432	0.469401

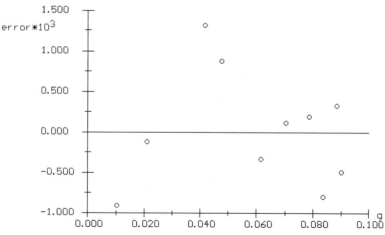

Figure 2–17 Residual Plot for Multiple Linear Regression for Margules Equations

The POLYMATH problem solution file for part (a) is found in the *Polynomial, Multiple Linear and Nonlinear Regression Program Library* located in directory CHAP2 with file named P2-08A.POL.

(b) Nonlinear Regression for Sum of Equations (2-30) and (2-31)   Let's create a column for the summation equation, $(\gamma_1 + \gamma_2)$, and call it gsum, as it will provide the function values during the nonlinear regression. The function for regression is given as follows in mathematical form and equivalent POLYMATH coding.

$$gsum = \exp[x_2^2(2B - A) + 2x_2^3(A - B)] + \exp[x_1^2(2A - B) + 2x_1^3(B - A)]$$

$$gsum = \exp(x2^2*(2*B-A)+2*x2^3*(A-B))+\exp(x1^2*(2*A-B)+2*x1^3*(B-A))$$

(2-34)

Note that this nonlinear objective function has approximately equal contributions from both γ_1 and γ_2 since these activity coefficients are close to unity throughout the data set, and thus the nonlinear regression will weigh both activity expressions approximately equally.

A nonlinear regression on the expression for gsum with the initial parameter estimates for A and B from Table 2–8 converges to the results shown in Figure 2–18. The residual plot is reproduced in Figure 2–19, and The confidence intervals are given in Table 2–9.

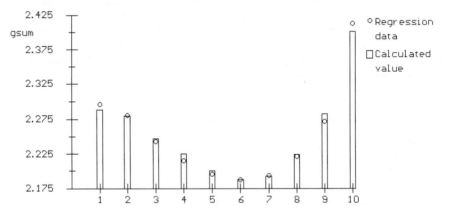

Model: gsum=exp(x2^2*(2*B-A)+2*x2^3*(A-B))+exp(x1^2*(2*A-B)+2*x1^3*(B-A))
B = 0.451573
A = 0.260765
4 positive residuals, 6 negative residuals.   Sum of squares = 0.000418584

Figure 2–18 Nonlinear Regression for Parameters of the Margules Equations

Table 2–9 Multiple Linear Regression Results for Margules Equation Parameters

Parameter	Value	95% Confidence Interval	Lower Limit	Upper Limit
B	0.451573	0.0156435	0.435929	0.467216
A	0.260765	0.0120537	0.248712	0.272819

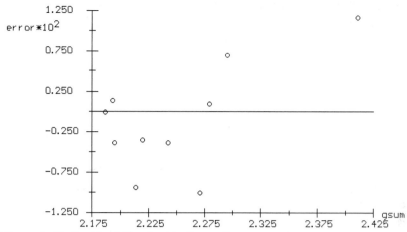

Figure 2–19 Residual Plot for Nonlinear Regression for Margules Equations

 The POLYMATH problem solution file for part (b) is found in the *Polynomial, Multiple Linear and Nonlinear Regression Library* located in directory CHAP2 with file named P2-08B.POL.

(c) Compare the Results of the Regressions in (a) and (b)   The basic results from the two regressions are quite similar as both correlations reproduce the input data fairly well, have narrow confidence intervals on the various parameters, and show fairly random distribution on the residual plots. A close examination, however, indicates that the multiple linear treatment has both a slightly better fit and lower confidence intervals.

Since the two regression expressions are quite different, the variance or sum of squares as calculated cannot be directly compared. A final comparison will involve an additional evaluation of the *sum of squares* of the errors for both activity coefficients as determined from the same form given by

$$SS = \sum_{i=1}^{N} [(\gamma_{1i(obs)} - \gamma_{1i(calc)})^2 + (\gamma_{2i(obs)} - \gamma_{2i(calc)})^2] \qquad \textbf{(2-35)}$$

where (obs) refers to the observed data values and (calc) refers to the calculated values from the regression. An evaluation of Equation (2-35) can be set up in POLYMATH by defining additional columns to calculate both gam1calc and gam2calc using the corresponding values of A and B from each regression. A final column definition providing the terms for each data point can be summed to yield SS (for multiple linear regression) = 1.222×10^{-3}, and SS (for nonlinear regression) = 7.993×10^{-4}

Thus both regressions provide highly accurate correlations for activity coefficients. While the nonlinear regression provides somewhat better values for the activity coefficients based on an evaluation of the sum of squares, the multiple linear regression parameter estimates have somewhat smaller confidence intervals. Residual plots for both show no definite trends.

2.9 MARGULES EQUATIONS FOR BINARY SYSTEMS CONTAINING TRICHLOROETHANE

2.9.1   Concepts Demonstrated

Estimation of parameters in the Margules equations for the correlation of binary activity coefficients.

2.9.2   Numerical Methods Utilized

Linear and nonlinear regression, transformation of data for regression, and calculations with comparisons involving confidence intervals, residual plots and sum of squares.

2.9.3   Problem Statement

The binary activity coefficient data for binary systems containing 1,1,1-trichloroethane are summarized in Tables B–21 through B–23.

(a) Use multiple linear regression of Equation (2-32) to determine A and B in the Margules equations for one of the binary systems containing 1,1,1-trichloroethane.

(b) Estimate A and B by employing nonlinear regression on a single equation that is the sum of Equations (2-30) and (2-31).

(c) Compare the results of the regressions in (a) and (b) using parameter confidence intervals, residual plots, and sums of squares of errors (least-squares summations calculated with both activity coefficients).

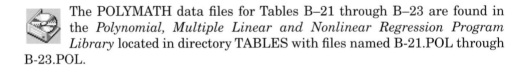 The POLYMATH data files for Tables B–21 through B–23 are found in the *Polynomial, Multiple Linear and Nonlinear Regression Program Library* located in directory TABLES with files named B-21.POL through B-23.POL.

2.10 RATE DATA ANALYSIS FOR A CATALYTIC REFORMING REACTION

2.10.1 Concepts Demonstrated

Evaluation of catalytic reaction rate expressions for experimental data.

2.10.2 Numerical Methods Utilized

Linear and nonlinear regression, transformation of data for regression, and calculations with comparisons involving confidence intervals, residual plots and sum of squares.

2.10.3 Problem Statement

Quanch and Rouleau[5] investigated different models for the reversible catalytic reforming reaction

$$CH_4 + 2H_2O \leftrightarrow CO_2 + 4H_2$$

The experimental results of reaction rate as function of partial pressure of the products at 350 °C are given in Table 2–10.

Table 2–10 Reaction Rate Data for Catalytic Reforming Reaction

Partial Pressure (atm)				Reaction Rate of CO_2
CH_4	H_2O	CO_2	H_2	(g-mol/hr · gm) $\times 10^3$
0.06298	0.23818	0.00420	0.01669	0.13717
0.03748	0.26315	0.00467	0.01686	0.15584
0.05178	0.29557	0.00542	0.02079	0.20028
0.04978	0.23239	0.00177	0.07865	0.05700
0.04809	0.29491	0.00655	0.02464	0.20150
0.03849	0.24171	0.00184	0.06873	0.07887
0.03886	0.26048	0.00381	0.01480	0.14983
0.05230	0.26286	0.05719	0.01635	0.15988
0.05185	0.33529	0.00718	0.02820	0.26194
0.06432	0.24787	0.00509	0.02055	0.14426
0.09609	0.28457	0.00652	0.02627	0.20195

One of several models for this catalytic reaction, in which methane is adsorbed on the catalyst surface, is given by

$$r_{CO_2} = \frac{k_s K_{CH_4}\left(P_{CH_4}P_{H_2O}^2 - \dfrac{P_{CO_2}P_{H_2}^4}{K_P}\right)}{1 + K_{CH_4}P_{CH_4}} \tag{2-36}$$

where the overall equilibrium constant is known from thermodynamic calculation to be $K_P = 5.051 \times 10^{-5}$ atm^2. Thus there are only two parameters which need to be evaluated for this catalytic rate expression.

Another simpler model is simply for a reversible reaction in which there is no component adsorption on the catalyst. This rate expression is given by

$$r_{CH_4} = k_1\left(P_{CH_4}P_{H_2O}^2 - \frac{P_{CO_2}P_{H_2}^4}{K_P}\right) \tag{2-37}$$

where there is only one parameter, k_1.

Note that the rate, r_{CO_2}, is the positive net generation of CO$_2$, and that the rate of CH$_4$ would be the negative of this same rate.

(a) Find the values of parameters k_s and K_{CH_4} using nonlinear regression on the data from Table 2–10.
(b) Determine the value of parameter k_1.
(c) Which of these two rate equations best represents the given data set? Justify your selection.

The POLYMATH problem data file is found in the *Polynomial, Multiple Linear and Nonlinear Regression Program Library* located in directory CHAP2 with file named P2-10.POL.

2.11 REGRESSION OF RATE DATA - CHECKING DEPENDENCY AMONG VARIABLES

2.11.1 Concepts Demonstrated

Correlation of reaction rate data with various reaction rate models.

2.11.2 Numerical Methods Utilized

Multiple linear regression with determination of parameter confidence intervals, residual plots, and identification of linear dependency among regression variables.

2.11.3 Problem Statement

Table 2–11 presents rate data for the reaction $A \leftrightarrow R$, as reported by Bacon and Downie[1]. They suggested fitting the rate data with two reaction rate models. An irreversible model has the form of a first-order reaction

$$r_R = k_0 C_A \tag{2-38}$$

and a reversible model has the form of reversible first-order reactions

$$r_R = k_1 C_A - k_2 C_R \tag{2-39}$$

where r_R is the rate of generation of component R (gm-mol/dm$^3 \cdot$ s); C_A and C_R are the respective concentrations of components A and R (gm-mol/dm^3); and k_0, k_1, and k_2 are reaction rate coefficients (s^{-1}).

(a) Calculate the parameters of both reaction rate expressions using the data in Table 2–11.
(b) Compare the two models and determine which one better correlates the rate data.
(c) Determine if the two variables, C_A and C_R, are correlated.
(d) Discuss the practical significance of any correlation among the regression variables.

Table 2–11 Reaction Rate Data from Bacon and Downie[1] (With permission.)

Run No.	r_R	C_A	C_R
	gm-mol/dm$^3 \cdot$ s $\times 10^8$	gm-mol/dm$^3 \times 10^4$	gm-mol/dm$^3 \times 10^4$
1	1.25	2.00	7.98
2	2.50	4.00	5.95
3	4.05	6.00	4.00

Table 2–11 Reaction Rate Data from Bacon and Downie[1] (With permission.)

Run No.	r_R $\text{gm-mol/dm}^3 \cdot \text{s} \times 10^8$	C_A $\text{gm-mol/dm}^3 \times 10^4$	C_R $\text{gm-mol/dm}^3 \times 10^4$
4	0.75	1.50	8.49
5	2.80	4.00	5.99
6	3.57	5.50	4.50
7	2.86	4.50	5.47
8	3.44	5.00	4.98
9	2.44	4.00	5.99

2.11.4   Solution

(a) Regression of Rate Expressions   Both rate expressions are in a standard form for linear regression, and the POLYMATH *Polynomial, Multiple Linear and Nonlinear Regression Program* has an option for regression without the free parameter. POLYMATH *hint*: For convenience, the data values can be entered from the table into columns without the power of 10, and subsequent transformation can be used to scale all the data to their proper values.

(b) Comparison of Rate Models   The linear regressions of both expressions are summarized in Table 2–12, where the two-parameter model is shown to

Table 2–12 Multiple Linear Regression Results

Parameter	Value	0.95 Conf. Interval	Variance
k_0	6.551×10^{-5}	2.73×10^{-6}	2.34×10^{-18}
k_1	6.867×10^{-5}	4.51×10^{-6}	1.728×10^{-18}
k_2	2.630×10^{-6}	3.18×10^{-6}	

have a lower variance than the one-parameter model. The residual plots for both equations appear to be randomly distributed. However, the confidence interval on k_2 is large and includes zero and negative values, so the two-parameter model is of questionable value relative to the one-parameter model.

(c) Checking for Correlation of Variables   A simple test for correlation among problem variables is to carry out linear regression to determine if one variable is linearly related to another. In this case, a regression of C_R versus C_A provides the correlation shown in Figure 2–20. This plot clearly shows that the two experimental variables, C_R and C_A, are linearly related.

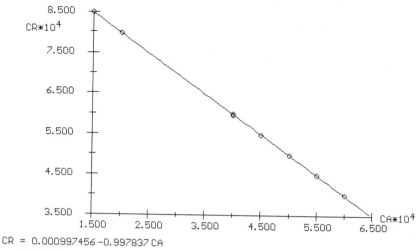

$$CR = 0.000997456 - 0.997837 \, CA$$

Figure 2–20 Regression of C_R versus C_A for the Reaction $A \leftrightarrow R$

(d) Significance of Any Correlation among the Regression Variables The discovery of this linear relationship means that the previous regression of Equation (2-39) is not valid because it assumes independence of all variables. In a new regression, the known relationship between C_R and C_A can be simplified to Equation (2-40) and then utilized in a new regression.

$$C_R = 0.001 - C_A \qquad\qquad (2\text{-}40)$$

Introducing Equation (2-40) into Equation (2-39) gives

$$r_R = k_1 C_A - k_2(0.001 - C_A) = (-0.001)k_2 + (k_1 + k_2)C_A \qquad (2\text{-}41)$$

which is a linear relationship. Denoting $a_0 = -0.001k_2$ and $a_1 = k_1 + k_2$ allows a linear regression to be done, which gives the results in Table 2–13 for the expression

$$r_R = a_0 + a_1 C_A \qquad\qquad (2\text{-}42)$$

The parameter values in the table indicate that a_0 is not significantly different from zero (the confidence interval is larger than the parameter value itself) so it can be removed from the correlation, yielding the irreversible model of Equation (2-38). Table 2–12 indicates that $k_0 = 6.551 \times 10^{-5} \pm 2.74 \times 10^{-6}$. The

Table 2–13 Regression Results for Equation (2-42)

Parameter	Value	0.95 Conf. Interval
a_0	-2.62633×10^{-09}	3.1673×10^{-09}
a_1	7.12978×10^{-05}	7.38046×10^{-06}

residual plot for this irreversible model showing random distribution of the errors is presented in Figure 2–21. All of this seems to support the Bacon and Downie conclusion[1] that the irreversible model is to be preferred.

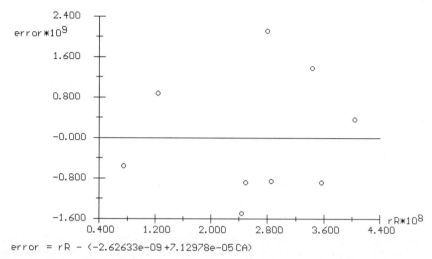

$$error = rR - (-2.62633e-09 + 7.12978e-05\,CA)$$

Figure 2–21 Residual Plot for Reaction Rate Data Represented by Equation (2-38)

Actually, no conclusion can be derived from the data and the regression results because of the linear dependency between C_R and C_A. Because of this dependency, no independent information on k_2 is available, and the calculation of k_2 from the relationship $a_0 = -0.001 k_2$ cannot provide a reasonable estimate for k_2. This is supported by the large confidence limits on a_0 as summarized in Table 2–13.

This problem brings out the most important consideration in the design of experiments, which is always to change the variables independent of one another. Otherwise, some important information will be lost due to linear or other dependencies between variables. In this example, the concentrations always sum to the same total, which will happen for binary gas mixtures at a given temperature and pressure. Certainly the use of a diluent would have been very appropriate during the experimental measurement of the reaction rates in this case.

The POLYMATH problem data file is found in the *Polynomial, Multiple Linear and Nonlinear Regression Program Library* located in directory CHAP2 with file named P2-11.POL.

2.12 REGRESSION OF HETEROGENEOUS CATALYTIC RATE DATA

2.12.1 Concepts Demonstrated

Correlation of heterogeneous catalytic reaction rate data with a rate expression.

2.12.2 Numerical Methods Utilized

Multiple linear and nonlinear regression with linearization of expressions and transformation of variables, and identification of possible dependency among regression variables.

2.12.3 Problem Statement

Table 2–14 presents reaction rate data for the heterogeneous catalytic reaction given by $A \rightarrow B$.

Table 2–14 Reaction Rate Data for Heterogeneous Catalytic Reaction

No.	P_A (atm)	P_B(atm)	$r \times 10^5$	No.	P_A (atm)	P_B(atm)	$r \times 10^5$
1	1	0	5.1	5	0.6	0.4	6
2	0.9	0.1	5.4	6	0.5	0.5	6.15
3	0.8	0.2	5.55	7	0.4	0.6	6.3
4	0.7	0.3	5.85	8	0.3	0.7	6.45

The following equation has been suggested to correlate the data:

$$r = \frac{k_1 P_A}{(1 + K_A P_A + K_B P_B)^2} \tag{2-43}$$

where k_1, K_A, and K_B are coefficients to be determined by regression. Note that Equation (2-43) can be linearized by rearranging, inverting, and taking the square root.

$$\left(\frac{P_A}{r}\right)^{1/2} = \frac{1}{\sqrt{k_1}} + \frac{K_A}{\sqrt{k_1}} P_A + \frac{K_B}{\sqrt{k_1}} P_B \tag{2-44}$$

(a) Determine how many parameters of Equations (2-43) and (2-44) should be estimated by regressing the data in Table 2–14.
(b) Calculate the parameters using linear and nonlinear regression and compare the results obtained by these two methods.

 The POLYMATH problem data file is found in the *Polynomial, Multiple Linear and Nonlinear Regression Program Library* located in directory CHAP2 with file named P2-12.POL.

2.13 VARIATION OF REACTION RATE CONSTANT WITH TEMPERATURE

2.13.1 Concepts Demonstrated

Correlation of the change in reaction rate constant with temperature using the Arrhenius equation.

2.13.2 Numerical Methods Utilized

Multiple linear and nonlinear regression with comparison of regression results using variances, confidence intervals, and residual plots.

2.13.3 Problem Statement

The catalytic hydrogenation of ethylene over copper magnesia catalyst has been studied in a continuous flow tubular reactor by Wynkoop and Wilhelm,[9] whose experiments were carried out at various temperatures. Some of the data are tabulated in Table B–24 as the reaction rate constant k with units of (g-mol/cm$^3 \cdot$ s$\cdot$ atm) versus the temperature T in (°C).

The change of k as a function of temperature can be expressed by the Arrhenius equation

$$k = A \exp[-E/(RT)] \qquad\qquad \textbf{(2-45)}$$

where typically T is the absolute temperature, R is the gas constant (1.987 cal/g-mol$\cdot$ K), E is the activation energy (typically with units of cal/g-mol), and A is the frequency factor with units of the rate constant.

A convenient alternative form of the Arrhenius expression for the rate constant is

$$k = k_0 \exp[E/R(1/T_0 - 1/T)] \qquad\qquad \textbf{(2-46)}$$

where T_0 is some arbitrary absolute temperature where $k = k_0$. Note that $k_0 = A \exp(-E/RT_0)$.

(a) Use both linear and nonlinear regression to find the Arrhenius parameters in Equations (2-45) and (2-46). Set $T_0 = 298$ K in Equation (2-46).
(b) Which equation and regression gives the most accurate correlation of the data? Explain your choice.
(c) Compare your most accurate correlation results with the values reported by Wynkoop and Wilhelm[9] as $A = 5960$ g mol/cm$^3 \cdot$ s$\cdot$ atm and $E = 13320$ cal/g-mol.

The POLYMATH data file for Table B–24 is found in the *Polynomial, Multiple Linear and Nonlinear Regression Program Library* located in directory TABLES with file named B-24.POL.

2.14   CALCULATION OF ANTOINE EQUATION PARAMETERS USING LINEAR REGRESSION

2.14.1   Concepts Demonstrated

Direct use of the Antoine equation to correlate vapor pressure versus temperature data.

2.14.2   Numerical Methods Utilized

Multiple linear regression with determination of the overall variance and confidence intervals of individual parameters.

2.14.3   Problem Statement

Multiple Linear Regression   Multiple linear regression can be defined as fitting a linear function of the form given in Equation (2-47)

$$y_{(calc)} = a_0 + a_1 x_1 + a_2 x_2 + \dots a_n x_n \qquad (2\text{-}47)$$

to N observed data points, where $x_1, x_2 \dots x_n$ are n independent variables, $a_0, a_1 \dots a_n$ are $n + 1$ parameters, $y_{(calc)}$ is the estimated value of the dependent variable, and $y_{(obs)}$ is the observed value of the dependent variable. The parameters of Equation (2-47) can be calculated by solving the following system of linear equations:

$$\mathbf{X^T X A} = \mathbf{X^T Y} \qquad (2\text{-}48)$$

where $\mathbf{X}$ is the matrix of the observed value of the independent variables, $\mathbf{A}$ is the vector of the parameters, and $\mathbf{Y}$ is the vector of observed values of the dependent variable. Thus

$$\mathbf{X} = \begin{bmatrix} 1 & x_{1,1} & x_{2,1} & \cdots & x_{n,1} \\ 1 & x_{1,2} & x_{2,2} & \cdots & x_{n,2} \\ \cdot & \cdot & \cdot & \cdot & \cdot \\ \cdot & \cdot & \cdot & \cdot & \cdot \\ 1 & x_{1,N} & x_{2,N} & \cdots & x_{n,N} \end{bmatrix} \quad \mathbf{A} = \begin{bmatrix} a_0 \\ a_1 \\ \cdot \\ \cdot \\ a_n \end{bmatrix} \quad \text{and} \quad \mathbf{Y} = \begin{bmatrix} y_{1(obs)} \\ y_{2(obs)} \\ \cdot \\ \cdot \\ y_{N(obs)} \end{bmatrix} \qquad (2\text{-}49)$$

Note that the first index in the elements of the matrix $\mathbf{X}$ is the variable number and the second index is the observed data point number. The total number of observed data points is N.

The variance σ^2 can be calculated from Equation (2-4), where $y_{i(obs)}$ is the observed and $y_{i(calc)}$ is the estimated value of the dependent variable and v is the degrees of freedom given by $[N - (n + 1)]$.

The exact parameter values of Equation (2-47) denoted by $\beta_0, \beta_1 \dots \beta_n$ should be located inside the interval

$$a_{i-1} - \sqrt{\alpha_{ii}}\sigma t_v \leq \beta_{i-1} < a_{i-1} + \sqrt{\alpha_{ii}}\sigma t_v \tag{2-50}$$

where α_{ii} are the diagonal elements of the $(\mathbf{X}^T\mathbf{X})^{-1}$ matrix. t_v is the statistical t distribution value corresponding to the degrees of freedom given by v at the desired percent confidence level and σ is the standard deviation (square root of the variance). The confidence interval is given by the term $\sqrt{\alpha_{ii}}\sigma t_v$.

The error, which is the difference between the observed and calculated values of the dependent variable, can be calculated from Equation (2-5).

Calculate the Antoine equation parameters of Equation (2-1) and the various statistical indicators for the propane vapor pressure data of Table B–5. Report the parameters for the vapor pressure in psia and the temperature in °F. (This problem is similar to Problem 2.1, but the parameters have different units.) The fundamental calculations for linear regression are to be carried out during the solution. The following sequence is to be used:

(a) Transform the data so that the Antoine equation parameters can be calculated using multiple linear regression.
(b) Find the matrices $\mathbf{X}^T\mathbf{X}$ and $\mathbf{X}^T\mathbf{Y}$.
(c) Solve system of equations to obtain the vector $\mathbf{A}$.
(d) Calculate the variance, the diagonal elements of $(\mathbf{X}^T\mathbf{Y})^{-1}$ and the 95% confidence intervals of the parameters (use the t distribution values provided in Table B–6).
(e) Prepare a residual plot (plot of ε_i versus $y_{i(\text{obs})}$).
(f) Assess the precision of the data and the appropriateness of the Antoine equation for correlation of the data.

2.14.4 Solution

Most of the steps of the solution can be carried out using the POLYMATH *Polynomial, Multiple Linear and Nonlinear Regression Program*. To start the solution, the data from Table B–5 must be entered into the program.

(a) The Antoine equation can be linearized as shown in Problem 1.7. An alternative linear form of the equation is

$$T \log P_v = (AC + B) + AT - C \log P_v \tag{2-51}$$

where log represents the logarithm to the base of 10. This form of the equation will be used in this example. The original data should be entered into POLYMATH and transformed into the variables (columns) shown in Table 2–15 as specified by $y = T \log P_V$, $x1 = T$ and $x2 = \log P_V$.

Table 2–15 Transformed Variables for the Antoine
Equation Regression

y	x1	x2
−60.7227	−70	0.867467
−59.26	−60	0.987666
−55.0185	−50	1.10037
−48.3806	−40	1.20952
−39.2249	−30	1.3075
−28.0967	−20	1.40483
−14.9693	−10	1.49693
0	0	1.58206
16.6276	10	1.66276
34.8859	20	1.74429
54.6454	30	1.82151
75.6838	40	1.89209
98.1421	50	1.96284
121.787	60	2.02979
146.54	70	2.09342
172.378	80	2.15473
199.336	90	2.21484
227.184	100	2.27184
256.122	110	2.32838
285.625	120	2.38021

(b) For the case of multiple linear regression with two independent variables $x1$ and $x2$ and one dependent variable y, the matrix $\mathbf{X}^T\mathbf{X}$ and the vector $\mathbf{X}^T\mathbf{Y}$ can be written

$$\mathbf{X}^T\mathbf{X} = \begin{bmatrix} N & \sum x_{1,i} & \sum x_{2,i} \\ \sum x_{1,i} & \sum x_{1,i}^2 & \sum x_{1,i} x_{2,i} \\ \sum x_{2,i} & \sum x_{1,i} x_{2,i} & \sum x_{2,i}^2 \end{bmatrix} \qquad \mathbf{X}^T\mathbf{Y} = \begin{bmatrix} \sum y_i \\ \sum x_{1,i} y_{i(\text{obs})} \\ \sum x_{2,i} y_{i(\text{obs})} \end{bmatrix} \qquad (2\text{-}52)$$

The sums can be calculated by summing the numbers in the respective columns. Five new columns should be defined in order to obtain the sums of $x_{1,i}$, $x_{1,i}^2$, $x_{2,i}$, $x_{2,i}^2$, $x_{1,i}y_{i(\text{obs})}$ and $x_{2,i}y_{i(\text{obs})}$. The terms, as they have been entered into the POLYMATH *Linear Equation Solver*, are as follows:

$$\begin{bmatrix} 20 & 500 & 34.5131 \\ 500 & 79000 & 1383.28 \\ 34.5131 & 1383.28 & 63.6854 \end{bmatrix} \begin{bmatrix} a_0 \\ a_1 \\ a_2 \end{bmatrix} = \begin{bmatrix} 1383.28 \\ 159281 \\ 3339.67 \end{bmatrix} \tag{2-53}$$

Note that in order to obtain accurate results, the numbers should be entered with at least six significant decimal digits.

(c) The solution to the general Equation (2-48) as given by Equation (2-53) is obtained using POLYMATH *Linear Equation Solver Program*. The results are

$$a_0 = 677.892 \qquad a_1 = 5.22878 \qquad a_2 = -428.502 \tag{2-54}$$

The inverse $\mathbf{X^T X}$ matrix can be calculated using the *Linear Equation Solver Program* with the same $\mathbf{X^T X}$ matrix but changing the vector of constants to $(1, 0, 0)$, $(0, 1, 0)$, and $(0, 0, 1)$ in turn. Thus one diagonal element of the inverse matrix is obtained for each solution. The diagonal elements obtained are $\alpha_{11} = 43.0499$, $\alpha_{22} = 0.00113993$, and $\alpha_{33} = 18.3651$.

(d) The variance, σ^2, can be calculated by defining a new column given by

$$\text{var} = (y - (677.892 + 5.22878 * x1 - 428.502 * x2))^2 \tag{2-55}$$

Summation of this column and division of the sum by $v = 20 - 3 = 17$ yields $\sigma^2 = 0.349557$. The respective t distribution value, from Table B–6, is $t = 2.1098$ (17 degrees of freedom, 95% confidence interval). Thus the respective confidence intervals as given by $\sqrt{\alpha_{ii}}\sigma t$ in Equation (2-50) are

for $a_0 = (43.0499 * 0.349557)^{1/2} 2.1098 = 8.184$;

for $a_1 = (0.00113993 * 0.349557)^{1/2} 2.1098 = 0.04211$

and for $a_2 = (18.3651 * 0.349557)^{1/2} 2.1098 = 5.3456$

(e) Each ε_i can be calculated by using Equation (2-55) without the power of 2. A residual plot can be obtained during the linear regression of ε versus y, as shown in Figure 2–22.

(f) The residual plot shows that the errors are larger for small or negative values of y that correspond to low vapor pressure. To obtain a more accurate cor-

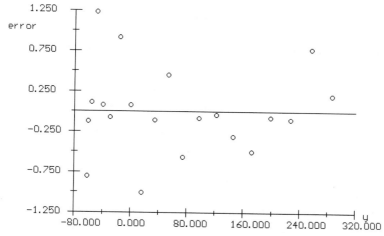

Figure 2–22 Residual Plot for Vapor Pressure Data Represented by Linearized
Antoine Equation

relation, more precise measurements should be carried out in the low-pressure
(low-temperature) range.

The even distribution of the errors around zero and the narrow confidence
intervals on the parameter indicate that the Antoine equation adequately correlates the vapor pressure data for propane in the region where measurements
were made.

The POLYMATH problem data file is found in the *Polynomial, Multiple
Linear and Nonlinear Regression Program Library* located in directory
CHAP2 with file named P2-14.POL.

REFERENCES

1. Bacon, D. W., and Downie, J. *Evaluation of Rate Data–III*, in Crynes, B. L., and Fogler, H. S., eds., *AICHEMI Modular Instruction: Series E, Kinetics,* Vol 2. New York: AICHE, 1981, pp. 65–74.

2. Dow, W. M., and Jacob, M. *Chem. Eng. Progr.*, *47*, 637 (1951).

3. Geankoplis, C. J. *Transport Processes and Unit Operations*, 3rd ed. Englewood Cliffs, NJ: Prentice Hall, 1993.

4. Perry, R. H., Green, D. W., and Malorey, J. D., eds. *Perry's Chemical Engineers Handbook*, 6th ed, New York: McGraw-Hill, 1984.

5. Quanch, Q. P., and Rouleau, J. *Appl. Chem. Biotechnol.*, *25*, 445 (1975).

6. Reid, R. C., Prausnitz, J. M., and Poling, B. F. *The Properties of Gases and Liquids*, 4th ed. New York: McGraw-Hill, 1987.

7. Sieder, E. N., and Tate, G. E. *Ind. and Eng. Chem.*, *28*, 1429 (1936).

8. Williams, R. B., and Katz, D. L. *Trans. ASME*, *74*, 1307–1320 (1952).

9. Wynkoop, R., and Wilhelm, R. H. *Chem. Eng. Progr.*, *46*, 300 (1950).

Advanced Techniques in Problem Solving

3.1 DEMONSTRATION OF ITERATIVE METHODS FOR SOLVING A NONLINEAR EQUATION

3.1.1 Concepts Demonstrated

Calculation of terminal velocity of solid particles falling in liquid or gas under gravity and additional forces.

3.1.2 Numerical Methods Utilized

Step by step use of the successive substitution method for solving one nonlinear equation and comparison with solution by graphical approximation.

3.1.3 Problem Statement

Terminal velocity of a solid spherical particle falling in fluid media can be expressed by the equation

$$v_t = \sqrt{\frac{4g(\rho_p - \rho)D_p}{3C_D\rho}} \qquad (3\text{-}1)$$

where v_t is the terminal velocity (m/s), g is gravitational acceleration (m/s^2), ρ_p is the particle's density (kg/m^3), D_p is the diameter of a spherical particle (m), and C_D is a dimensionless drag coefficient.

The drag coefficient varies with the Reynolds number (Re) as (see Perry[10])

$$C_D = \frac{24}{Re} \qquad \text{for} \qquad Re < 0.1 \qquad (3\text{-}2)$$

$$C_D = \frac{24}{Re}(1 + 0.14Re^{0.7}) \qquad \text{for} \qquad 0.1 \le Re \le 1000 \qquad (3\text{-}3)$$

where $Re = v_t D_p \rho / \mu$, ρ is the density of the fluid (kg/m³) and μ is the fluid viscosity (Pa·s).

Note that Equation (3-1) is an implicit equation for v_t (it includes v_t in both sides) because the C_D term that appears on the right-hand side of the equation is a function of Re and Re is a function of v_t. Thus Equation (3-1) can be rewritten as

$$v_t = F(v_t) \tag{3-4}$$

where $F(v_t) = \sqrt{\dfrac{4g(\rho_p - \rho)D_p}{3C_D \rho}}$.

Consider uniform spherical iron pellets with $D_p = 0.5$ mm, $\rho_p = 7860$ kg/m³ that are falling in air ($\rho = 1.23$ kg/m³, $\mu = 1.79 \times 10^{-5}$ Pa·s) at terminal velocity.

> (a) Plot $F(v_t)$ and v_t versus v_t in the range $0.01 \leq v_t \leq 10$ m/s and locate the approximate root of Equation (3-1).
> (b) Find the root of this equation using the successive substitution method, starting from an initial estimate of $v_{t,0} = 2.0$ with an error tolerance given by $\varepsilon_d = 10^{-3}$.

3.1.4   Solution

Simple Graphical Solution

A very simple way to solve Equation (3-4) is to construct a graph of $F(v_t)$ and v_t versus v_t, and the solution will be where the $F(v_t)$ and v_t functions cross. This is an excellent way to estimate the solution for this problem when numerical methods with higher accuracy are needed for solution.

(a) In order to plot $F(v_t)$ versus v_t in the indicated range, v_t should be continuously changed from the initial value of 0.01 to the final value of 10. This can be accomplished by setting the independent variable, t, in an ordinary differential equation (ODE) solver equal to v_t and integrating the differential equation

$$\frac{d(v_t)}{dt} = 1.0$$

from $t_0 = 0.01$; $v_{t0} = 0.01$ up to $t_f = 10.0$. During the integration, $F(v_t)$ can be calculated and subsequently plotted. Note that this is not the typical use of the POLYMATH *Simultaneous Differential Equation Solver*, but it is helpful here because POLYMATH has no separate plotting capability.

It can be expected that during integration the value of Re will pass through the two regions indicated in Equations (3-2) and (3-3). In order to calculate the appropriate C_D term, an if ... then ... else ... " statement supported by POLYMATH can be used within the program to control the logic for C_D.

```
CD=if (Re<0.1) then (24/Re) else (24*(1+0.14*Re^0.7)/Re)
```

The complete set of equations as entered into the POLYMATH *Simultaneous Differential Equation Solver* is as follows:

```
Equations:
d(vt)/d(t)=1.0
g=9.80665
rhop=7860
rho=1.23
Dp=0.5e-3
vis=1.79e-5
Re=vt*Dp*rho/vis
CD=if (Re<0.1) then (24/Re) else (24*(1+0.14*Re^0.7)/Re)
Fvt=sqrt(4*g*(rhop-rho)*Dp/(3*CD*rho))
Initial Conditions:
t(0)=0.01
vt(0)=0.01
Final Value:
t(f)=10
```

Figure 3–1 shows the plot of $F(v_t)$ and v_t versus v_t. The root of the equation

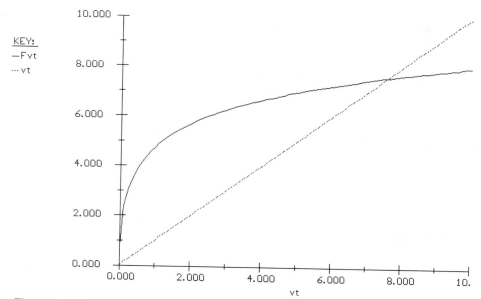

Figure 3–1 Plot of $F(v_t)$ and v_t versus v_t

is where the straight line of v_t crosses the $F(v_t)$ curve at about 7.5 m/s. A closer examination of the tabular results in POLYMATH indicates the root to be approximately 7.57 m/s.

The POLYMATH problem solution file for part (a) is found in the *Simultaneous Differential Equation Solver Library* located in directory CHAP3 with files named P3-01A.POL.

Successive Substitution Method

The successive substitution method for solving Equation (3-4) first requires an initial estimate for the solution to be provided, which in this case will be designated by $v_t = v_{t0}$. This estimate is then introduced into the right-hand side of Equation (3-4) and a new estimate, v_{t1}, is calculated from the equation. This procedure is continued until the difference between two consecutive values of v_t becomes smaller than a desired error tolerance, ε_d. Generalizing

$$v_{t,k} = F(v_{t,k-1}); \quad k = 1, 2, 3 \ldots \tag{3-5}$$

where k is the iteration number. During the solution the error is estimated from the absolute value of the difference between two consecutive values of v_t. Thus

$$\varepsilon_{k-1} = \left| v_{t,k} - v_{t,k-1} \right| \tag{3-6}$$

and the iterations are stopped when $\varepsilon_{k-1} < \varepsilon_d$.

(b) The POLYMATH *Simultaneous Algebraic Equation Solver* can be used to carry out the iterations defined by Equation (3-5) with the error criteria of Equation (3-6). Note that POLYMATH is used here to only solve explicit equations to demonstrate the successive substitution method. Equation (3-1) can, of course, be solved simply by converting it to an implicit equation of the form $f(v_t) = v_t - F(v_t)$. This approach is demonstrated in Problem 1.1.

From iteration to iteration the variables v_t, Re, and C_D will change. An index representing the number of iterations can be added to these variables, and several sets of equations representing those variables can be entered to carry out several iterations at one time. (Note that the copy option from the equation entry menu in POLYMATH allows convenient copying of equations.) The set of equations for two iterations that calculate v_{t1}, v_{t2}, ε_0 and ε_1 is given by

```
Equations:
g=9.80665
rhop=7860
rho=1.23
Dp=0.5e-3
vis=1.79e-5
vt0=2
Re0=vt0*Dp*rho/vis
CD0=if (Re0<0.1) then (24/Re0) else (24*(1+0.14*Re0^0.7)/Re0)
vt1=sqrt(4*g*(rhop-rho)*Dp/(3*CD0*rho))
err0=abs(vt1-vt0)
Re1=vt1*Dp*rho/vis
CD1=if (Re1<0.1) then (24/Re1) else (24*(1+0.14*Re1^0.7)/Re1)
vt2=sqrt(4*g*(rhop-rho)*Dp/(3*CD1*rho))
err1=abs(vt2-vt1)
```

Additional iterations can be carried out by either replacing the value of v_{t0} with the value obtained for v_{t2} after solving the preceding equation set, or by adding expressions for Re2, CD2, vt3, etc. to the set. Table 3–1 summarizes the

Table 3–1 Successive Substitution Iterations for Free-Falling Velocity

Iteration Number	v_t	Re	C_D	ε
0	2	68.72	1.294	3.682
1	5.682	195.2	0.8135	1.484
2	7.166	246.2	0.7416	0.3393
3	7.505	257.9	0.7283	0.06813
4	7.573	260.2	0.7257	0.01334
5	7.587	260.7	0.7252	0.002596
6	7.589	260.7	0.7251	0.00050
7	7.590	260.8	0.7251	

results obtained (numbers rounded to four digits). The results indicate that after six iterations the estimated error becomes smaller than the desired error tolerance of 10^{-3}, and the terminal velocity is 7.590 m/s.

 The POLYMATH problem solution file for part (b) is found in the *Simultaneous Algebraic Equation Solver Library* located in directory CHAP3 with file named P3-01B.POL.

3.2 SOLUTION OF STIFF ORDINARY DIFFERENTIAL EQUATIONS

3.2.1   Concepts Demonstrated

Simulation of chemical or biological reactions in a batch process that can lead to very high reaction rates with very low reactant concentrations.

3.2.2   Numerical Methods Utilized

Solution of systems of ordinary differential equations that become stiff during the course of the integration.

3.2.3   Problem Statement

A biological process involves the growth of biomass from substrate as studied by Garritsen.[4] The material balances on this batch process yield

$$\frac{dB}{dt} = \frac{kBS}{(K+S)} \tag{3-7}$$

$$\frac{dS}{dt} = \frac{0.75kBS}{(K+S)} \tag{3-8}$$

where B and S are the respective biomass and substrate concentrations. The reaction kinetics are such that $k = 0.3$ and $K = 10^{-6}$ in consistent units.

(a)   Solve this set of differential equations starting at $t_0 = 0$, where $S = 5.0$ and $B = 0.05$ to a final time given by $t_f = 20$. Assume consistent units.
(b)   Plot S and B with time for the conditions of part (a).

3.2.4   Solution

This set of equations can be introduced into the POLYMATH *Simultaneous Differential Equation Solver* as presented previously. If the equations are entered correctly, the integration will proceed in the usual way up to $t = 16.34$. At this point the arrow on the screen that indicates the progress toward t_f stops moving ahead and the following message appears:

```
Solution process halted because it was not going anywhere.

Do you want to try again using STIFF algorithm? (Y/N)
```

If you choose not to try again using the stiff algorithm, the table of partial results will be displayed up to the point at which the integration was stopped. At this point both S and B have begun to change extremely rapidly with t. Also, the substrate amount has attained a very small, negative value that is not physically possible. These are all indications that *the equation set has become "stiff" at this particular point.* Detailed discussions of stiff systems and the methods that can be used to solve them are outside the scope of this book. However, most textbooks of numerical analysis (such as Press et al.[11]) contain a detailed description of this subject.

In order to determine mathematically whether a particular system is stiff, the matrix of partial derivatives of the differential equations with respect to each of the dependent variables must be calculated. If among the eigenvalues of this matrix there is at least one that is negative and has a large absolute value, then the system is referred to as stiff.

For this problem, the matrix of partial derivatives may be evaluated by first rewriting Equations (3-7) and (3-8) using simplified notation and introducing the known kinetic constants.

$$f_1 = \frac{dB}{dt} = \frac{0.3BS}{10^{-6} + S} \tag{3-9}$$

$$f_2 = \frac{dS}{dt} = -\frac{0.225BS}{10^{-6} + S} \tag{3-10}$$

Differentiation of these equations with respect to B and S yields

$$\frac{\partial f_1}{\partial B} = \frac{0.3S}{(10^{-6} + S)} \qquad \frac{\partial f_1}{\partial S} = \frac{0.3 \times 10^{-6} B}{(10^{-6} + S)^2} \tag{3-11}$$

$$\frac{\partial f_2}{\partial B} = -\frac{0.225S}{(10^{-6} + S)} \qquad \frac{\partial f_2}{\partial S} = -\frac{0.225 \times 10^{-6} B}{(10^{-6} + S)^2} \tag{3-12}$$

Two of the preceding partial derivatives have small negative values as long as S does not become small. But when S reaches small values, the derivative $\partial f_2/\partial S$ is negative and can reach very large absolute values (if $S = 10^{-6}$ then the order of magnitude of the derivative is 10^6). At this point the system becomes stiff, and the default integration algorithm (fifth-order Runge-Kutta-Fehlberg) in POLYMATH and most general packages cannot solve this problem.

Normally additional stiff integration algorithms are provided, and POLY-MATH allows the user to use an algorithm that solves stiff systems. In this biomass problem, the use of this option results in the biomass and substrate concentration profiles shown in Figure 3–2, where at $t \sim 16.3$ the amount of substrate becomes essentially zero and there is no growth in the amount of the biomass. So for this example problem, the stiff integration algorithm was required in order to achieve the desired results.

In general, it is usually advisable to use the stiff algorithm whenever a default integration algorithm either fails to solve a problem giving error messages with strange results or progresses very slowly toward a solution. If both the default and stiff integration algorithms fail to solve an ODE problem, then

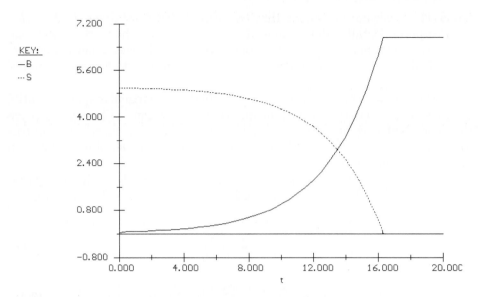

Figure 3–2 Change of Substrate and Biomass Concentration

there are probably mistakes in the model equations, the values of the constants, or the initial values.

3.3 STIFF ORDINARY DIFFERENTIAL EQUATIONS IN CHEMICAL KINETICS

3.3.1   Concepts Demonstrated

Solution of a reaction kinetics model for an unsteady-state batch process and model validation for reaction rate equations.

3.3.2   Numerical Methods Utilized

Solution of a stiff system of simultaneous ordinary differential equations.

3.3.3   Problem Statement

Gear,[6] who has developed well-known methods for solving stiff systems of ODEs, presented the following problem (which he entitled "Chemistry Problem") for testing software to solve stiff ODEs.

$$\frac{dy_1}{dt} = -0.013y_1 - 1000y_1y_3$$

$$\frac{dy_2}{dt} = -2500y_2y_3 \tag{3-13}$$

$$\frac{dy_3}{dt} = -0.013y_1 - 1000y_1y_3 - 2500y_2y_3$$

The initial conditions are $y_1(0) = 1$, $y_2(0) = 1$, and $y_3(0) = 0$. These equations are usually integrated from $t_0 = 0$ up to $t_f = 50$.

Since its introduction, this example has been frequently used and cited in the literature (see, for example, p. 734 in Press et al.[11]).

(a)   Solve the system defined by Equation (3-13) with the given initial conditions.

(b)   Assuming that y_1, y_2, and y_3 represent concentrations of different species, does the solution obtained make sense and seem feasible?

(c)   If the system defined by Equation (3-13) represents reaction rate equations, what typographical error could cause infeasible results? Suggest a reaction sequence that can be represented by the system, correct the typographical error, and resolve the equations in their correct form.

3.4 MULTIPLE STEADY STATES IN A SYSTEM OF ORDINARY DIFFERENTIAL EQUATIONS

3.4.1   Concepts Demonstrated

Dynamic material and energy balances for the unsteady-state model of a fluidized bed reactor leading to possible multiple steady states.

3.4.2   Numerical Methods Utilized

Determination of all solutions of a system of nonlinear algebraic equations, solution of stiff ODE systems, and effect of round-off errors in ill-conditioned algebraic and differential systems.

3.4.3   Problem Statement

Luss and Amundson[9] studied a simplified model for dynamics of a catalytic fluidized bed in which an irreversible gas phase reaction $A \to B$ was assumed to occur. The mass and energy conservation equations along with the kinetic rate constant for this system are given as Equations (3-14), which will be referred to as Set I.

$$\frac{dP}{d\tau} = P_e - P + H_g(P_p - P)$$

$$\frac{dT}{d\tau} = T_e - T + H_T(T_p - T) + H_W(T_W - T)$$

$$\frac{dP_p}{d\tau} = \frac{H_g}{A}[P - P_p(1 + K)] \qquad\qquad \textbf{(3-14)}$$

$$\frac{dT_p}{d\tau} = \frac{H_T}{C}[(T - T_p) + FKP_p]$$

$$K = 0.0006 \exp(20.7 - 15000/T_p)$$

where T = absolute temperature of reactant in fluid (°R), P = partial pressure of the reactant in the fluid (atm), T_p = temperature of the reactant at the catalyst surface (°R), P_p = partial pressure of the reactant at the catalyst surface (atm), K = dimensionless reaction rate constant, τ = dimensionless time, and the subscript e indicates entrance conditions. The dimensionless constants were given as H_g = 320, T_e = 600, H_T = 266.67, H_W = 1.6, T_W = 720, F = 8000, A = 0.17142, C = 205.74, and P_e = 0.1.

Aiken and Lapidus[1] subsequently used Set I as a test example for a program they had developed, but they rewrote the equations by introducing the numerical values into the system and rounded some of the coefficients. The result is the set of Equations (3-15), which will be referred to as Set II.

$$\frac{dP}{d\tau} = 0.1 + 320P_p - 321P$$

$$\frac{dT}{d\tau} = 1752 - 269T + 267T_p$$

$$\frac{dP_p}{d\tau} = 1.88\times10^3[P - P_p(1 + K)] \tag{3-15}$$

$$\frac{dT_p}{d\tau} = 1.3(T - T_p) + 1.04\times10^4 KP_p$$

$$K = 0.0006 \exp(20.7 - 15000/T_p)$$

(a) Introduce the values provided by Luss and Amundson into Set I and observe the differences between Set I and Set II.

(b) Find all the steady-state solutions for both Set I and Set II in the range of $500°R \le T \le 1300°R$.

(c) Solve both Set I and Set II using the initial values at $\tau = 0$ given by $P = 0.1$, $T = 600$, $P_p = 0$, and $T_p = 761$. The final value is to be $\tau = 1500$.

(d) Explain the differences in steady-state and dynamic solutions obtained while using the original Set I and the modified Set II.

3.4.4   Solution (Partial)

(a) The steady-state solutions are easily determined by setting the time derivatives in the four differential equations of Sets I and II equal to zero. Each set can then be reformulated to give a single implicit nonlinear algebraic equation, which should be equal to zero, while the rest of the variables can be calculated from explicit expressions. Introducing the numerical values of the constants into Set I and reformulating yields

$$f(T) = 1.296(T - T_p) + 10369KP_p$$

$$T_p = (269.267T - 1752)/266.667$$

$$P_p = 0.1/(1 + 321K) \tag{3-16}$$

$$P = (320P_p + 0.1)/321$$

$$K = 0.0006 \exp(20.7 - 15000/T_p)$$

Converting a system of nonlinear algebraic equations into a form where there is only one implicit equation is the simplest way to find all of the solutions for such a system in a particular interval.

3.5 SINGLE-VARIABLE OPTIMIZATION

3.5.1   Concepts Demonstrated

Heat transfer in a one-dimensional slab which involves both conduction with variable thermal conductivity and radiation to the surroundings at one surface.

3.5.2   Numerical Methods Utilized

Numerical solution of an ordinary differential equation where a variable describing the system must be optimized using the secant method and the method of false position.

3.5.3   Problem Statement

Heat conduction is occurring within a one-dimensional slab of variable thermal conductivity as shown in Figure 3–3. One surface of the slab is maintained at

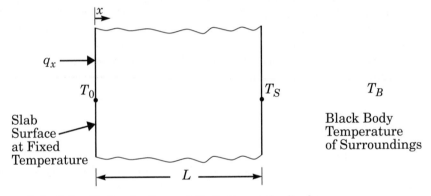

Figure 3–3 Slab with Conduction and Radiation at the Surface

temperature T_0, and the other surface at temperature T_S has radiative heat transfer with the surroundings that act as a black body at temperature T_B. The slab thickness is given by L. There is negligible convection because a vacuum is maintained between the slab and the surroundings. Details on various modes of heat transfer can be found in Geankoplis[5] and Thomas.[14] Chapter 6 also presents the essential equations, boundary conditions, and typical units.

    A steady-state energy balance on a differential element within the slab indicates that the heat flux is constant since there is no generation of heat within the slab. Application of Fourier's law in the x direction therefore gives

$$\frac{dT}{dx} = -\left(\frac{q_x}{A}\right)/k \qquad (3\text{-}17)$$

where T is in K, q_x is the heat transfer in the x direction in W or J/s, A is the cross-sectional area that is normal to the direction of heat conduction in m^2, k is the thermal conductivity of the medium in $W/m \cdot K$, and x is the distance in m.

For this problem, the thermal conductivity of the medium is temperature dependent and given by

$$k = 30(1 + 0.002T) \tag{3-18}$$

The radiation from the slab surface is given by the Stefan-Boltzmann law for a black body with an emissivity of unity and a view factor of unity. The resulting heat flux at the slab surface (or at any position within the slab) is

$$\frac{q_x}{A}\bigg|_{x=L} = \sigma(T_S^4 - T_B^4)\big|_{x=L} \tag{3-19}$$

where σ is the Stefan-Boltzmann constant with a value of 5.676×10^{-8} $W/m^2 \cdot K^4$.

The surface of the slab is maintained at $T_0 = 290$ K and the black body temperature of the surroundings is $T_B = 1273$ K. $L = 0.2$ m.
(a)   Calculate and plot the temperature profile within the slab using the secant method to optimize the constant heat flux within the slab. What is the corresponding value of T_S?
(b)   Repeat part (a) using the method of false position.

3.5.4   Solution

This problem requires the solution of the ODE given by Equation (3-17) during which the thermal conductivity must be calculated by Equation (3-18). The initial condition for T is T_0, which is known. The final condition is given by Equation (3-19).

This problem will be solved by optimizing the value of the heat flux, q_x/A, so that the final condition is satisfied. In this case, an objective function representing the error at the final condition can be expressed by

$$\varepsilon(q_x/A) = \varepsilon(Q_x) = [Q_x - \sigma(T^4 - T_B^4)]\big|_{x=L} \tag{3-20}$$

where the heat flux is designated by Q_x and T is the final value from the numerical integration at $x = L$. Equation (3-20) should approach zero when the correct value of heat flux has been determined.

A number of different techniques can be used to accomplish this one variable optimization.

(a) Secant method The secant method (see Himmelblau[8] or Hanna and Sandall[7]) can be applied by applying the Newton formula

$$Q_{x,new} = Q_x - \varepsilon(Q_x)/\varepsilon'(Q_x) \tag{3-21}$$

with the derivative approximated by

$$\varepsilon'(Q_x) \cong \frac{\varepsilon(Q_x + \delta Q_x) - \varepsilon(Q_x)}{\delta Q_x} \tag{3-22}$$

where δ is 0.0001 to provide a small increment in the current value of Q_x. The equations necessary to determine $Q_{x,\text{new}}$ can be calculated *simultaneously with the numerical ODE solution* for $\varepsilon(Q_x)$, thereby allowing the approximation of $\varepsilon'(Q_x)$ from Equation (3-22) and a *new estimate* for $Q_{x,\text{new}}$ from Equation (3-21) at the conclusion of each iteration.

The POLYMATH equation set for carrying out the first iteration using the secant method with an arbitrary initial estimate of $Q_x = -100000$ is given by

```
Equations:
d(T1)/d(x)=-Qx1/k1
d(T)/d(x)=-Qx/k
Qx=-100000
k=30*(1+0.002*T)
k1=30*(1+0.002*T1)
TB=1273
delta=0.0001
Qx1=(1+delta)*Qx
err=Qx-5.676e-8*(T^4-TB^4)
err1=Qx1-5.676e-8*(T1^4-TB^4)
derr=(err1-err)/(delta*Qx)
QxNEW=Qx-err/derr
Initial Conditions:
x(0)=0
T1(0)=290
T(0)=290
Final Value:
x(f)=0.2
```

The final value for QxNEW at the end of the integration for the first iteration is −133,944 with units of J/m²·s. The negative sign here indicates that the heat flux is in the negative x direction. The second iteration gives a QxNEW value of −133.014, and the third iteration yields −133.013, indicating that a converged solution has been obtained. The final value of T, which is T_S, is 729.17 K. The temperature distribution from the POLYMATH solution is given in Figure 3–4, where the nonlinear profile is a result of the temperature dependency of the thermal conductivity.

The secant method converges very rapidly for this problem, but initial estimates should be fairly accurate. The major disadvantage is the need to carry the equations for calculation of the derivative expression during the solution. Several trial solutions or some simplified calculations can be used to obtain the initial estimate of the heat flux Q_x.

 The POLYMATH problem solution file for part (a) is found in the *Simultaneous Differential Equation Solver Library* located in directory CHAP3 with file named P3-05A.POL.

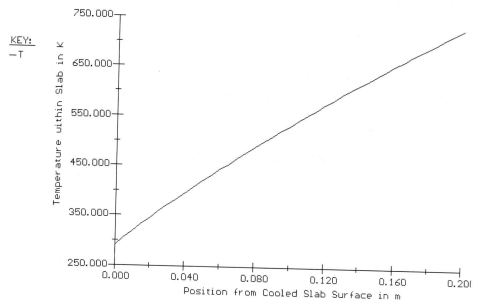

Figure 3–4 Temperature Profile in the Slab

(b) False position method This method for the optimization of a single variable is initiated by trial solutions that determine two points that bracket the solution at which the values of the function to be minimized are of opposite sign. (See Himmelblau[8] or Carnahan et al.[2] for more details.) For this problem, let the negative functional value be given by $FQ_{xN} = \varepsilon(Q_xN)$ and the positive functional value by $FQ_{xP} = \varepsilon(Q_xP)$, where Q_{xN} and Q_{xP} lead to negative and positive values, respectively.

The new estimate of the variable for the next step is given by

$$Q_{x,\text{new}} = Q_{xN} - \frac{(Q_{xN} - Q_{xP})(FQ_{xN})}{(FQ_{xN} - FQ_{xP})} \qquad (3\text{-}23)$$

and the function is determined to be FQ_{xNEW}. The new variable estimate replaces either Q_{xN} or Q_{xP} and the new function replaces either FQ_{xN} or FQ_{xP} depending on the sign of FQ_{xNEW}.

Initial trials determined that $Q_x = -100{,}000$ resulted in $\varepsilon(Q_xP) = 39{,}764$ and that $Q_x = -150{,}000$ resulted in $\varepsilon(Q_xN) = -21{,}355$. These values bracketed the solution and were used to start the iterations utilizing Equation (3-23). The POLYMATH equation set for carrying out the first iteration using the false position method is given by

```
Equations:
d(T)/d(x)=-QxNEW/k
QxN=-150000
k=30*(1+0.002*T)
```

```
QxP=-100000
TB=1273
FQxN=-21355
FQxP=39764
QxNEW=QxN-(QxN-QxP)*(FQxN)/(FQxN-FQxP)
FQxNEW=QxNEW-5.676e-8*(T^4-TB^4)
Initial Conditions:
x(0)=0
T(0)=290
Final Value:
x(f)=0.2
```

The first three iterations of the false position method are summarized in Table 3–2, where the converged values of iteration 3 agree with part (a) and the temperature profile is identical.

Table 3–2 Iterations for False Position Method

Iteration	QxP	FQxP	QxN	FQxN	QxNEW	FQxNEW	T_S
1	−100000	39764	−150000	−21355	−132530	597.849	727.86
2	−132530	597.849	−150000	−21355	−133006	8.7448	729.15
3	−133006	8.7448	−150000	−21355	−133013	−0.1614	729.17

This method has the advantage that derivatives are not needed in the calculations and thus the POLYMATH equation set is simpler for complex problems. The method may not converge typically when the initial starting points are not close to the solution and the function is very nonlinear.

 The POLYMATH problem solution file for part (b) is found in the *Simultaneous Differential Equation Solver Library* located in directory CHAP3 with file named P3-05B.POL.

3.6 SHOOTING METHOD FOR SOLVING TWO-POINT BOUNDARY VALUE PROBLEMS

3.6.1 Concepts Demonstrated

Methods for solving second-order ordinary differential equations with two-point boundary values typically used in transport phenomena and reaction kinetics.

3.6.2 Numerical Methods Utilized

Conversion of a second-order ordinary differential equation into a system of two first-order differential equations, a shooting method for solving ODEs, and use of the secant method to solve two-point boundary value problems.

3.6.3 Problem Statement

The diffusion and simultaneous first-order irreversible chemical reaction in a single phase containing only reactant A and product B results in a second-order ordinary differential equation given by

$$\frac{d^2 C_A}{dz^2} = \frac{k}{D_{AB}} C_A \tag{3-24}$$

where C_A is the concentration of reactant A (kg-mol/m^3), z is the distance variable (m), k is the homogeneous reaction rate constant (s^{-1}), and D_{AB} is the binary diffusion coefficient (m^2/s). A typical geometry for Equation (3-24) is that of a one- dimensional layer that has its surface exposed to a known concentration and allows no diffusion across its bottom surface. Thus the initial and boundary conditions are

$$C_A = C_{A0} \qquad \text{for } z = 0 \tag{3-25}$$

$$\frac{dC_A}{dz} = 0 \qquad \text{for } z = L \tag{3-26}$$

where C_{A0} is the constant concentration at the surface ($z = 0$) and there is no transport across the bottom surface ($z = L$) so the derivative is zero.

This differential equation has an analytical solution given by

$$C_A = C_{A0} \frac{\cosh[L(\sqrt{k/D_{AB}})(1 - z/L)]}{\cosh(L\sqrt{k/D_{AB}})} \tag{3-27}$$

(a) Numerically solve Equation (3-24) with the boundary conditions of (3-25) and (3-26) for the case where $C_{A0} = 0.2$ kg-mol/m^3, $k = 10^{-3}$ s^{-1}, D_{AB} = 1.2 10^{-9} m^2/s, and $L = 10^{-3}$ m. This solution should utilize an ODE solver with a shooting technique and employ the secant method for converging on the boundary condition given by Equation (3-26).

(b) Compare the concentration profiles over the thickness as predicted by the numerical solution of (a) with the analytical solution of Equation (3-27).

(c) Obtain a numerical solution for a second-order reaction that requires the C_A term on the right side of Equation (3-24) to become squared. The second-order rate constant is given by $k = 0.02$ m^3/(kg-mol·s).

3.6.4   Solution (Partial)

Solving Higher Order Ordinary Differential Equations

Most mathematical software packages can solve only systems of first-order ODEs. Fortunately, the solution of an nth-order ODE can be accomplished by expressing the equation as a series of simultaneous first-order differential equations each with a boundary condition. This is the approach that is typically used for the integration of higher-order ODEs.

(a) Equation (3-24) is a second-order ODE, but it can be converted into a system of first-order equations by substituting new variables for the higher order derivatives. In this particular case, a new variable y can be defined that represents the first derivative of C_A with respect to z. Thus Equation (3-24) can be written as

$$\frac{dC_A}{dz} = y$$

$$\frac{dy}{dz} = \frac{k}{D_{AB}}C_A$$

(3-28)

This set of first-order ODEs can be entered into the *POLYMATH Simultaneous Differential Equation Solver* for solution, but initial conditions for both C_A and y are needed.   Since the initial condition of y is not known, an iterative method (also referred to as a shooting method) can be used to find the correct initial value for y that will yield the boundary condition given by Equation (3-26).

Shooting Method—Trial and Error

The shooting method is used to solve a boundary value problem by iterating on the solution of an initial value problem. Known initial values are utilized while unknown initial values are optimized to achieve the corresponding boundary conditions. Either trial-and-error or variable optimization techniques are used to achieve convergence on the boundary conditions.

For this problem, a first trial-and-error value for the initial condition of y (for example, $y_0 = -150$) is used to carry out the integration and calculate the error for the boundary condition designated by ε. Thus the difference between the calculated and desired final value of y at $z = L$ is given by

$$\varepsilon(y_0) = y_{f, \text{calc}} - y_{f, \text{desired}} \qquad (3\text{-}29)$$

Note that for this example, $y_{f, \text{desired}} = 0$ and thus $\varepsilon(y_0) = y_{f, \text{calc}}$ only because this desired boundary condition is zero.

The equations as entered in the POLYMATH *Simultaneous Differential Equation Solver* for an initial trial-and-error solution are

```
Equations:
d(CA)/d(z)=y
d(y)/d(z)=k*CA/DAB
k=0.001
DAB=1.2E-9
err=y
Initial Conditions:
z(0)=0
CA(0)=0.2
y(0)=-150
Final Value:
z(f)=0.001
```

The calculation of err in the POLYMATH equation set that corresponds to Equation (3-29) is only valid at the end of the ODE solution. Repeated reruns of this POLYMATH equation set with different initial conditions for y can be used in a trial-and-error mode to converge upon the desired boundary condition for y_0, where $\varepsilon(y_0)$ or err $\cong 0$. Some results are summarized in Table 3–3 for various val-

Table 3–3 Trial Boundary Conditions for Equation Set (3-28) in Problem Part (a)

y_0 $(z = 0)$	−120	−130	−140	−150
$y_{f,\text{calc}}$ $(z = L)$	17.23	2.764	−11.70	−26.16
$\varepsilon(y_0)$	17.23	2.764	−11.70	−26.16

ues of y_0. The desired initial value for y_0 lies between −130 and −140. This trial-and-error approach can be continued to obtain a more accurate value for y_0, or an optimization technique can be applied.

 The POLYMATH problem solution file is found in the *Simultaneous Differential Equation Solver Library* located in directory CHAP3 with file named P3-06A1.POL.

Secant Method for Boundary Condition Convergence

A very useful method for optimizing the proper initial condition is to consider this determination to be a problem in finding the zero of a function. In the notation of this problem, the variable is y_0 and the function is $\varepsilon(y_0)$. The secant method, an effective method for optimizing a single variable, has been described in Problem 3.5 and will be applied here.

According to this method, an improved estimate for y_0 can be calculated using the equation

$$y_{0,\,new} = y_0 - \varepsilon(y_0)/\varepsilon'(y_0) \tag{3-30}$$

where $\varepsilon'(y_0)$ is the derivative of ε at $y = y_0$. The derivative, $\varepsilon'(y_0)$, can be estimated using a finite difference approximation:

$$\varepsilon'(y_0) \cong \frac{\varepsilon(y_0 + \delta y_0) - \varepsilon(y_0)}{\delta y_0} \tag{3-31}$$

where δy_0 is a small increment in the value of y_0. It is very convenient that $\varepsilon(y_0 + \delta y_0)$ can be calculated *simultaneously with the numerical ODE solution* for $\varepsilon(y_0)$, thereby allowing calculation of $\varepsilon'(y_0)$ from Equation (3-31) and a *new estimate* for y_0 from Equation (3-30).

Using $\delta = 0.0001$ for this example, the POLYMATH equation set for carrying out the first step in the secant method procedure is given by

```
Equations:
d(CA)/d(z)=y
d(y)/d(z)=k*CA/DAB
d(CA1)/d(z)=y1
d(y1)/d(z)=k*CA1/DAB
k=0.001
DAB=1.2E-9
err=y-0
err1=y1-0
y0=-130
L=.001
delta=0.0001
CAanal=0.2*cosh(L*(k/DAB)^.5*(1-z/L))/(cosh(L*(k/DAB)^.5))
derr=(err1-err)/(.0001*y0)
ynew=y0-err/derr
Initial Conditions:
z(0)=0
CA(0)=0.2
y(0)=-130
CA1(0)=0.2
y1(0)=-130.013
Final Value:
z(f)=0.001
```

This set of equations yields the results summarized in Table 3–4, where the new estimate for y_0 is the final value of the POLYMATH variable `ynew` or −131.911. Another iteration of the secant method can be obtained by starting with the new estimate and modifying the initial conditions for `y` and `y1` and the value of `y0` in the POLYMATH equation set. The second iteration indicates that the `err` is approximately 3.e−4 and that `ynew` is unchanged, indicating that convergence has been obtained.

Table 3–4 Partial Results for Selected Variables during First Secant Method Iteration

Variable	Initial Value	Maximum Value	Minimum Value	Final Value
z	0	0.001	0	0.001
y	−130	2.76438	−130	2.76438
CA	0.2	0.2	0.140428	0.140461
err	−130	2.76438	−130	2.76438
y1	−130.013	2.74558	−130.013	2.74558
CA1	0.2	0.2	0.140446	−0.142229
err1	−130.013	2.74558	−130.013	2.74558
derr	1	1.44642	1	1.44642
ynew	−5.22675e−11	−5.22675e−11	−131.911	−131.911

This problem solution emphasizes the value of first obtaining an approximate solution for a split boundary value problem. Application of the secant method procedure from a reasonable starting point will usually converge very efficiently to a solution; however, unreasonable starting points can lead to numerical difficulties and often do not yield a solution.

The POLYMATH problem solution file is found in the *Simultaneous Differential Equation Solver Library* located in directory CHAP3 with file named P3-06A2.POL. This problem is also solved with Excel, Maple, MathCAD, MATLAB, Mathematica, and POLYMATH as problem 7 in the Set of Ten Problems discussed in Appendix F.

3.7 EXPEDITING THE SOLUTION OF SYSTEMS OF NONLINEAR ALGEBRAIC EQUATIONS

3.7.1 Concepts Demonstrated

Complex chemical equilibrium calculations.

3.7.2 Numerical Methods Utilized

Solution of systems of nonlinear algebraic equations, and techniques useful for effective solutions and for examining possible multiple solutions of such systems.

3.7.3 Problem Statement

The following reactions are taking place in a constant volume, gas-phase batch reactor:

$$A + B \leftrightarrow C + D$$
$$B + C \leftrightarrow X + Y$$
$$A + X \leftrightarrow Z$$

A system of algebraic equations describes the equilibrium of the preceding reactions. The nonlinear equilibrium relationships utilize the thermodynamic equilibrium expressions, and the linear relationships have been obtained from the stoichiometry of the reactions.

$$K_{C1} = \frac{C_C C_D}{C_A C_B} \qquad K_{C2} = \frac{C_X C_Y}{C_B C_C} \qquad K_{C3} = \frac{C_Z}{C_A C_X}$$

$$C_A = C_{A0} - C_D - C_Z \qquad C_B = C_{B0} - C_D - C_Y$$

$$C_C = C_D - C_Y \qquad C_Y = C_X + C_Z$$

(3-32)

In this equation set C_A, C_B, C_C, C_D, C_X, C_Y and C_Z are concentrations of the various species at equilibrium resulting from initial concentrations of only C_{A0} and C_{B0}. The equilibrium constants K_{C1}, K_{C2}, and K_{C3} have known values.

Solve this system of equations when $C_{A0} = C_{B0} = 1.5$, $K_{C1} = 1.06$, $K_{C2} = 2.63$, and $K_{C3} = 5$ starting from three sets of initial estimates.
(a) $C_D = C_X = C_Z = 0$
(b) $C_D = C_X = C_Z = 1$
(c) $C_D = C_X = C_Z = 10$

3.7.4 Solution

The equation set (3-32) can be entered into the POLYMATH *Simultaneous Algebraic Equation Solver*, but the nonlinear equilibrium expressions must be written as functions that are equal to zero at the solution. A simple transformation of the equilibrium expressions to the required functional form yields

$$f(C_D) = \frac{C_C C_D}{C_A C_B} - K_{C1}$$

$$f(C_X) = \frac{C_X C_Y}{C_B C_C} - K_{C2} \qquad \text{(3-33)}$$

$$f(C_Z) = \frac{C_Z}{C_A C_X} - K_{C3}$$

The POLYMATH equation set utilizing the transformed nonlinear equations is as follows:

```
Equations:
f(CD)=CC*CD/(CA*CB)-KC1
f(CX)=CX*CY/(CB*CC)-KC2
f(CZ)=CZ/(CA*CX)-KC3
KC1=1.06
CY=CX+CZ
KC2=2.63
KC3=5
CA0=1.5
CB0=1.5
CC=CD-CY
CA=CA0-CD-CZ
CB=CB0-CD-CY
Initial Estimates:
CD(0)=0
CX(0)=0
CZ(0)=0
```

When the preceding equations are used in POLYMATH, the Newton-Raphson algorithm will fail to solve the problem for all sets of initial estimates specified in (a), (b), and (c). Either the error message *"one of the expressions is undefined at the starting point"* [for part (a)] or the error message *"singular matrix was encountered"* [parts (b) and (c)] will be displayed. The failure of POLYMATH and most other programs for solving nonlinear equations is that division by unknowns makes the equations very nonlinear or sometimes undefined. The solution methods that are based on linearization (such as the Newton-Raphson method) may diverge for highly nonlinear systems or cannot continue for undefined functions.

Expediting the Solution of Nonlinear Equations

A simple transformation of the nonlinear function can make many functions much less nonlinear and easier to solve by simply eliminating division by the unknowns. In this case, the equation set (3-33) can be modified to

$$f(C_D) = C_C C_D - K_{C1} C_A C_B$$
$$f(C_X) = C_X C_Y - K_{C2} C_B C_C \qquad\qquad (3\text{-}34)$$
$$f(C_Z) = C_Z - K_{C3} C_A C_X$$

Using the modified nonlinear equations in POLYMATH produces the solutions summarized in Table 3–5 for the three sets of initial conditions in parts (a),

Table 3–5 Multiple Solutions of the Chemical Equilibrium Problem

Variable	Part (a)	Part (b)	Part (c)
CD	0.705334	0.0555561	1.0701
CX	0.177792	0.59722	−0.322716
CZ	0.373977	1.08207	1.13053
CA	0.420689	0.36237	−0.700638
CB	0.242897	−0.234849	−0.377922
CC	0.153565	−1.62374	0.262286
CY	0.551769	1.67929	0.807818

(b), and (c). Note that the initial conditions for problem part (a) converged to all positive concentrations. However, the initial conditions for parts (b) and (c) converged to some negative values for some of the concentrations. Thus a "reality check" on Table 3–5 for physical feasibility reveals that the negative concentrations in parts (b) and (c) are the basis for rejecting these solutions as not representing a physically valid situation.

This problem illustrates the desirability of entering nonlinear functions in a way in which the unknown variable will not lead to highly nonlinear behavior or division by zero. Another option to alleviate the solution of systems of algebraic equations is to convert them to a system where there is only one implicit equation, and the rest of the variables can be calculated from explicit expressions. This approach is demonstrated in Problem 3.4 although it cannot be applied to this chemical equilibrium problem.

Additionally, this problem shows that a correct numerical solution to a properly posed problem may be an infeasible solution, and thus it should be rejected as an unrealistic solution (for example negative concentrations). *In all cases where the solutions of simultaneous nonlinear equations are required, it is very important to specify initial estimates inside the feasible region and as close to the ultimate solution as possible.*

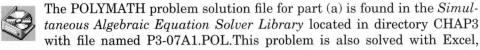

The POLYMATH problem solution file for part (a) is found in the *Simultaneous Algebraic Equation Solver Library* located in directory CHAP3 with file named P3-07A1.POL. This problem is also solved with Excel, Maple, MathCAD, MATLAB, Mathematica, and POLYMATH as problem 4 in the Set of Ten Problems discussed in Appendix F.

3.8 SOLVING DIFFERENTIAL ALGEBRAIC EQUATIONS

3.8.1 Concepts Demonstrated

Batch distillation of an ideal binary mixture.

3.8.2 Numerical Methods Utilized

Solution of a system of equations comprising of ordinary differential and implicit algebraic equations using the controlled integration technique.

3.8.3 Problem Statement

For a binary distillation process involving two components designated 1 and 2, the moles of liquid remaining, L, as a function of the mole fraction of the component 2, x_2, can be expressed by the following equation:

$$\frac{dL}{dx_2} = \frac{L}{x_2(k_2 - 1)} \tag{3-35}$$

where k_2 is the vapor liquid equilibrium ratio for component 2. If the system may be considered ideal, the vapor liquid equilibrium ratio can be calculated from $k_i = P_i/P$, where P_i is the vapor pressure of component i and P is the total pressure.

A common vapor pressure correlation is the Antoine equation, which utilizes three parameters A, B, and C for component i as given next, where T is the temperature in °C.

$$P_i = 10^{\left(A + \frac{B}{T + C}\right)} \tag{3-36}$$

The temperature in the batch still follows the bubble point curve. The bubble point temperature is defined by the implicit algebraic equation, which can be written using the vapor liquid equilibrium ratios as

$$k_1 x_1 + k_2 x_2 = 1 \tag{3-37}$$

Consider a binary mixture of benzene (component 1) and toluene (component 2) that is to be considered as ideal. The Antoine equation constants for benzene are $A_1 = 6.90565$, $B_1 = -1211.033$, and $C_1 = 220.79$. For toluene $A_2 = 6.95464$, $B_2 = -1344.8$, and $C_2 = 219.482$ (Dean[3]). P is the pressure in mm Hg and T the temperature in °C.

The batch distillation of benzene (component 1) and toluene (component 2) mixture is being carried out at a pressure of 1.2 atm. Initially, there are 100 mol of liquid in the still, comprised of 60% benzene and 40% toluene (mole fraction basis). Calculate the amount of liquid remaining in the still when concentration of toluene reaches 80% using the two approaches discussed in the following section and compare the results.

3.8.4   Solution (Partial)

This problem requires the simultaneous solution of Equation (3-35) while the temperature is calculated from the bubble point considerations implicit in Equation (3-37). A system of equations comprising differential and implicit algebraic equations is called differential algebraic and referred to as DAEs. There are several numerical methods for solving DAEs. Most problem-solving software packages including POLYMATH do not have the specific capability for DAEs.

Approach 1   The first approach will be to use the *controlled integration technique* proposed by Shacham et al.[13] Using this method, the nonlinear Equation (3-37) is rewritten with an error term given by

$$\varepsilon = 1 - k_1 x_1 - k_2 x_2 \tag{3-38}$$

where the ε calculated from this equation provides the basis for keeping the temperature of the distillation at the bubble point. This is accomplished by changing the temperature in proportion to the error in a manner analogous to a proportion controller action. Thus this can be represented by another differential equation:

$$\frac{dT}{dx_2} = K_c \varepsilon \tag{3-39}$$

where a proper choice of the proportionality constant K_c will keep the error below a desired error tolerance.

A simple trial-and-error procedure is sufficient for the calculation of K_c. At the beginning K_c is set to a small value (say $K_c = 1$), and the system is integrated. If ε is too large, then K_c must be increased and the integration repeated. This trial-and-error procedure is continued until ε becomes smaller than a desired error tolerance throughout the entire integration interval.

The temperature at the initial point is not specified in the problem, but it is necessary to start the problem solution at the bubble point of the initial mixture. This separate calculation can be carried out on Equation (3-37) for $x_1 = 0.6$ and $x_2 = 0.4$ and the Antoine equations using the POLYMATH *Simultaneous Algebraic Equation Solver*. The resulting initial temperature is found to be $T_0 = 95.5851$.

The system of equations for the batch distillation as they are introduced into the POLYMATH *Simultaneous Differential Equation Solver* using $K_c = 0.5 \times 10^6$ is as follows and the partial results from the solution are summarized in Table 3–6.

Equations:
```
d(L)/d(x2)=L/(k2*x2-x2)
d(T)/d(x2)=Kc*err
Kc=0.5e6
k2=10^(6.95464-1344.8/(T+219.482))/(760*1.2)
x1=1-x2
k1=10^(6.90565-1211.033/(T+220.79))/(760*1.2)
err=(1-k1*x1-k2*x2)
```
Initial Conditions:
```
x2(0)=0.4
L(0)=100
T(0)=95.5851
```
Final Value:
```
x2(f)=0.8
```

Table 3–6 Partial Results for DAE Binary Distillation Problem

Variable	Initial Value	Maximum Value	Minimum Value	Final Value
x2	0.4	0.8	0.4	0.8
L	100	100	14.0456	14.0456
T	95.5851	108.569	95.5851	108.569
k2	0.532535	0.785753	0.532535	0.785753
x1	0.6	0.6	0.2	0.2
k1	1.31164	1.8566	1.31164	1.8566
err	−3.64587e−07	7.75023e−05	−3.64587e−07	7.75023e−05

The final values from the table indicate that 14.05 mol of liquid remain in the column when the concentration of toluene reaches 80%. During the distillation the temperature increases from 95.6 °C to 108.6 °C. The error calculated from Equation (3-38) increases from about -3.6×10^{-7} to 7.75×10^{-5} during the numerical solution, but it is still small enough for the solution to be considered accurate.

 The POLYMATH data file for Approach 1 is found in the *Simultaneous Differential Equation Solver Library* located in directory CHAP3 with file named P3-08A.POL. This problem is also solved with Excel, Maple, MathCAD, MATLAB, Mathematica, and POLYMATH as problem 8 in the Set of Ten Problems discussed in Appendix F.

Approach 2   A different approach for solving this problem can be used because Equation (3-37) can be differentiated with respect to x_2 to yield

$$\frac{dT}{dx_2} = \frac{(k_2 - k_1)}{\ln(10)\left[x_1 k_1 \dfrac{-B_1}{(C_1 + T)^2} + x_2 k_2 \dfrac{-B_2}{(C_2 + T)^2}\right]} \qquad \textbf{(3-40)}$$

This equation can be integrated simultaneously with Equation (3-35) to provide the bubble point temperature during the problem solution.

3.9 METHOD OF LINES FOR PARTIAL DIFFERENTIAL EQUATIONS

3.9.1   Concepts Demonstrated

Unsteady-state heat conduction in one-dimensional slab with one face insulated and constant thermal diffusivity.

3.9.2   Numerical Methods Utilized

Application of the numerical method of lines to solve a partial differential equation which involves the solution of simultaneous ordinary differential equations and explicit algebraic equations.

3.9.3   Problem Statement[*]

Unsteady-state heat transfer in a slab in the x direction is described by the partial differential equation (see Geankoplis[5] for derivation)

$$\frac{\partial T}{\partial t} = \alpha \frac{\partial^2 T}{\partial x^2} \qquad\qquad\qquad (3\text{-}41)$$

where T is the temperature in K, t is the time in s, and α is the thermal diffusivity in m²/s given by $k/\rho c_p$. In this treatment, the thermal conductivity k in W/m·K, the density ρ in kg/m³, and the heat capacity c_p in J/kg·K are all considered to be constant.

Consider an example problem posed by Geankoplis[5] where a slab of material with a thickness 1.00 m is supported on nonconducting insulation. This slab is shown in Figure 3–5. For a numerical problem solution, the slab is divided into N sections with $N + 1$ node points. The slab is initially at a uniform temperature of 100 °C. This gives the initial condition that all the internal node temperatures are known at time $t = 0$.

$$T_n = 100 \text{ for } n = 2 \ldots (N+1) \text{ at } t = 0 \qquad\qquad (3\text{-}42)$$

If at time zero the exposed surface is suddenly held constant at a temperature of 0 °C, this gives the boundary condition at node 1:

$$T_1 = 0 \quad \text{for } t \geq 0 \qquad\qquad\qquad (3\text{-}43)$$

Another boundary condition is that the insulated boundary at node $N + 1$ allows no heat conduction. Thus

$$\frac{\partial T_{N+1}}{\partial x} = 0 \quad \text{for } t \geq 0 \qquad\qquad (3\text{-}44)$$

Note that this problem is equivalent to having a slab of twice the thickness exposed to the initial temperature on both faces.

[*] Adapted from Geankoplis[5], pp. 353–56, with permission.

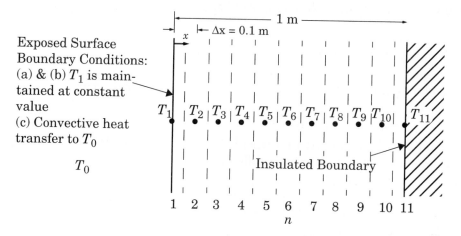

Figure 3–5 Unsteady-State Heat Conduction in a One-Dimensional Slab

When convection is considered as the only mode of heat transfer to the surface of the slab, an energy balance can be made at the interface that relates the energy input by convection to the energy output by conduction. Thus at any time for transport normal to the slab surface in the x direction

$$h(T_0 - T_1) = -k\frac{\partial T}{\partial x}\bigg|_{x=0} \tag{3-45}$$

where h is the convective heat transfer coefficient in W/m$^2 \cdot$K and T_0 is the ambient temperature.

(a) Numerically solve Equation (3-41) with the initial and boundary conditions of (3-42), (3-43), and (3-44) for the case where $\alpha = 2 \times 10^{-5}$ m^2/s and the slab surface is held constant at $T_1 = 0$ °C. This solution should utilize the numerical method of lines with 10 sections. Plot the temperatures T_2, T_3, T_4, and T_5 as functions of time to 6000 s.

(b) Repeat part (a) with 20 sections. Compare results with part (a) to verify that solution for part (a) is accurate.

(c) Repeat parts (a) and (b) for the case where heat convection is present at the slab surface. The heat transfer coefficient is $h = 25.0$ W/m$^2 \cdot$K, and the thermal conductivity is k = 10.0 W/m$\cdot$K.

The Numerical Method of Lines

The method of lines is a general technique for solving partial differential equations (PDEs) by typically using finite difference relationships for the spatial derivatives and ordinary differential equations for the time derivative, as discussed by Schiesser.[12] For this problem with $N = 10$ sections of length $\Delta x = 0.1$ m, Equation (3-41) can be rewritten using a central difference formula for the second derivative [Appendix A, Equation (A-9)] as

$$\frac{\partial T_n}{\partial t} = \frac{\alpha}{(\Delta x)^2}(T_{n+1} - 2T_n + T_{n-1}) \quad \text{for } (2 \leq n \leq 10) \tag{3-46}$$

The boundary condition given in Equation (3-43) can be expressed as

$$T_1 = 0 \tag{3-47}$$

The second boundary condition represented by Equation (3-44) can be written using a second-order backward finite difference [Appendix A, Equation (A-7)] as

$$\frac{\partial T_{11}}{\partial x} = \frac{3T_{11} - 4T_{10} + T_9}{2\Delta x} = 0 \tag{3-48}$$

which can be solved for T_{11} to yield

$$T_{11} = \frac{4T_{10} - T_9}{3} \tag{3-49}$$

Surface Boundary Condition

The energy balance at the slab surface given by Equation (3-45) can be used to determine a relationship between the slab surface temperature T_1, the ambient temperature T_0, and the temperatures at internal node points. In this case, the second-order forward difference equation for the first derivative [Appendix A, Equation (A-5)] can be applied to T_1

$$\frac{\partial T_1}{\partial x} = \frac{(-T_3 + 4T_2 - 3T_1)}{2\Delta x} \tag{3-50}$$

and can be substituted into Equation (3-45) to yield

$$h(T_0 - T_1) = -k\frac{\partial T}{\partial x}\bigg|_{x=0} = -k\frac{(-T_3 + 4T_2 - 3T_1)}{2\Delta x} \tag{3-51}$$

The preceding equation can by solved for T_1 to give

$$T_1 = \frac{2hT_0\Delta x - kT_3 + 4kT_2}{3k + 2h\Delta x} \tag{3-52}$$

and this can be used to calculate T_1 during the method of lines solution.

3.9.4   Solution

(a) The problem then requires the solution of Equations (3-46), (3-47), and (3-49) which results in nine simultaneous ordinary differential equations and two explicit algebraic equation for the 11 temperature nodes. This set of equations can be entered into the POLYMATH *Simultaneous Differential Equation*

Solver. Note that there is the capability to duplicate equations in POLYMATH that is a very helpful option for this problem. The resulting equation set is

```
Equations:
d(T2)/d(t)=alpha/deltax^2*(T3-2*T2+T1)
d(T3)/d(t)=alpha/deltax^2*(T4-2*T3+T2)
d(T4)/d(t)=alpha/deltax^2*(T5-2*T4+T3)
d(T5)/d(t)=alpha/deltax^2*(T6-2*T5+T4)
d(T6)/d(t)=alpha/deltax^2*(T7-2*T6+T5)
d(T7)/d(t)=alpha/deltax^2*(T8-2*T7+T6)
d(T8)/d(t)=alpha/deltax^2*(T9-2*T8+T7)
d(T9)/d(t)=alpha/deltax^2*(T10-2*T9+T8)
d(T10)/d(t)=alpha/deltax^2*(T11-2*T10+T9)
alpha=2.e-5
T1=0
T11=(4*T10-T9)/3
deltax=.10
Initial Conditions:
t(0)=0
T2(0)=100
T3(0)=100
T4(0)=100
T5(0)=100
T6(0)=100
T7(0)=100
T8(0)=100
T9(0)=100
T10(0)=100
Final Value:
t(f)=6000
```

The plots of the temperatures in the first four sections, node points 2 ... 5, are shown in Figure 3–6. The transients in temperatures show an approach to steady state. The numerical results are compared to the hand calculations of a finite difference solution by Geankoplis[5](pp. 471–3) at the time of 6000 s in Table 3–7. These results indicate that there is general agreement regarding the prob-

Table 3–7 Results for Unsteady-State Heat Transfer in One-Dimensional Slab at t = 6000 s

Distance from Slab Surface in m	Geankoplis[5] $\Delta x = 0.20$ m		Method of Lines (a) $\Delta x = 0.10$ m		Method of Lines (b) $\Delta x = 0.05$ m	
	n	*T* in °C	*n*	*T* in °C	*n*	*T* in °C
0	1	0.0	1	0.0	1	0.0
0.2	2	31.25	3	31.71	5	31.68
0.4	3	58.59	5	58.49	9	58.47
0.6	4	78.13	7	77.46	13	77.49
0.8	5	89.84	9	88.22	17	88.29
1.0	6	93.75	11	91.66	21	91.72

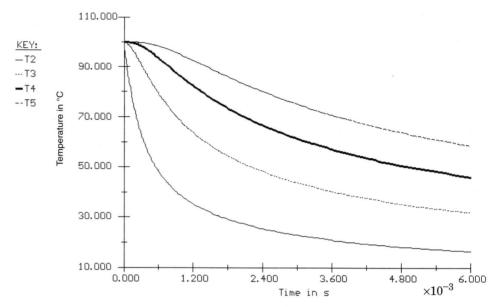

Figure 3–6 Temperature Profiles for Unsteady-State Heat Conduction in a One-Dimensional Slab

lem solution, but differences between the temperatures at the nodes increase as the nodes approach the insulated boundary of the slab.

 The POLYMATH problem solution file for part (a) is found in the *Simultaneous Differential Equation Solver Library* located in directory CHAP3 with file named P3-09A.POL.

(b) The accuracy of the numerical solution can be investigated by doubling the number of sections for the numerical method of lines solution. This just involves adding an additional 10 equations given by the relationship in Equation (3-46) and modifying Equation (3-49) to calculate T_{21}. The results for this change in the POLYMATH equation set are also summarized in Table 3–7. Here the numerical solution is only slightly changed from the previous solution in part (a), which gives reassurance to the first choice of 10 sections for this problem. The temperature profiles are virtually unchanged.

 The POLYMATH problem solution file for part (b) is found in the *Simultaneous Differential Equation Solver Library* located in directory CHAP3 with file named P3-09B.POL.

(c) The calculation of T_1 at node 1 is required by the convection boundary condition for this case, and Equation (3-52) can be entered into the equation set used in part (a) along with an equation for the ambient temperature T_0. This equation set should indicate a somewhat slower response of the temperatures within the slab because of the additional resistance to heat transfer.

A comparison with the approximate hand calculations by Geankoplis[5] is summarized in Table 3–8. In this case, the simplified hand calculations give results that have some error relative to the numerical method of lines solutions, which are in good agreement with each other.

Table 3–8 Unsteady-State Heat Transfer with Convection in One-Dimensional Slab at t = 1500 s

Distance from Slab Surface in m	Geankoplis[5] $\Delta x = 0.20$ m		Method of Lines (a) $\Delta x = 0.10$ m		Method of Lines (b) $\Delta x = 0.05$ m	
	n	*T* in °C	*n*	*T* in °C	*n*	*T* in °C
0	1	64.07	1	64.40	1	64.99
0.2	2	89.07	3	88.13	5	88.77
0.4	3	98.44	5	97.38	9	97.73
0.6	4	100.00	7	99.61	13	99.72
0.8	5	100.00	9	99.96	17	99.98
1.0	6	100.00	11	100.00	21	100.00

The POLYMATH problem solution files for part (c) are found in the *Simultaneous Differential Equation Solver Library* located in directory CHAP3 with files named P3-09C1.POL and P3-09C2.POL.

REFERENCES

1. Aiken, R. C., and Lapidus, L. "An Effective Integration Method for Typical Stiff Systems," *AIChE J.*, *20* (2), 368 (1974).
2. Carnahan, B., Luther, H. A., and Wilkes, J. O. *Applied Numerical Methods*, New York: Wiley, 1969.
3. Dean, A. (Ed.), *Lange's Handbook of Chemistry*, New York: McGraw-Hill, 1973.
4. Garritsen, A.W., University of Technology, Delft, Netherlands, personal communication, 1992.
5. Geankoplis, C. J. *Transport Processes and Unit Operations*, 3rd ed. Englewood Cliffs, NJ: Prentice Hall, 1993.
6. Gear, C. W., "The Automatic Integration of Stiff Ordinary Differential Equations," *Proc. of the IP68 Conf.*, North-Holland, Amsterdam, 1969.
7. Hanna, O. T., and Sandall, O. C. *Computational Methods in Chemical Engineering*, Englewood Cliffs, NJ: Prentice Hall, 1995.
8. Himmelblau, D. M., *Basic Principles and Calculations in Chemical Engineering*, 6th ed., Englewood Cliffs, NJ: Prentice Hall, 1996.
9. Luss, D., and Amundson, N. R. "Stability of Batch Catalytic Fluidized Beds," *AIChE J.*, *14* (2), 211 (1968).
10. Perry, R. H., Green, D. W., and Malorey, J. D., Eds. *Perry's Chemical Engineers Handbook*, 6th ed. New York: McGraw-Hill, 1984.
11. Press, W. H., Teukolsky, S. A., Vetterling, W. T., and Flannery, B. P. *Numerical Recipes,* 2nd ed. Cambridge, MA: Cambridge Univ. Press., 1992.
12. Schiesser, W. E. *The Numerical Method of Lines*, San Diego, CA: Academic Press, 1991.
13. Shacham, M., Brauner, N., and Pozin, M. *Computers Chem Engng.*, *20*, Suppl., pp. S1329–S1334 (1996).
14. Thomas, L. C. *Heat Transfer*, Englewood Cliffs, NJ: Prentice Hall, 1992.

Thermodynamics

4.1 COMPRESSIBILITY FACTOR VARIATION FROM VAN DER WAALS EQUATION

4.1.1 Concepts Demonstrated

Compressibility factor variation with reduced pressure from the van der Waals equation of state at various reduced temperatures.

4.1.2 Numerical Methods Utilized

Solution of a single nonlinear equation and integration of an ordinary differential equation to permit continuous plotting of various functions.

4.1.3 Problem Statement

The van der Waals equation of state can be used to calculate the compressibility factors for gases, as has been discussed in Problem 1.1. Thermodynamic textbooks often provide charts of compressibility factors for various reduced temperatures, T_r, as a function of reduced pressure, P_r. An example is given by Kyle.[6]

Consider the calculations necessary for a general compressibility factor chart for CO_2 whose critical temperature is $T_C = 304$ K and whose critical pressure is $P_C = 72.9$ atm.

(a) Plot the compressibility factor for CO_2 over the reduced pressure range of $0.1 \leq P_r \leq 10$ for a constant reduced temperature of $T_r = 1.1$.

(b) Repeat part (a) at $T_r = 1.3$.

(c) Repeat part (a) at $T_r = 2.0$.

(d) Plot the results of (a), (b), and (c) on a single figure and compare your results with those from a generalized compressibility chart, as given in textbooks such as Kyle[6] and Sandler.[8]

4.1.4   Solution

(a)–(d) Approach 1 The first approach is a direct solution of the van der Waals equation for a variety of reduced pressures at the desired reduced temperature. Thus the POLYMATH *Simultaneous Algebraic Equation Solver* can be used, as discussed in Problem 1.1. For desired values of T_r and P_r, the molar volume V and the compressibility factor Z can be calculated. This approach requires repetitive solution of a nonlinear equation in order to generate continuous curves of Z. While this approach is straightforward, it requires a number of calculations plus the creation of a figure from the results.

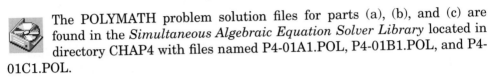 The POLYMATH problem solution files for parts (a), (b), and (c) are found in the *Simultaneous Algebraic Equation Solver Library* located in directory CHAP4 with files named P4-01A1.POL, P4-01B1.POL, and P4-01C1.POL.

(a)–(d) Approach 2 Another more efficient approach to generating the continuous curves of compressibility factor is to convert the van der Waals equation into a differential equation to describe how the specific volume changes with reduced pressure. This differential equation, along with the definitions of reduced pressure and compressibility factor, can be used to generate a continuous curve for compressibility, as outlined next.

The van der Waals equation (see Problem 1.1) can be rewritten as

$$P = \frac{RT}{V-b} - \frac{a}{V^2} \tag{4-1}$$

Differentiating Equation (4-1) with respect to V yields

$$\frac{dP}{dV} = -\frac{RT}{(V-b)^2} + \frac{2a}{V^3} \tag{4-2}$$

The variation of V with P_r can be determined using

$$\frac{dV}{dP_r} = \frac{dV}{dP}\frac{dP}{dP_r} \tag{4-3}$$

By definition, $P_r = P/P_c$; therefore,

$$\frac{dP}{dP_r} = P_c \tag{4-4}$$

The derivative from the preceding equation and the inverse of Equation (4-2) can be introduced into Equation (4-3) to yield the variation of V with P_r. Thus

$$\frac{dV}{dP_r} = \left[-\frac{RT}{(V-b)^2} + \frac{2a}{V^3} \right]^{-1} P_c \tag{4-5}$$

Differential Equation (4-5) can be solved simultaneously using the POLY-MATH *Simultaneous Differential Equation Solver* for a single value of T_r while the independent variable P_r is changed over the desired range $0.1 \leq P_r \leq 10$. In order to obtain the initial value (at $P_r = 0.1$) for Equation (4-5), the POLYMATH *Simultaneous Algebraic Equation Solver* can be used to solve the van der Waals equation for V when P_r is set to 0.1 and T_r is known. For example, when P_r is 0.1 and $T_r = 1.1$, the solution is $V = 3.71998$ liter/g-mol.

The equations for solution of the van der Waals equation for a range of P_r values can be entered into the POLYMATH *Simultaneous Differential Equation Solver* as

```
Equations:
d(V)/d(Pr)=Pc/(-R*T/(V-b)^2+2*a/V^3)
Tc=304.2
Pc=72.9
R=0.08206
Tr=1.1
P=Pc*Pr
T=Tc*Tr
b=R*Tc/(8*Pc)
a=27*R^2*Tc^2/(Pc*64)
Z=P*V/(R*T)
Initial Conditions:
Pr(0)=0.1
V(0)=3.71998
Final Value:
Pr(f)=10
```

Solution of this set of equations generates the plot of Z versus P_r for $T_r = 1.1$ that is given in Figure 4–1.

This plot can be compared with similar curves generated from experimental data as presented in Kyle[6] and Sandler.[8] Deviations are found at higher reduced temperatures and pressures where the van der Waals equation becomes less accurate. Similar plots for other T_r values can be made for other reduced temperatures when the initial value for V is changed as a result of solving the van der Waals equation using the POLYMATH *Simultaneous Algebraic Equation Solver*.

The POLYMATH problem solution files for parts (a), (b), and (c) are found in the *Simultaneous Differential Equation Solver Library* located in directory CHAP4 with files named P4-01A2.POL, P4-01B2.POL, and P4-01C2.POL.

In part (d) it is desirable to have the compressibility for all three T_r's on the same plot. This can be accomplished in POLYMATH by modifying the equation set with indices 1, 2, and 3 for reduced temperatures of 1.1, 1.3, and 2.0, respectively. The respective initial conditions as determined from separate solutions of the van der Waals equations are V1 = 3.71998, V2 = 4.4344, and V3 = 6.89988. The POLYMATH equation set is as follows:

KEY:
−Z

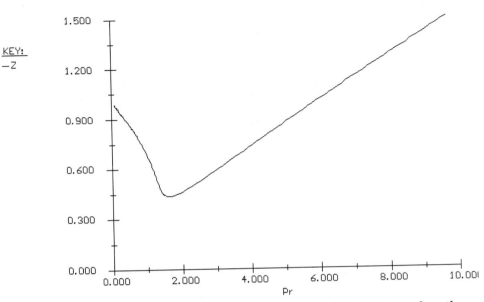

Figure 4–1 Compressibility Factor Variation as Function of P_r at $T_r = 1.1$, from the van der Waals Equation

Equations:
d(V1)/d(Pr)=Pc/(-R*T1/(V1-b)^2+2*a/V1^3)
d(V2)/d(Pr)=Pc/(-R*T2/(V2-b)^2+2*a/V2^3)
d(V3)/d(Pr)=Pc/(-R*T3/(V3-b)^2+2*a/V3^3)
Tc=304.2
Pc=72.9
R=0.08206
Tr1=1.1
Tr2=1.3
Tr3=2.0
P=Pc*Pr
T1=Tc*Tr1
T2=Tc*Tr2
T3=Tc*Tr3
b=R*Tc/(8*Pc)
a=27*R^2*Tc^2/(Pc*64)
Z1=P*V1/(R*T1)
Z2=P*V2/(R*T2)
Z3=P*V3/(R*T3)
Initial Conditions:
Pr(0)=0.1
V1(0)=3.71998
V2(0)=4.4344
V3(0)=6.89988
Final Value:
Pr(f)=10

The plot for all three T_r's is shown in Figure 4–2.

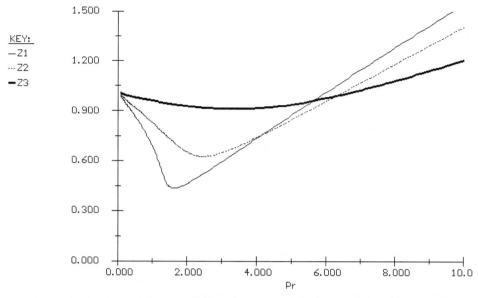

Figure 4–2 Compressibility Factors from the van der Waals Equation at $T_r = 1.1$ (Z1), 1.3 (Z2), and 2.0 (Z3)

 The POLYMATH problem solution file for part (d) is found in the *Simultaneous Differential Equation Solver Library* located in directory CHAP4 with file named P4-01D.POL.

4.2 COMPRESSIBILITY FACTOR VARIATION FROM VARIOUS EQUATIONS OF STATE

4.2.1   Concepts Demonstrated

Use of Virial, Soave-Redlich-Kwong, Peng-Robinson, and Beattie-Bridgeman equations of state for calculation of compressibility factors, and plotting compressibility factors versus reduced pressure at constant reduced temperatures.

4.2.2   Numerical Methods Utilized

Solution of a single nonlinear algebraic equation and conversion of a nonlinear equation to a differential equation for generation of a continuous solution.

4.2.3   Problem Statement

(a)   Select a component from Table 4–1 and plot the compressibility factor over the reduced pressure range of $0.5 \leq P_r \leq 11$ for a constant reduced temperature of $T_r = 1.3$. Calculations should be based on one of the following equations of state: Soave-Redlich-Kwong, Peng-Robinson, Beattie-Bridgeman or Virial.

(b)   Repeat part (a) at $T_r = 2.0$.

(c)   Repeat part (a) at $T_r = 10.0$.

(d)   Plot the results of parts (a), (b), and (c) on a single figure.

Additional Information and Data

The Soave-Redlich-Kwong, Peng-Robinson and Beattie-Bridgeman equations of state are described in Problem 1.8.

The Virial Equation of State The Virial equation of state[7] is normally expressed as

$$P = \frac{RT}{V - B} \tag{4-6}$$

where

P = pressure in atm

V = molar volume in liters/g-mol

T = temperature in K

R = gas constant   ($R = 0.08206$ atm·liter/g-mol·K)

B = the second Virial coefficient in liter/g-mol

Various techniques have been suggested to calculate B. The method outlined by Smith and Van Ness[9] utilizes

$$B = \frac{RT_c}{P_c}[B^{(0)} + \omega B^{(1)}]$$ (4-7)

where

T_c = the critical temperature (K)

P_c = the critical pressure (atm)

ω = the acentric factor

and

$$B^{(0)} = 0.083 - \frac{0.422}{T_r^{1.6}}$$ (4-8)

$$B^{(1)} = 0.139 - \frac{0.172}{T_r^{4.2}}$$ (4-9)

The preceding treatment of the virial equation of state is only accurate for $T_r > 4$. Table 4–1 summarizes values of T_c, P_c and ω for selected substances. The

Table 4–1 Critical Constants of Selected Substances[a]

Substance		Critical Temperature T_c (K)	Critical Pressure P_c (atm)	Acentric Factor ω
Hydrogen	H_2	33.3	12.8	−0.218
Nitrogen	N_2	126.2	33.5	0.039
Oxygen	O_2	154.8	50.1	0.025
Carbon Dioxide	CO_2	304.2	72.9	0.239
Carbon Monoxide	CO	133	34.5	0.066
Nitrous Oxide	N_2O	309.7	71.7	0.165
Water	H_2O	647.4	218.3	0.344
Ammonia	NH_3	405.5	111.3	0.250
Methane	CH_4	191.1	45.8	0.011
Ethane	C_2H_6	305.5	48.2	0.099
Propane	C_3H_8	370.0	42.0	0.153
n-Butane	C_4H_{10}	425.2	37.5	0.199

[a]A more extensive table is given by Kyle[6]

constants for the Beattie-Bridgeman equation for selected substances are given in Table 4–2.

Table 4–2 Constants for the Beattie-Bridgeman Equation

Gas	A_0	a	B_0	b	$c \times 10^{-4}$
Hydrogen	0.1975	−0.00506	0.02096	−0.04359	0.0504
Nitrogen	1.3445	0.02617	0.05046	−0.00691	4.20
Oxygen	1.4911	0.02562	0.04624	0.004208	4.80
Air	1.3012	0.01931	0.04611	−0.001101	4.34
Carbon Dioxide	5.0065	0.07132	0.10476	0.07235	66.00
Ammonia	2.3930	0.17031	0.03415	0.19112	476.87
Carbon Monoxide	1.3445	0.02617	0.05046	−0.00691	4.20
Nitrous Oxide	5.0065	0.07132	0.10476	0.07235	66.00
Methane	2.2769	0.01855	0.05587	−0.01587	12.83
Ethane	5.8800	0.05861	0.09400	0.01915	90.00
Propane	11.9200	0.07321	0.18100	0.04293	120.00
n-Butane	17.7940	0.12161	0.24620	0.094620	350.00

4.2.4   Solution (Suggestions)

See Section 4.1.4 for methods of solution.

4.3 ISOTHERMAL COMPRESSION OF GAS USING REDLICH-KWONG EQUATION OF STATE

4.3.1 Concepts Demonstrated

Work and volume change during isothermal compression of an ideal and real gas.

4.3.2 Numerical Methods Utilized

Solution of an ordinary differential equation.

4.3.3 Problem Statement

Carbon dioxide is being compressed isothermally at 50 °C to 1/100th of its initial volume. The initial pressure is 1.0 atm.

(a) Calculate the work per g-mol of gas required using the ideal gas assumption and employing the Redlich Kwong equation of state.

(b) Check whether the variation of the pressure P and volume V during compression follows the expected path of PV = constant for the ideal and nonideal cases.

(c) Repeat (a) and (b) for the cases when the gas to be compressed is (1) air and (2) methane. Explain the differences in the behavior of the three gases.

Additional Information and Data

Redlich-Kwong Equation of State The Redlich-Kwong equation of state is given by

$$P = \frac{RT}{(V-b)} - \frac{a}{V(V+b)\sqrt{T}}$$

(4-10)

where

$$a = 0.42747\left(\frac{R^2 T_c^{5/2}}{P_c}\right)$$

(4-11)

$$b = 0.08664\left(\frac{RT_c}{P_c}\right)$$

(4-12)

P = pressure in atm
V = molar volume in liters/g-mol
T = temperature in K
R = gas constant (R = 0.08206 (atm·liter/g-mol·K))
T_c = critical temperature in K
P_c = critical pressure in atm

Critical properties of CO_2 and methane can be found in Table 4–1.

4.3.4   Solution (Partial)

(a)   For a closed system in a reversible isothermal compression from V_1 to V_2, the work W done on the system can be calculated from

$$\frac{dW}{dV} = -P \tag{4-13}$$

For an ideal gas, this equation can be integrated from V_1 to V_2, yielding

$$W = -RT \ln\frac{V_2}{V_1} \tag{4-14}$$

For the nonideal case, Equation (4-13) should be numerically integrated using Equation (4-10) to calculate the pressure as a function of the volume.   The initial volume, V_1, should be calculated using the Redlich-Kwong equation of state given in Equation (4-10) by solving this nonlinear equation for V_1 at 1.0 atm and 50 °C. This solution for V_1 can be obtained using the POLYMATH *Simultaneous Algebraic Equation Solver.* The result is V_1 = 26.4134 liters so that V_2 = 0.2641 liters since the problem states that V_2 is simply $0.01\ V_1$.

The POLYMATH problem solution file for the initial volume estimate of part (a) is found in the *Simultaneous Algebraic Equation Solver Library* located in directory CHAP4 with file named P4-03A1.POL.

Change of Variable to Allow Decreasing Independent Variable in ODEs
Most numerical integration programs, including POLYMATH, proceed with increasing values of the independent variable. In problems where this is not the case, a convenient new independent variable can be introduced. In this case, a simple linear relationship of this new variable Y to variable V can be employed:

$$V = 26.4134 - Y \tag{4-15}$$

so that as this new (dummy) variable Y progresses from an initial condition of Y = 0.0 to a final value of Y = 26.1493, then V goes from V_1 = 26.4134 to V_2 = 0.2641. The only major change to the original problem equations is that Equation (4-13) must be rewritten using the independent variable Y as

$$\frac{dW}{dY} = P \tag{4-16}$$

since $dV = -\ dY$ from Equation (4-15). Also, algebraic Equation (4-15) must be introduced into the ODE solution.

(b)   If the rule, PV = constant, holds for the process, then a plot of $\ln(P)$ versus $\ln(V)$ should give a straight line with slope of -1.0.

The necessary calculations can be obtained during a single numerical solution by using Equations (4-10) to (4-12) and Equations (4-14) to (4-16) with the initial/final conditions as discussed for Y to solve part (a) of this problem. Part (b) can also be solved during the same numerical ODE solution by simply creating the necessary plotting variables. The equation set for the POLYMATH *Simultaneous Differential Equation Solver* is as follows:

```
Equations:
d(W)/d(Y)=P
R=0.08206
T=50+273.15
Tc=304.2
Pc=72.9
V=26.4134-Z
lnV=ln(V)
b=0.08664*R*Tc/Pc
a=0.42747*R^2*Tc^(5/2)/Pc
Pideal=R*T/V
Wideal=-R*T*ln(V/26.4134)
lnPideal=ln(Pideal)
P=R*T/(V-b)-a/(V*(V+b)*sqrt(T))
lnP=ln(P)
Initial Conditions:
Y(0)=0
W(0)=0
Final Value:
Y(f)=26.1493
```

Numerical results for the calculated values of P, Pideal, W and Wideal for different volumes during the compression are summarized in Table 4–3.

Table **4–3** Partial Tabular Results for Isothermal Compression of Carbon Dioxide

V	P	Pideal	W	Wideal
26.4134	1.0000015	1.0039483	0	0
23.79847	1.1093993	1.1142602	2.7530198	2.7644734
21.18354	1.2456727	1.2518063	5.8253164	5.8510483
18.56861	1.420112	1.4280923	9.3007788	9.3448054
15.95368	1.6513613	1.6621675	13.30143	13.36974
13.33875	1.9725694	1.988019	18.014738	18.116835
10.72382	2.4489014	2.4727839	23.750825	23.903152
8.10889	3.228489	3.2701996	31.080157	31.315016
5.49396	4.7360817	4.8266986	41.24304	41.638663
2.87903	8.8831912	9.2106331	57.928225	58.774284
0.2641	67.419813	100.40776	112.66436	122.12188

Here the suffix "ideal" refers to the value calculated using the ideal gas assumption. The results in the preceding table indicate that during the initial compression, the ideal and the real work are nearly equivalent as the system is nearly ideal. However, as the compression proceeds toward very small volume and high pressure, the differences between the ideal and real pressure and the ideal and real work become very substantial as the system becomes very nonideal.

Figure 4–3 presents the plot of ln(P) and ln(Pideal) versus ln(V). This fig-

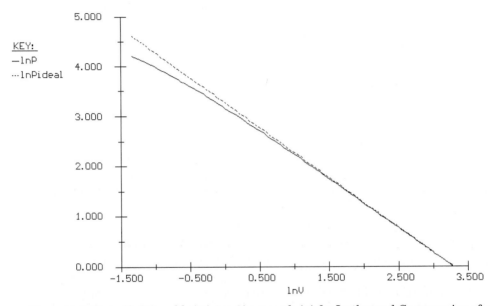

Figure 4–3 Plot of ln(P) and ln(Pideal) versus ln(V) for Isothermal Compression of Carbon Dioxide

ure provides a clear indication that ln(Pideal) versus ln(V) for the ideal gas indeed gives a straight line with slope of negative one as expected, whereas the slope of ln(P) versus ln(V) for the real CO_2 calculations changes considerably for pressures higher than $P \sim 12$ atm as the system becomes nonideal.

Similar calculations using the Redlich-Kwong equation of state for air and methane require only modifications to T_c, P_c, and the initial/final conditions in the POLYMATH program; however, the initial/final conditions will first require solving the equation of state as a nonlinear equation for the starting temperature and pressure.

 The problem solution file for parts (a) and (b) is found in the *Simultaneous Differential Equation Solver Library* located in directory CHAP4 with files named P4-03A2.POL.

4.4 THERMODYNAMIC PROPERTIES OF STEAM FROM REDLICH-KWONG EQUATION

4.4.1   Concepts Demonstrated

Calculation of compressibility factor, molar volume, liquid enthalpy, and vapor enthalpy from an equation of state and ideal gas enthalpy change.

4.4.2   Numerical Methods Utilized

Solution of a nonlinear algebraic equation.

4.4.3   Problem Statement

If you have ever used a thermodynamic diagram (such as pressure enthalpy diagram) to try to find the specific volume, compressibility factor or enthalpy of a pure substance, you probably noticed how difficult it is to get an accurate value from the diagram. You probably concluded that with access to a computer "there ought to be a better way."

This problem requires the development of a simple numerical computation that will provide thermodynamic properties of steam to approximately three digits accuracy with calculations utilizing the Redlich-Kwong equation of state.

(a)   Outline a general calculational procedure to use the Redlich-Kwong equation of state to calculate the molar volume (m^3/kg-mol), specific volume (m^3/kg), compressibility factor, and enthalpy of water vapor at a specified temperature and pressure.

(b)   Compare your results with data found in the steam tables or thermodynamics textbooks for saturated steam at 100, 200, and 300 °C. Please pay particular attention to the possibility of multiple solutions to the R-K equation in some regions. Typically the highest molar volume represents the correct value, while other solutions are extraneous.

(c)   Observe the behavior of the R-K equation in three different regions: (1) below the critical point, (2) close to the critical point, and (3) above the critical point.

Additional Information and Data

The Redlich-Kwong equation of state is described in Problem 4.3, and the critical properties for water can be found in Table 4–1.

The molar specific heats of gases in the ideal gas state are typically given in the literature as a polynomial in temperature.

$$C_p^o = a_0 + a_1 T + a_2 T^2 + a_3 T^3 \tag{4-17}$$

where

T = temperature in K

C_p = molar specific heat in cal/g-mol·K

$a_0, a_1, a_2,$ and a_3 = constants specific for the substance. For water vapor, a_0 = 7.700, $a_1 = 0.04594 \times 10^{-2}$, $a_2 = 0.2521 \times 10^{-5}$, and $a_3 = -0.8587 \times 10^{-9}$.

The usual reference state for the enthalpy of liquid water is $H_l^o = 0$ kJ/kg at $T_0 = 0$ °C = 273.15 K, where the enthalpy of vaporization is $\Delta H_{vap}^o = 2501.3$ kJ/kg. Thus the enthalpy of water vapor as an ideal gas at temperature T can be calculated from

$$H_v^o = H_l^o + \Delta H_{vap}^o + \int_{T_0}^{T} C_p^o dT \qquad (4\text{-}18)$$

where

$$\int_{T_0}^{T} C_p^o dT = a_0(T-T_0) + \frac{a_1}{2}(T^2 - T_0^2) + \frac{a_2}{3}(T^3 - T_0^3) + \frac{a_3}{4}(T^4 - T_0^4) \qquad (4\text{-}19)$$

and the temperatures are in K.

The isothermal enthalpy departure can be expressed in the integral equation

$$(H_v - H_v^o) = \int_{\infty}^{V} \left[T\left(\frac{\partial P}{\partial T}\right)_V + V\left(\frac{\partial P}{\partial V}\right)_T \right] dV \qquad (4\text{-}20)$$

The Redlich-Kwong equation of state, which accounts for nonidealities, can be introduced into Equation (4-20) and the integration can be carried out analytically. The integrated form of the equation can be found in Edmister.[2]

$$(H_v - H_v^o) = RT\left[Z - 1 - \frac{1.5a}{bRT^{1.5}} \ln\left(1 + \frac{b}{V}\right) \right] \qquad (4\text{-}21)$$

Thus the enthalpy of steam (water in the vapor phase) at T and P can be calculated from

$$H_v = H_l^o + \Delta H_{vap}^o + \int_{T_0}^{T} C_p^o dT + RT\left[Z - 1 - \frac{1.5a}{bRT^{1.5}} \ln\left(1 + \frac{b}{V}\right) \right] \qquad (4\text{-}22)$$

using the usual reference states and any conversion factors to maintain consistent units.

4.4.4   Solution (Partial)

(a) For specified values of T and P, the Redlich-Kwong equation can be solved for the molar volume using the POLYMATH *Simultaneous Algebraic*

Equation Solver. In the two-phase region this equation may have up to three roots; the smallest one is an approximate solution for liquid, the largest is accurate for vapor, and the intermediate has no physical meaning.

After calculation of the molar volume and then specific volume, the compressibility factor can be calculated from its definition:

$$Z = \frac{PV}{RT}$$

(4-23)

Additional calculations follow from the previous discussion and lead to the final calculation of H_v by evaluation of Equation (4-22). All of the preceding equations can be solved with the POLYMATH *Simultaneous Algebraic Equation Solver.*

(a) The general equation set for this problem when T in °C and P in atm are provided is given as

```
Equations:
f(V)=P-(R*T/(V-b)-a/(V*(V+b)*sqrt(T)))
P=1
R=0.08206
Tc=647.4
Pc=218.3
Vsp=V/18
TC=100
T0=273.15
T=273.15+TC
b=0.08664*R*Tc/Pc
a=0.42747*R^2*Tc^(5/2)/Pc
Hv0=7.7*(T-T0)+0.04594e-2*(T^2-T0^2)/2+0.2521e-5*(T^3-T0^3)/3-
    0.8587e-9*(T^4-T0^4)/4
Z=P*V/(R*T)
Hdep=24.218*R*T*(Z-1-1.5*a*ln(1+b/V)/(b*R*T^1.5))
Hv=2501.3+4.1868*(Hv0+Hdep)/18
V(min)=20, V(max)=40
```

Note that the conversion factor 24.218 has been added to the equation for Hdep to convert from liter·atm to cal. The conversion factor 4.1868/18 is added to the equation for H_v to convert cal/g-mol to kJ/kg.

(b) The case of saturated steam at $T = 100$ °C where $P = 1$ atm has been entered in the equation set. The solution yields $V = 30.4027$ liter/g-mol, $v = 1.68904$ m³/kg, $Z = 0.99288$, and $H_v = 2686.09$ kJ/kg. Steam tables give $v = 1.6729$ m³/kg and $H_v = 2676.1$ kJ/kg for saturated steam at the same condition. Only the temperature and pressure in the POLYMATH equation set need to be changed in order to obtain the specific volume and saturated steam enthalpy. Modification for other pure components can easily be made.

 The POLYMATH problem solution files for part (b) are found in the *Simultaneous Algebraic Equation Solver Library* located in directory CHAP4 with files named P4-04B1.POL, P4-04B2.POL, and P4-04B3.POL.

4.5 ENTHALPY AND ENTROPY DEPARTURE USING THE REDLICH-KWONG EQUATION

4.5.1   Concepts Demonstrated

Use of the Redlich-Kwong equation to calculate isothermal enthalpy and entropy departure versus reduced temperature and pressure.

4.5.2   Numerical Methods Utilized

Solution of a nonlinear algebraic equation and integration of an ordinary differential equation.

4.5.3   Problem Statement

(a) Select a component from Table 4–1 and plot the enthalpy departure function over the reduced pressure range of $0.5 \leq P_r \leq 30$ for constant reduced temperatures of $T_r = 1.2$, 1.4, and 3.0. Please place all results on a single plot.

(b) Repeat part (a) for the entropy departure function.

(c) Modify the output of parts (a) and (b) to allow a direct comparison of your plots with generalized plots available in the literature. Typically these plots require $(\Delta H^*)/T_c$ and ΔS^* versus $\ln(P_r)$.

(d) Compare your calculated results with generalized correlations and experimental data wherever available (data from Mollier diagrams, etc.).

Additional Information and Data

The enthalpy and entropy departure expressions for the Redlich-Kwong equation are presented by Edmister[2] as

$$\frac{\Delta H^*}{RT} = \frac{3a}{2bRT^{1.5}} \ln\left(1 + \frac{b}{V}\right) - (Z - 1) \tag{4-24}$$

and

$$\frac{\Delta S^*}{R} = \frac{a}{2bRT^{1.5}} \ln\left(1 + \frac{b}{V}\right) - \ln\left(Z - \frac{Pb}{RT}\right) \tag{4-25}$$

where

$$\Delta H^* = H^\circ - H \qquad \Delta S^* = S^\circ - S$$

H° = ideal gas enthalpy at temperature T in cal/g-mol

H = gas enthalpy at temperature T in cal/g-mol

S° = ideal gas entropy at temperature T in cal/g-mol·K

S = gas entropy at temperature T in cal/g-mol·K

P = pressure in atm

V = molar volume in liters/g-mol

T = temperature in K

R = gas constant (R = 0.08206 liter · atm/g-mol · K) (R = 1.9872 cal/g-mol · K)

Z = compressibility factor

and a and b are the Redlich Kwong equation constants calculated from critical pressure and temperature as discussed in Problem 4.3.

4.5.4   Solution (Partial)

Calculational Approach Since continuous curves for enthalpy and entropy departure are desired, the approach discussed in Section 4.1.4 will be used to develop a differential equation for the Redlich-Kwong equation. The enthalpy and entropy departure equations will then be calculated and plotted during the numerical solution.

The Redlich-Kwong equation, which is discussed in Section 4.3.3,

$$P = \frac{RT}{V - b} - \frac{a}{V(V + b)\sqrt{T}} \qquad \text{(4-26)}$$

can be differentiated with respect to V to yield

$$\frac{dP}{dV} = -\frac{RT}{(V - b)^2} + \frac{a}{\sqrt{T}}\left[\frac{2V + b}{V^2(V + b)^2}\right] \qquad \text{(4-27)}$$

The introduction of Equations (4-27) and (4-4) into Equation (4-3) yields

$$\frac{dV}{dP_r} = \left\{-\frac{RT}{(V - b)^2} + \frac{a}{\sqrt{T}}\left[\frac{2V + b}{V^2(V + b)^2}\right]\right\}^{-1} P_c \qquad \text{(4-28)}$$

Equation (4-28) can be numerically integrated using the POLYMATH *Simultaneous Differential Equation Solver* for a desired value of T_r while the independent variable P_r is changed over the desired range ($0.5 \le P_r \le 30$). During the integration, the enthalpy and entropy departure functions given by Equations (4-24) and (4-25) can be calculated for plotting.

(a) and (b) The component selected for this example is water so that a comparison can be made with data from steam tables. The initial condition for V must be separately calculated using the POLYMATH *Simultaneous Algebraic Equation Solver* and the Redlich-Kwong equation of state. This calculation, which is similar to that presented in Section 4.4.4, gives $V = 0.523726$ liters for the initial conditions where $P_r = 0.5$ and $T_r = 1.2$.

The POLYMATH problem solution file for the molar volume calculation is found in the *Simultaneous Algebraic Equation Solver Library* located in directory CHAP4 with file named P4-05A1.POL.

The POLYMATH equation set for both the enthalpy departure function (Hdep) and the entropy departure function (Sdep) is

```
Equations:
d(V)/d(Pr)=Pc/(-R*T/((V-b)^2)+a*(2*V+b)/((V^2*(V+b)^2)*sqrt(T)))
Tc=647.4
Pc=218.3
R=0.08206
lnPr=ln(Pr)
Tr=1.2
P=Pc*Pr
T=Tc*Tr
b=0.08664*R*Tc/Pc
a=0.42747*R^2*Tc^(5/2)/Pc
Z=P*V/(R*T)
Hdep=(3*a/(2*b*R*T^1.5))*ln(1+b/V)-(Z-1)
Sdep=(a/(2*b*R*T^1.5))*ln((1+b/V))-ln(Z-P*b/(R*T))
Initial Conditions:
Pr(0)=0.5
V(0)=0.523726
Final Value:
Pr(f)=30
```

A plot of POLYMATH variables Hdep and Sdep versus Pr for Tr = 1.2 is shown in Figure 4–4, where both functions go through a maximum with Pr.

Since it is desirable to have the Hdep plots for all three Tr's on the same plot, the POLYMATH equation set can be modified to solve all three cases simultaneously. The Tr's can be coded with indices 1, 2, and 3 for reduced temperatures of 1.2, 1.4, and 3.0, respectively, in a similar manner to the method discussed in Section 4.1.4. The respective initial conditions as determined from the Redlich-Kwong equation solutions are V1 = 0.523726, V2 = 0.639885, and V3 = 1.46182. The results for all three reduced temperatures are given in Figure 4–5.

The POLYMATH problem solution files for parts (a) and (b) are found in the *Simultaneous Differential Equation Solver Library* located in directory CHAP4 with files named P4-05A2.POL and P4-05A3.POL.

(c)   Most of the plots in the literature require $(\Delta H^*)/T_c$ and ΔS^* versus P_r on a logarithmic scale. Thus these variables can also be calculated during the numerical solution. Typical results are presented in Figure 4–6.

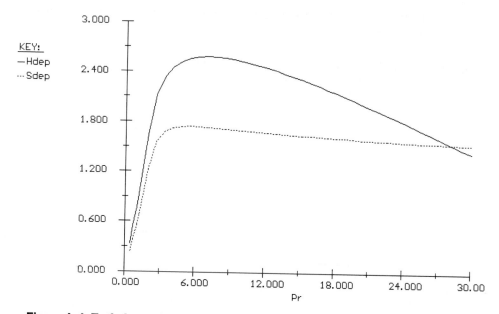

Figure 4–4 Enthalpy and Entropy Departures for Water Vapor at $T_r = 1.2$

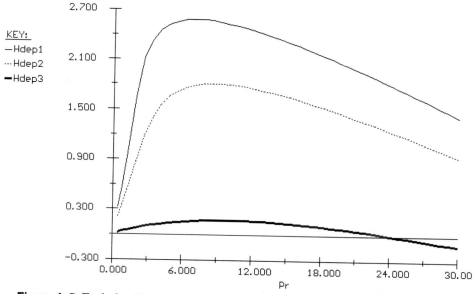

Figure 4–5 Enthalpy Departures for Water Vapor at $T_r = 1.2, 1.4,$ and 3.0

(d) Most thermodynamic textbooks (for example, Kyle[6] and Sandler[8]) present figures for generalized enthalpy and entropy departure for ideal gas behavior. The numerical results for $T_r = 1.2$ and $P_r = 10$ are calculated to be

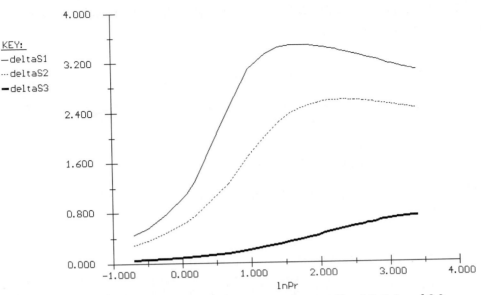

KEY:
—deltaS1
··· deltaS2
— deltaS3

Figure 4–6 Entropy Departure Curves for Water Vapor at T_r = 1.2, 1.4, and 3.0

$(\Delta H^*)/T_c$ = 6.031 cal/g-mol·K and ΔS^* = 3.374 cal/g-mol·K. These values are slightly lower than what is shown in the generalized figures (pp. 100 and 101 of Kyle[6]), but the generalized figures are for substances with Z_c = 0.27 while for water Z_c = 0.23.

4.6 FUGACITY COEFFICIENTS OF PURE FLUIDS FROM VARIOUS EQUATIONS OF STATE

4.6.1   Concepts Demonstrated

Use of various equations of state to calculate fugacity coefficients for pure substances in the liquid or gaseous state.

4.6.2   Numerical Methods Utilized

Solution of a single nonlinear algebraic equation and conversion of a nonlinear equation to a differential equation.

4.6.3   Problem Statement

(a) Select a component from Table 4–1 and use the van der Waals equation of state to plot the calculated fugacity coefficient over the reduced pressure range of $0.5 \leq P_r \leq 30$ for constant reduced temperatures of $T_r = 1.2, 1.4$ and 3.0. Please place all results on a single plot, and compare your results with those from a generalized compressibility chart, as given in textbooks such as Kyle[6] and Sandler.[8]

(b) Repeat part (a) using the Redlich-Kwong equation of state.

(c) Repeat part (a) using the Peng-Robinson equation of state.

(d) Compare your calculated results with experimental data wherever available (data from Mollier diagrams, etc.).

Additional Information and Data

The fugacity coefficient is defined by $\phi = f/P$, where f is the fugacity of the component with units of pressure and P is the pressure. The fugacity coefficients of pure substances can be calculated from a number of equations of state which are summarized by Walas[11].

van der Waals

$$\ln\phi = Z - 1 - \frac{a}{RTV} - \ln\left[Z\left(1 - \frac{b}{V}\right)\right]$$ (4-29)

Redlich-Kwong

$$\ln\phi = Z - 1 - \ln\left[Z\left(1 - \frac{b}{V}\right)\right] - \frac{a}{bRT^{1.5}}\ln\left(1 + \frac{b}{V}\right)$$ (4-30)

Peng-Robinson

$$\ln\phi = Z - 1 - \ln(Z - B) + \left(-\frac{A}{2\sqrt{2}B}\right)\ln\frac{(Z + 2.414B)}{(Z - 0.414B)} \tag{4-31}$$

where $A = 0.45724\alpha P_r/T_r^2$, $\alpha = [1 + (0.37464 + 1.54226\omega - 0.26992\omega^2)(1 - T_r^{0.5})]^2$, and $B = 0.07780P_r/T_r$.

Equations (4-29) to (4-31) are applicable to both the vapor and liquid phases, according to the root selected in solving the equation of state for Z (or for V). The smallest root for Z is for the liquid phase, the largest for the vapor phase.

4.6.4   Solution (Comment)

The equation of state of interest can be solved while continuously changing the pressure along an isotherm for the compressibility factor as demonstrated in Problem 4.5. After the molar volume and compressibility factor have been calculated, the fugacity coefficient can be determined.

Critical properties and the acentric factor for water can be found in Table 4–1. The van der Waals equation and the appropriate constants are discussed in Problem 1.1. The Redlich-Kwong equation of state is described in Problem 4.3.

4.7 FUGACITY COEFFICIENTS FOR AMMONIA—EXPERIMENTAL AND PREDICTED

4.7.1   Concepts Demonstrated

Determination of fugacity from experimental data and calculation of fugacity coefficients from various equations of state.

4.7.2   Numerical Methods Utilized

Integration of experimental data and solution of nonlinear algebraic equations.

4.7.3   Problem Statement

Experimental measurements have yielded the compressibility factors at various pressures for ammonia at 100 °C summarized in Table 4–4.

(a) Calculate the fugacity coefficients from the experimental data given in Table 4–4 and summarize these values in a table for $100 \leq P \leq 1100$ in increments of 100 atm.

(b) Calculate the fugacity coefficients from the van der Waals equation of state and enter these values in the table created in part (a).

(c) Repeat (b) for the Redlich-Kwong equation of state.

(d) Repeat (b) for the Peng-Robinson equation of state.

(e) Discuss any conclusions you can make about the accuracy of the equations of state that you have considered.

Table 4–4 Experimental Data for Ammonia[a]

Pressure atm	Z	Pressure atm	Z	Pressure atm	Z
1.374	0.9928	30.47	0.8471	300	0.3212
3.537	0.9828	33.21	0.831	400	0.4145
5.832	0.9728	36.47	0.8111	500	0.506
8.632	0.9599	40.41	0.7864	600	0.5955
11.352	0.9468	45.19	0.7538	700	0.6828
14.567	0.9315	51.09	0.7102	800	0.7684
19.109	0.9085	58.28	0.6481	900	0.8507
22.84	0.889	100	0.1158	1000	0.9333
26.12	0.8714	200	0.2221	1100	1.014

[a]Data are from Walas[11] where the original reference is given as Gmelin, *Handbook der anorganischen Chemie, 5,* 426 (1935).

Additional Information and Data

The fugacity coefficient, either from experimental data or from an equation of state, may be calculated from

$$\ln\phi = \int_0^P \frac{Z-1}{P} dP \tag{4-32}$$

The use of a cubic spline is recommended for evaluation of the integral in Equation (4-32) when experimental data are available for Z as a function of P.

4.7.4   Solution (Suggestions)

The formulas for the calculation of fugacity coefficients from various equations of state are given by Equations (4-29) to (4-31).

 The POLYMATH data file for Table 4–4 is found in the *Polynomial, Multiple Linear and Nonlinear Regression Library* located in directory CHAP4 with file named P4-07.POL.

4.8 FLASH EVAPORATION OF AN IDEAL MULTICOMPONENT MIXTURE

4.8.1 Concepts Demonstrated

Calculation of bubble point and dew point temperatures and associated vapor and, and, and liquid compositions for flash evaporation of an ideal multicomponent mixture.

4.8.2 Numerical Methods Utilized

Solution of a single nonlinear algebraic equation.

4.8.3 Problem Statement

A flash evaporator must separate ethylene and ethane from a feed stream which also contains propane and n-butane. A diagram of the evaporator is given in Figure 4–7.

Figure 4–7 Flash Evaporator

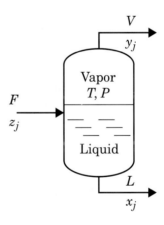

where

F = total feed flow rate in lb-mol/hr

V = total vapor flow rate in lb-mol/hr

L = total liquid flow rate in lb-mol/hr

z_j = mole fraction of component j in the feed stream

y_j = mole fraction of component j in the vapor stream

x_j = mol fraction of component j in the liquid stream

n_c = total number of components

The feed composition entering the evaporator is given in Table 4–5. The evaporator will operate under high pressure, between 15 and 25 atm, with a feed

Table 4–5 Liquid Composition and Antoine Equation Constants [a]

| | Mole | | | |
Component	Fraction	A	B	C
Ethylene	0.1	6.64380	395.74	266.681
Ethane	0.25	6.82915	663.72	256.681
Propane	0.5	6.80338	804.00	247.04
n-Butane	0.15	6.80776	935.77	238.789

[a]Thermodynamics Research Center API44 Hydrocarbon
Project, *Selected Values of Properties of Hydrocarbon and
Related Compounds*, Texas A&M University, College Station,
Texas (1978) with permission.

stream at 50 °C.

> (a) Calculate the percent of the total feed at 50 °C that is evaporated and
>   the corresponding mole fractions in the liquid and vapor streams for
>   the following pressures: $P = 15, 17, 19, 21, 23$, and 25 atm.
> (b) Determine the dew and the bubble point temperatures of the feed
>   stream.

Additional Information and Data

The vapor pressure of the individual components can be calculated from the
Antoine equation:

$$P_j = 10^{\left[A_j - \left(\frac{B_j}{C_j + T}\right)\right]} \tag{4-33}$$

where P_j is the vapor pressure of component j. The constants A_j, B_j, and C_j are
specific for component j, and T is the temperature in °C. The Antoine equation
constants for the hydrocarbons of this problem are presented in Table 4–5, where
P_j has units of mm Hg.

4.8.4   Solution (Partial)

The material balance and phase equilibrium equations for a flash evaporator can
be formulated into a single nonlinear algebraic equation (Henley and Rosen[4], p.
341) given by

$$f(\alpha) = \sum_{j=1}^{n_c} (x_j - y_j) = \sum_{j=1}^{n_c} \frac{z_j(1 - k_j)}{1 + \alpha(k_j - 1)} = 0 \tag{4-34}$$

where α is the vapor to feed ratio, $\alpha = V/F$, and k_j is the vapor liquid equilibrium ratio for component j. Once α has been determined, the vapor and liquid mole fractions can be calculated from

$$x_j = \frac{z_j}{1 + \alpha(k_j - 1)} \tag{4-35}$$

and

$$y_j = k_j x_j \tag{4-36}$$

For ideal systems, the vapor liquid equilibrium ratio can be calculated from

$$k_j = \frac{P_j}{P} \tag{4-37}$$

where P_j is the vapor pressure of component j and P is the total pressure in the evaporator. The k_j's needed in this problem can be calculated at a particular pressure and temperature by using the Antoine correlation, Equation (4-33), for each vapor pressure P_j in Equation (4-37).

   (a) This problem requires the solution of the single nonlinear Equation (4-34) along with Equations (4-33) and (4-35) through (4-37) for each component j. This set of equations can be solved for α, the vapor to feed ratio, and the resulting mole fractions. The interval for the solution is $0 \le \alpha \le 1$ since there is no physical meaning to $V/F > 1$ or $V/F < 0$. The equations as they are entered into POLYMATH *Simultaneous Algebraic Equation Solver* for the case of $P = 20$ atm and $T = 50\ °C$ are given as follows:

```
Equations:
f(alpha)=x1*(1-k1)+x2*(1-k2)+x3*(1-k3)+x4*(1-k4)
P=20*760
TC=50
k1=10^(6.6438-395.74/(266.681+TC))/P
k2=10^(6.82915-663.72/(256.681+TC))/P
k3=10^(6.80338-804/(247.04+TC))/P
k4=10^(6.80776-935.77/(238.789+TC))/P
x1=0.1/(1+alpha*(k1-1))
x2=0.25/(1+alpha*(k2-1))
x3=0.5/(1+alpha*(k3-1))
x4=0.15/(1+alpha*(k4-1))
y1=k1*x1
y2=k2*x2
y3=k3*x3
y4=k4*x4
alpha(min)=0, alpha(max)=1
```

   The solution obtained for this case is $\alpha = 0.6967$, indicating that at this temperature 69.67% of the feed has evaporated at a pressure of 20 atm. The corresponding mole fractions in the vapor and in the liquid streams are as follows:

Component Mole Fractions	Ethylene	Ethane	Propane	n-Butane
Feed	0.1	0.25	0.5	0.15
Vapor	0.1398	0.313904	0.46917	0.077126
Liquid	0.00857431	0.103203	0.570821	0.317401

 The POLYMATH problem solution file for part (a) ($P = 20$ atm and $T = 50$ °C) is found in the *Simultaneous Algebraic Equation Solver Library* located in CHAP4 with file named P4-08A.POL.

(b) For $\alpha = 0$ and constant pressure, Equation (4-34) reduces to

$$f(T) = 1 - \sum_{j=1}^{n_c} z_j k_j \qquad (4\text{-}38)$$

Solution of this equation for T gives the bubble point temperature.
For $\alpha = 1$ and constant pressure, Equation (4-34) reduces to

$$f(T) = \sum_{j=1}^{n_c} \frac{z_j}{k_j} - 1 \qquad (4\text{-}39)$$

Solution of Equation (4-39) for T yields the dew point temperature.

4.9 FLASH EVAPORATION OF VARIOUS HYDROCARBON MIXTURES

4.9.1 Concepts Demonstrated

Calculation of bubble point and dew point temperatures and associated vapor and liquid compositions for flash evaporation of an ideal multicomponent mixture.

4.9.2 Numerical Methods Utilized

Solution of a single nonlinear algebraic equation.

4.9.3 Problem Statement

Complete Problem 4.8 for a different three- or four-component mixture of hydrocarbons. Select the components from Table 4–6 and assign mole fractions in the feed that sum to one.

Table 4–6 Antoine Equation Constants for Various Hydrocarbons[a]

Substance	A	B	C
Methane	6.64380	395.74	266.681
Ethane	6.82915	663.72	256.681
Propane	6.80338	804.00	247.04
n-Butane	6.80776	935.77	238.789
n-Pentane	6.85296	1064.84	232.012
n-Hexane	6.87601	1171.17	224.408
n-Heptane	6.89677	1264.90	216.544
n-Octane	6.91868	1351.99	209.155
n-Nonane	6.93893	1431.82	202.011
n-Decane	6.94363	1495.17	193.858

[a]Thermodynamics Research Center API44 Hydrocarbon Project, *Selected Values of Properties of Hydrocarbon and Related Compounds*, Texas A&M University, College Station, Texas (1978).

4.9.4 Solution (Suggestion)

The pertinent equations are given in Problem 4.8.

4.10 CORRELATION OF ACTIVITY COEFFICIENTS WITH THE VAN LAAR EQUATIONS

4.10.1 Concepts Demonstrated

Estimation of parameters in the Van Laar equations for the correlation of binary activity coefficients.

4.10.2 Numerical Methods Utilized

Linear and nonlinear regression, transformation of data for regression, calculation and comparisons of confidence intervals, residual plots, and sum of squares.

4.10.3 Problem Statement

The Van Laar equations for correlation of binary activity coefficients are

$$\gamma_1 = \exp\left\{ A/[1 + (x_1/x_2)(A/B)]^2 \right\} \tag{4-40}$$

$$\gamma_2 = \exp\left\{ B/[1 + (x_2/x_1)(B/A)]^2 \right\} \tag{4-41}$$

where x_1 and x_2 are the mole fractions of components 1 and 2, respectively, and γ_1 and γ_2 are the activity coefficients. Parameters A and B are constant for a particular binary mixture.

Equations (4-40) and (4-41) can be combined to give the excess Gibbs energy expression:

$$g = G_E/RT = x_1 \ln \gamma 1 + x_2 \ln \gamma 2 = ABx_1 x_2/(Ax_1 + Bx_2) \tag{4-42}$$

Activity coefficients at various mole fractions are available from Table 2–7 for the benzene and n-heptane system from which the g in Equation (4-42) can be calculated. Linear regression can be used to estimate the parameter values of A and B. An alternate method is to sum Equations (4-40) and (4-41) and to utilize nonlinear regression on this sum to determine the values of both A and B.

(a) Use linear regression on Equation (4-42) with the data of Table 2–7 to determine A and B in the Van Laar equations for the benzene and n-heptane binary system.

(b) Estimate A and B by employing nonlinear regression on Equation (4-42) and a single equation that is the sum of Equations (4-40) and (4-41).

(c) Compare the results of the regressions in (a) and (b) using parameter confidence intervals, residual plots, and sums of squares of errors (least-squares summations calculated with both activity coefficients).

(d) Comment on the best choice to correlate this data set between the Margules equations of Problem 2.8 and the Van Laar equations of this problem.

4.10.4 Solution (Suggestions)

The approaches to this problem are similar to those that were used in Problem 2.8 to determine the Margules equation parameters.

Equation (4-42) can be rewritten in a linearized form for the determination of A and B using linear regression as

$$\frac{x_1}{x_1 \ln \gamma_1 + x_2 \ln \gamma_2} = \frac{1}{A} + \frac{1}{B}\frac{x_1}{x_2} \tag{4-43}$$

 The POLYMATH problem data file is found in the *Polynomial, Multiple Linear and Nonlinear Regression Library* located in directory CHAP2 with file named P2-08A.POL.

4.11 VAPOR LIQUID EQUILIBRIUM DATA FROM TOTAL PRESSURE MEASUREMENTS I

4.11.1   Concepts Demonstrated

Calculation of vapor composition from liquid composition and total pressure data using the Gibbs-Duhem equation, and consistency testing for activity coefficients.

4.11.2   Numerical Methods Utilized

Differentiation and integration of tabular data, solution of ordinary differential equations, and use of l'Hôpital's rule for functions undefined at the initial point of the solution interval.

4.11.3   Problem Statement

Data of liquid composition versus total pressure for the system benzene (1) and acetic acid (2) are presented in Table 4–7 .

(a)   Calculate the composition of the vapor phase as a function of liquid phase composition using the Gibbs-Duhem equation.
(b)   Calculate the activity coefficients for benzene and acetic acid in the liquid phase.
(c)   Check the consistency of the activity coefficients using the Gibbs-Duhem equation.

Table 4–7   Total Pressure Data at 50 °C for the System Benzene (1) Acetic Acid (2)[a]

x_1	P (mm Hg)	x_1	P (mm Hg)
0.0	57.52	0.8286	250.20
0.0069	58.2	0.8862	259.00
0.1565	126.00	0.9165	261.11
0.3396	175.30	0.9561	264.45
0.4666	189.50	0.9840	266.53
0.6004	224.30	1.0	271.00
0.7021	236.00'	0.8286	250.20
0.8286	250.20		

[a]*International Critical Tables*, 1st ed., Vol III, McGraw Hill Book Co., New York, 1928, p. 287.

Additional Information and Data

Vapor liquid equilibrium data for binary systems are very important for comput-
ing multicomponent and multiphase equilibrium. A simple and economical
experimental method for obtaining such data is to measure the total vapor pressure of the binary mixture as a function of composition at constant temperature.
The total pressure and liquid composition data can then be used to calculate the
vapor phase composition and activity coefficients in the liquid phase.

The Gibbs-Duhem equation is used as the basis for calculating the vapor
phase composition and the activity coefficients. This equation can be written for
a binary mixture as

$$x_1\left(\frac{\partial \ln \gamma_1}{\partial x_1}\right)_{P,T} + x_2\left(\frac{\partial \ln \gamma_2}{\partial x_1}\right)_{P,T} = 0 \tag{4-44}$$

where x_i is the mole fraction of the component i in the liquid phase and γ_i is the
liquid phase activity coefficient of component i.

The activity coefficients are defined as

$$\gamma_i = \frac{y_i P}{x_i P_i^v} \tag{4-45}$$

where y_i is the mole fraction of component i in the vapor phase, P is the total
pressure and P_i^v is the vapor pressure of component i.

Introduction of Equation (4-45) into Equation (4-44) with some manipula-
tions (see p. 448 in Balzhiser et al.[1]) leads to

$$\frac{dy_1}{dx_1} = \frac{y_1(1-y_1)}{y_1 - x_1}\frac{d \ln P}{dx_1} \tag{4-46}$$

If P versus x data are available, Equation (4-46) can be used to calculate y_1
versus x_1. This calculation requires an expression for the derivative $d \ln P/dx_1$
as a function of x_1. The generation of y_1 then requires the numerical solution of
Equation (4-46) integrated from $x_1 = 0$, where $y_1 = 0$ to $x_1 = 1.0$.

4.11.4   Solution

For this problem, the POLYMATH *Polynomial, Multiple Linear and Nonlinear
Regression Program* can be used to obtain a continuous representation of the
total pressure versus composition data from Table 4–7. The third-order polynomial that results is given by

$$P = 57.6218 + 463.137x_1 - 419.736x_1^2 + 169.871x_1^3 \tag{4-47}$$

This is the highest-degree polynomial in which all the coefficients are significantly different from zero. Consequently, this polynomial can be used to represent the variation of P as a function of x_1.

The POLYMATH program solution file for Table 4–7 is found in the *Polynomial, Multiple Linear and Nonlinear Regression Program Library* located in directory CHAP4 with file named P4-11A.POL.

Application of Equation (4-46) requires values for $d\ln P/dx_1$, and these can be obtained from Equation (4-47) as

$$\frac{d\ln(P)}{dx_1} = \frac{1}{P}\frac{dP}{dx_1} = \frac{1}{P}[463.137 - 2(419.736)x_1 + 3(169.871)x_1^2] \tag{4-48}$$

The initial condition on the integration of Equation (4-46) poses a difficulty because the equation is undefined when $x_1 = 0$ and $y_1 = 0$. The application of L'Hopital's rule can be used to determine the initial value of the derivative as

$$\frac{dy_1}{dx_1} = 1 + \frac{d\ln(P)}{dx_1}\bigg|_{x_1 = 0} \tag{4-49}$$

The use of a finite difference expression for the left side of Equation (4-49) allows a value of y_1 to be calculated for a small value of x_1. Thus for $x_1 = 0.00001$

$$y_1\big|_{x_1 = 0.00001} = \left(1 + \frac{d\ln(P)}{dx_1}\bigg|_{x_1 = 0}\right)10^{-5} = (1 + 8.0375)10^{-5} = 9.0375 \times 10^{-5} \tag{4-50}$$

From these initial values of x_1 and y_1, Equation (4-46) can be integrated, but it becomes undefined once more for $x_1 = 1$ as $y_1 = 1$. The practical solution to this difficulty is to stop the integration when x_1 becomes 0.99999, giving reasonable results.

During the course of the integration, the activity coefficients can be calculated from

$$\gamma_1 = \frac{y_1 P}{x_1 P_1^v} \quad \text{and} \quad \gamma_2 = \frac{(1-y_1)P}{(1-x_1)P_2^v} \tag{4-51}$$

and P can be evaluated from Equation (4-47). The pure component vapor pressures are known from Table 4–7 to be $P_1^v = 271.00$ mm Hg and $P_2^v = 57.52$ mm Hg.

The Gibbs-Duhem equation can also be used to check the quality and the consistency of the results based on experimental data. For a binary mixture the Gibbs-Duhem equation requires that the integral given by

$$I = \int_0^1 \ln\frac{\gamma_1}{\gamma_2}dx_1 \tag{4-52}$$

is equal to zero.

Equation (4-52) can be written as

$$\frac{dI}{dx_1} = \ln\frac{\gamma_1}{\gamma_2} \tag{4-53}$$

and this differential equation can be evaluated during the problem solution since γ_1 and γ_2 are calculated during the numerical integration. The initial condition on I is zero.

Thus the problem consists of ordinary differential equations (4-46) and (4-53) to be solved simultaneously with algebraic equations (4-47), (4-48), and (4-51).

This set of equations can be entered into the POLYMATH *Simultaneous Differential Equation Solver* as follows:

```
Equations:
d(y1)/d(x1)=dy1dx1
d(I)/d(x1)=ln(gamma1/gamma2)
P=57.6218+463.137*x1-419.736*x1^2+169.871*x1^3
gamma1=y1*P/(x1*271)
gamma2=(1-y1)*P/((1-x1)*57.52)
dlnpdx=(463.137-2*419.736*x1+3*169.871*x1^2)/P
dy1dx1=y1*(1-y1)*dlnpdx/(y1-x1)
Initial Conditions:
x1(0)=1e-05
y1(0)=9.029e-05
I(0)=0
Final Value:
x1(f)=0.999999
```

The calculated mole fractions and activity coefficient values are summarized in Table 4–8.

Table 4–8 Calculated y_1, γ_1, and γ_2 Values for the Benzene Acetic Acid System

x1	y1	gamma1	gamma2	P
1e–05	9.029e–05	1.9199596	1.0017699	57.626431
0.1000089	0.47856451	1.7641996	1.0063723	99.911431
0.2000078	0.6512689	1.6199525	1.021749	134.82119
0.3000067	0.74097647	1.4889822	1.051022	163.3749
0.4000056	0.7967668	1.3714763	1.098806	186.59175
0.5000045	0.83594805	1.2677358	1.172166	205.49094
0.6000034	0.86655125	1.1782667	1.2823652	221.09166
0.7000023	0.89339823	1.1039724	1.4481365	234.4131
0.8000012	0.92059699	1.0466019	1.7012284	246.47445
0.9000001	0.95357238	1.0098518	2.0848454	258.29491
0.999999	0.99999949	0.99960812	2.3923907	270.89367

Figure 4–8 shows a plot of the integral given by $I = \int_0^{x_1} \ln\frac{\gamma_1}{\gamma_2}dx_1$ versus x_1 from $x_1 = 0.00001$ up to $x_1 = 0.99999$.

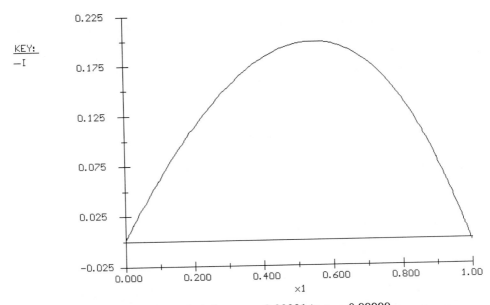

Figure 4–8 Integral of $\ln(\gamma_1/\gamma_2)$ from $x_1 = 0.00001$ to $x_1 = 0.99999$

While the integration cannot be carried out up to exactly $x_1 = 1$ because Equation (4-46) is undefined at this point, it is clearly seen that the integral gets very close to zero (value is -2.2×10^{-3}) for $x_1 = 0.99999$, indicating that the activity coefficient calculations are consistent.

The determination of the polynomial that represents the pressure P has a critical influence on this problem. This has a particularly major effect on the activity coefficients. Numerical treatment of experimental data to obtain derivative information is always challenging.

 The POLYMATH problem solution file for the *Simultaneous Differential Equation Solver Library* is located in directory CHAP4 with file named P4-11ABC.POL.

4.12 VAPOR LIQUID EQUILIBRIUM DATA FROM TOTAL PRESSURE MEASUREMENTS II

4.12.1 Concepts Demonstrated

Calculation of vapor composition from liquid composition and total pressure data using the Gibbs-Duhem equation, and consistency testing for activity coefficients.

4.12.2 Numerical Methods Utilized

Differentiation and integration of tabular data, solution of ordinary differential equations, and use of l'Hôpital's rule for functions undefined at the initial point of the solution interval.

4.12.3 Problem Statement

Data from the *International Critical Tables*[5] (pp. 286–290) of total pressure versus liquid composition are available for four different systems in the Appendix: water-methyl alcohol (Table C–1), ethyl ether-chloroform (Table C–2), toluene-acetic acid (Table C–3), and chloroform-acetone (Table C–4).

Select one of the four systems with total pressure versus liquid composition data.

(a) Calculate the composition of the vapor phase as function of liquid phase composition using the Gibbs-Duhem equation.
(b) Calculate the activity coefficients for both components in the liquid phase.
(c) Check the consistency of the activity coefficients using the Gibbs-Duhem equation.

4.12.4 Solution (Suggestions)

See Section 4.11.4 for method of solution. Note that Equation (4-46) must be integrated in the direction of increasing pressure (see Van Ness[10]). Note that systems that form azeotropes (such as toluene-acetic acid) must be divided into two parts at the azeotropic composition. The integration should be carried out in the two parts in opposite directions.

4.13 COMPLEX CHEMICAL EQUILIBRIUM

4.13.1   Concepts Demonstrated

Calculation of equilibrium concentrations for a complex reacting system in a constant volume batch reactor and use of the van't Hoff equation to estimate changes in equilibrium constants with temperature.

4.13.2   Numerical Methods Utilized

Effective solution of systems of nonlinear algebraic equations, and useful techniques for examining the possible multiple solutions of such systems.

4.13.3   Problem Statement

Problem 3.7 involves a gas mixture of 50% A and 50% B charged to a constant volume batch reactor in which equilibrium is quickly achieved. The initial total concentration is 3.0 g-mol/dm^3. Three independent reactions are known to occur.

Reaction 1: $A + B \leftrightarrow C + D$ $K_{C1}(350 \text{ K}) = 1.06$

Reaction 2: $C + B \leftrightarrow X + Y$ $K_{C2}(350 \text{ K}) = 2.63$

Reaction 3: $A + X \leftrightarrow Z$ $K_{C3}(350 \text{ K}) = 5.0 \text{ dm}^3/\text{g-mol}$

At 330 K, the equilibrium constants based on concentrations are $K_{C1}(330 \text{ K}) = 0.7$, $K_{C2}(330 \text{ K}) = 4.0$, and $K_{C3}(330 \text{ K}) = 5.0 \text{ dm}^3/\text{g-mol}$.

(a)   Calculate the equilibrium concentrations of all reaction components at 330 K.

(b)   Repeat part (a) at 370 K.

4.13.4   Solution (Suggestions)

(a) See Section 3.7.4 for solution assistance.

(b) The equilibrium constants at 370 K can be estimated using the van't Hoff equation with the assumption of constant heat of reaction, ΔH_R.

$$\frac{d \ln K_P}{dT} = \frac{\Delta H_R}{RT^2} \tag{4-54}$$

The relationship between K_C and K_P is discussed by Fogler.[3]

4.14 REACTION EQUILIBRIUM AT CONSTANT PRESSURE OR CONSTANT VOLUME

4.14.1 Concepts Demonstrated

Calculation of equilibrium concentrations for both constant pressure and constant volume conditions using a combination of elemental balances and equilibrium expressions.

4.14.2 Numerical Methods Utilized

Solution of simultaneous nonlinear and linear algebraic equations.

4.14.3 Problem Statement

At high temperatures and low pressures, hydrogen sulfide and sulfur dioxide undergo the following reversible reactions:

$$H_2S \leftrightarrow H_2 + \frac{1}{2}S_2 \qquad K_{P1} = 0.45 \text{ atm}^{1/2} \qquad \text{(4-55)}$$

$$2H_2S + SO_2 \leftrightarrow 2H_2O + \frac{3}{2}S_2 \qquad K_{P2} = 28.5 \text{ atm}^{1/2} \qquad \text{(4-56)}$$

An initial gas mixture contains 45 mol % H_2S, 25 mol % SO_2, and the balance inert N_2 at a pressure of 1.2 atm.

(a) Calculate the equilibrium mole fractions of all reaction components when the preceding reactions are at equilibrium at a constant pressure of 1.2 atm.

(b) Repeat (a) for equilibrium at constant volume when the initial pressure is 1.2 atm.

4.14.4 Solution (Suggestions)

Material balances on the various species involved in this system (H, S, O, N) can yield linear algebraic equations that can be solved with Equations (4-55) and (4-56). The POLYMATH *Simultaneous Algebraic Equation Solver* can be used to solve all the equations. It is advisable to enter all equations as nonlinear equations for simultaneous solution. Please apply the suggestions given in Section 3.7.4 regarding the expediting of solutions of nonlinear equations.

REFERENCES

1. Balzhiser, R. E., Samuels, M. R., and Eliassen, J. D. *Chemical Engineering Thermodynamics*, Englewood Cliffs, NJ: Prentice Hall, 1972.
2. Edmister, W. C. *Hydrocarb. Process.*, **47** (10), 145–149 (1968).
3. Fogler, H. S. *Elements of Chemical Reaction Engineering*, 2nd ed., Englewood Cliffs, NJ: Prentice Hall, 1992.
4. Henley, E. J., and Rosen, E. M. *Material and Energy Balance Computation*. New York: Wiley, 1969.
5. *International Critical Tables*, 1st ed., Vol III, New York: McGraw-Hill, 1928.
6. Kyle, B. G. *Chemical and Process Thermodynamics*, 2nd ed., Englewood Cliffs, NJ: Prentice Hall, 1992.
7. Modell, M., Reid, C. *Thermodynamics and its Applications*, 2nd ed., Englewood Cliffs, NJ: Prentice Hall, 1983.
8. Sandler, S. I. *Chemical and Engineering Thermodynamics*, New York: Wiley, 1989.
9. Smith, J. M., and Van Ness, H. C. *Introduction to Chemical Engineering Thermodynamics*, 3rd ed., New York: McGraw-Hill, 1975.
10. Van Ness, H. C. *AIChE J.* **16** (1), 18 (1970).
11. Walas, S. M. *Phase Equilibrium in Chemical Engineering*, Stoneham, MA: Butterworth, 1985.

Fluid Mechanics

5.1 LAMINAR FLOW OF A NEWTONIAN FLUID IN A HORIZONTAL PIPE

5.1.1 Concepts Demonstrated

Solution of momentum balance to obtain shear stress and velocity profiles for a Newtonian fluid in laminar flow in a horizontal pipe and comparison of numerical and analytical solutions.

5.1.2 Numerical Methods Utilized

Solution of simultaneous first order ordinary differential equations employing a shooting technique to converge on the desired boundary conditions, use of combined variables, and avoidance of division by zero in calculating expressions.

5.1.3 Problem Statement

An incompressible Newtonian fluid is flowing at steady state inside a horizontal circular pipe at constant temperature. This flow is fully developed so that the velocity profile does not vary in the direction of flow. Figure 5–1 presents a schematic of this laminar flow for the coordinate system where the flow is in the x direction and the inside pipe radius is R in m.

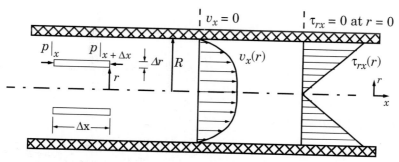

Figure 5–1 Differential Element for Laminar Flow Treatment in a Horizontal Pipe

159

A shell momentum balance of a control volume within the pipe gives the differential equation

$$\frac{d}{dr}(r\tau_{rx}) = \left(\frac{\Delta p}{L}\right)r \qquad (5\text{-}1)$$

where r is the radius in m, τ_{rx} is the shear stress at radius r in kg/m·s^2 or Pa, and Δp is the pressure drop in kg/m·s^2 or Pa over length L in m.

For a Newtonian fluid, the shear stress (or momentum flux) is linearly related to the velocity gradient by

$$\tau_{rx} = -\mu\frac{dv_x}{dr} \qquad (5\text{-}2)$$

where μ = the viscosity in kg/m·s or Pa·s.

The boundary conditions for Equations (5-1) and (5-2) are that

$$\tau_{rx} = 0 \qquad \text{at } r = 0 \qquad (5\text{-}3)$$

$$v_x = 0 \qquad \text{at } r = R \qquad (5\text{-}4)$$

The analytical solution with the two boundary conditions yields

$$\tau_{rx} = \left(\frac{\Delta p}{2L}\right)r \qquad (5\text{-}5)$$

$$v_x = \frac{\Delta p}{4\mu L}R^2\left[1 - \left(\frac{r}{R}\right)^2\right] \qquad (5\text{-}6)$$

where $\Delta p = p_0 - p_L$.

The average velocity $v_{x,\text{av}}$ is calculated from

$$v_{x,\text{av}} = \frac{1}{\pi R^2}\int_0^R v_x 2\pi r\, dr \qquad (5\text{-}7)$$

to give the analytical solution

$$v_{x,\text{av}} = \frac{(p_0 - p_L)R^2}{8\mu L} = \frac{(p_0 - p_L)D^2}{32\mu L} \qquad (5\text{-}8)$$

(a) Numerically solve Equations (5-1) and (5-2) with the boundary conditions given by Equations (5-3) and (5-4) for water at 25 °C with $\mu = 8.937 \times 10^{-4}$ kg/m·s, $\Delta p = 500$ Pa, $L = 10$ m, and $R = 0.009295$ m. This solution should utilize an ODE solver with a shooting technique and should employ some technique for converging on the boundary condition given by Equation (5-4).

(b) Compare the calculated shear stress and velocity profiles with the analytical solutions given by Equations (5-5) and (5-6).

(c) Modify your solution to part (a) to include calculation of the average velocity given by Equation (5-7) and compare your solution with the analytical solution of Equation (5-8).

5.1.4 Solution

This set of differential equations is similar to the set solved in Problem 3.6, and the same techniques can be used.

(a) The differential equations to be solved in this problem consist of Equations (5-1) and (5-2). Equation (5-2) can be rearranged as

$$\frac{dv_x}{dr} = -\frac{\tau_{rx}}{\mu} \tag{5-9}$$

There is no need to use the chain rule from differential calculus to work on the derivative in Equation (5-1) to separate out the derivative $d\tau_{rx}/dr$. Instead in this case, and in most problems, it is highly recommended that the combined variable such as $r\tau_{rx}$ should be retained, and an algebraic equation should be used to calculate the individual variables. In this case the algebraic equation is

$$\tau_{rx} = \frac{r\tau_{rx}}{r} \tag{5-10}$$

The simultaneous solution of the two ordinary differential equations, Equations (5-1) and (5-10), requires that an initial condition for v_x be assumed at $r = 0$ so that the boundary condition given by $v_x = 0$ will be satisfied at $r = R$. The shooting method discussed in Section 3.6.4 will be used to determine this initial condition. A first trial initial condition for v_x, say $v_{x0} = 2.0$, can be used to integrate the equations and calculate the error in the boundary condition for v_x at $r = R$. This error can be defined as ε which is the difference between the numerical solution and the desired final value of v_x at $r = R$.

$$\varepsilon(v_{x0}) = v_{x,\,calc} - v_{x,\,desired} \tag{5-11}$$

In this problem, $v_{x,desired} = 0$ indicates no velocity at the wall of the pipe.

The equation set as entered into the POLYMATH *Simultaneous Differential Equation Solver* for the first step of the shooting method is

```
Equations:
d(rTAUrx)/d(r)=deltaP*r/L
d(Vx)/d(r)=-TAUrx/mu
TAUrx=rTAUrx/r
mu=8.937e-4
deltaP=500
L=10
err=Vx-0
R=.009295
Initial Conditions:
r(0)=0
rTAUrx(0)=0
Vx(0)=1.20842
Final Value:
r(f)=0.009295
```

The numerical solution of the preceding equation set cannot be carried out because of the division by r in the calculation of TAUrx, because r is zero when the integration is initiated.

Division by Zero
A simple way to avoid division by zero is to use the "if ... then ... else ... " capability in POLYMATH. In this case the algebraic equation for calculation of τ_{rx} in POYMATH can be modified to

```
TAUrx=if(r>0)then(rTAUrx/r)else(0)
```

and the numerical solution of the initial trial solution can be achieved.

Boundary Condition Convergence
Once the initial trial solution has been obtained, additional integrations can be made in order to minimize the error as calculated by Equation (5-11). This can utilize either trial and error or some simple convergence logic, such as the secant or false position methods, which are discussed in Problem 3.5.

The final solution of this two-point boundary value problem for part (a) is presented in Table 5–1.

Table 5–1 Partial Results for Selected Variables

Variable	Initial Value	Maximum Value	Minimum Value	Final Value
r	0	0.009295	0	0.009295
rTAUrx	0	0.00215993	0	0.00215993
Vx	1.20842	1.20842	2.39622e-06	2.39622e-06
TAUrx	0	0.232375	0	0.232375
mu	0.0008937	0.0008937	0.0008937	0.0008937

Table 5–1 Partial Results for Selected Variables

Variable	Initial Value	Maximum Value	Minimum Value	Final Value
deltaP	500	500	500	500
L	10	10	10	10
err	1.20842	1.20842	2.39622e-06	2.39622e-06
R	0.009295	0.009295	0.009295	0.009295

 The POLYMATH problem solution file for part (a) is found in the *Simultaneous Differential Equation Solver Library* located in directory CHAP5 with file named P5-01A.POL.

(b) The analytical solutions for the shear stress and velocity profiles given in Equations (5-5) and (5-6) can be entered directly into the POLYMATH equation set for evaluation as

```
TAUrxANAL=(deltaP/(2*L))*r
VxANAL=(deltaP*R^2/(4*mu*L))*(1-(r/R)^2)
```

Evaluation of the preceding equations can easily be compared with the numerical solution by requesting POLYMATH to print a table for results at 10 intervals during the integration. This is shown in Table 5–2, where the numerical and analytical solutions are shown to agree to approximately six significant figures.

Table 5–2 Comparisons of Numerical and Analytical Solutions for τ_{rx} and v_x

r	TAUrx	TAUrxANAL	Vx	VxANAL
0	0	0	1.20842	1.2084176
0.0009295	0.0232375	0.0232375	1.1963358	1.1963334
0.001859	0.046475	0.046475	1.1600833	1.1600809
0.0027885	0.0697125	0.0697125	1.0996624	1.09966
0.003718	0.09295	0.09295	1.0150732	1.0150708
0.0046475	0.1161875	0.1161875	0.9063156	0.9063132
0.005577	0.139425	0.139425	0.77338966	0.77338727
0.0065065	0.1626625	0.1626625	0.61629537	0.61629298
0.007436	0.1859	0.1859	0.43503273	0.43503034
0.0083655	0.2091375	0.2091375	0.22960174	0.22959934
0.009295	0.232375	0.232375	2.396218e-06	0

(c) The average velocity in the pipe is calculated from Equation (5-7) which is an integral equation. This equation can be differentiated with respect to r to yield the differential equation

$$\frac{d}{dr}(v_{x,\,\mathrm{av}}) = \frac{v_x 2r}{R^2} \tag{5-12}$$

whose initial condition is zero. This equation can be entered as another differential equation in the POLYMATH equation set and integrated to $r = R$. Note that $v_{x,\,\mathrm{av}}$ is determined only at the end of the integration. The analytical solution for average velocity is given by Equation (5-8), which can be calculated during the numerical solution. The additional statements for POLYMATH are

```
d(Vxav)/d(r)=Vx*2*r/R^2
VxavANAL=deltaP*R^2/(8*mu*L)
```

The numerical and analytical solutions for the average v_x are 0.604211 m/s and 0.604209 m/s, respectively, which are in good agreement to five significant figures.

 The POLYMATH problem solution file for this complete problem is found in the *Simultaneous Differential Equation Solver Library* located in directory CHAP5 with file named P5-01ABC.POL.

5.2 LAMINAR FLOW OF NON-NEWTONIAN FLUIDS IN A HORIZONTAL PIPE

5.2.1 Concepts Demonstrated

Solution of momentum balance to obtain shear stress and velocity profiles for power law fluids in laminar flow in a horizontal pipe., and comparison of numerical and analytical solutions.

5.2.2 Numerical Methods Utilized

Solution of simultaneous first order ordinary differential equations employing a shooting technique to converge on the desired boundary conditions and use of combined variables in problem solution.

5.2.3 Problem Statement

The shear stress for a number of non-Newtonian fluids can be described by

$$\tau_{rx} = -K\left|\frac{dv_x}{dr}\right|^{n-1}\frac{dv_x}{dr} \tag{5-13}$$

where parameter K has units of $N \cdot s^n/m^2$ and exponent n is the flow index. For $n < 1$ the fluid is *pseudoplastic* and for $n > 1$ the fluid is *dilatant*. When $n = 1$, the fluid is Newtonian.

Application of Equation (5-13) in numerical solutions typically requires the differential equation for velocity.

$$\frac{dv_x}{dr} = -\left(\frac{\tau_{rx}}{K}\right)^{1/n} \quad \text{(if } \tau_{rx} > 0\text{)} \tag{5-14}$$

$$\frac{dv_x}{dr} = \left(\frac{-\tau_{rx}}{K}\right)^{1/n} \quad \text{(if } \tau_{rx} \le 0\text{)} \tag{5-15}$$

The analytical velocity profile for these fluids flowing in a horizontal cylindrical pipe is given by

$$v_x = \frac{n}{n+1}\left(\frac{\Delta p}{2KL}\right)^{1/n}R^{(n+1)/n}\left[1-\left(\frac{r}{R}\right)^{(n+1)/n}\right] \tag{5-16}$$

(a)   A dilatant fluid is flowing in a horizontal pipe where $K = 1.0 \times 10^{-6}$, $n = 2$, $\Delta p = 100$ Pa, $L = 10$ m, and $R = 0.009295$ m. The shear stress τ_{rx} and velocity profile for v_x should be plotted versus the radius r of the pipe. Calculate the average velocity in the pipe $v_{x,\text{av}}$.

(b)   Compare the calculated velocity profile with the analytical solution given by Equation (5-16).

(c)   Repeat part (a) for a pseudoplastic fluid with $K = 0.01$ and $n = 0.5$.

(d)   Repeat part (b) for the pseudoplastic fluid of part (c).

5.2.4   Solution (Suggestions)

(a) The numerical solution requires ordinary differential equations for both τ_{rx} and v_x. The equation for τ_{rx} in the pipe is obtained from a shell momentum balance in the pipe. The result is Equation (5-1), which is independent of the fluid properties. Note that use of the combined variable $r\tau_{rx}$ is strongly suggested in Problem 5.1. This requires a separate algebraic expression, Equation (5-10), to yield τ_{rx} during the numerical solution. The POLYMATH "if ... then ... else ... " statement can be used to select the correct option given by Equations (5-14) and (5-15). Convergence to the boundary conditions is discussed in Section 5.1.4.

5.3 VERTICAL LAMINAR FLOW OF A LIQUID FILM

5.3.1   Concepts Demonstrated

Shear stress, velocity profiles and average velocity calculations for laminar flow of Newtonian and Bingham fluids down a vertical surface.

5.3.2   Numerical Methods Utilized

Solution of simultaneous first order ordinary differential equations employing a shooting technique to converge on the desired boundary conditions.

5.3.3   Problem Statement

A liquid is flowing down a vertical surface as a fully established film where the velocity profile in the film does not vary in the direction of flow. This is illustrated in Figure 5–2. This situation is discussed in more detail by Geankoplis[4] and Bird et al.[1]

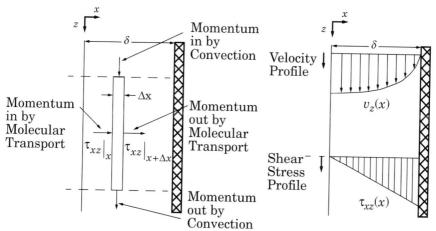

Figure 5–2 Differential Element for Vertical Flow of a Liquid Film

A shell momentum balance yields a differential equation for the shear stress (momentum flux)

$$\frac{d\tau_{xz}}{dx} = \rho g \qquad (5\text{-}17)$$

where the boundary condition for the free surface is

$$\tau_{xz} = 0 \qquad \text{at } x = 0 \qquad (5\text{-}18)$$

For a Newtonian fluid, the shear stress is related to the velocity gradient by

$$\tau_{xz} = -\mu \frac{dv_z}{dx} \tag{5-19}$$

where the boundary condition at the wall is

$$v_z = 0 \quad \text{at } x = \delta \tag{5-20}$$

The average velocity in the film can be calculated from

$$v_{z,\,av} = \frac{1}{\delta} \int_0^{\delta} v_z \, dz \tag{5-21}$$

The analytical solution for the velocity distribution is

$$v_z = \frac{\rho g \delta^2}{2\mu} \left[1 - \left(\frac{x}{\delta} \right)^2 \right] \tag{5-22}$$

and the analytical solution for the average velocity is

$$v_{z,\,av} = \frac{\rho g \delta^2}{3\mu} \tag{5-23}$$

For non-Newtonian fluids, the relationship between shear stress and velocity gradient varies from the linear expression given in Equation (5-19). A Bingham fluid is represented by the following relationships:

$$\tau_{xz} = -\mu_0 \frac{dv_z}{dx} \pm \tau_0 \quad (\text{if } |\tau_{xz}| > \tau_0) \tag{5-24}$$

$$\frac{dv_z}{dx} = 0 \quad (\text{if } |\tau_{xz}| \leq \tau_0) \tag{5-25}$$

where the positive sign on τ_0 is used when τ_{xz} is positive and the negative sign is used when τ_{xz} is negative.

For numerical solutions where the differential equation for velocity is needed, Equations (5-24) and (5-25) can be expressed as

$$\frac{dv_z}{dx} = \frac{(\tau_0 - \tau_{xz})}{\mu_0} \qquad \text{if } \tau_{xz} > \tau_0$$

$$\frac{dv_z}{dx} = 0 \qquad \text{if } \left|\tau_{xz}\right| \leq \tau_0 \qquad \textbf{(5-26)}$$

$$\frac{dv_z}{dx} = \frac{(\tau_0 + \tau_{xz})}{-\mu_0} \qquad \text{if } \tau_{xz} < (-\tau_0)$$

with constants τ_0 and μ_0 having known values for the particular fluid.

(a) Numerically solve Equations (5-17) and (5-19) with the boundary conditions given by Equations (5-18) and (5-20) for an oil that is a Newtonian fluid with $\mu = 0.15$ kg/m·s and $\rho = 840.0$ kg/m^3. The film thickness is $\delta = 0.002$ m. The shear stress τ_{xz} and velocity profile for v_z should be plotted versus the location x in the film. Calculate the average velocity in the film $v_{z,av}$.

(b) Compare the calculated velocity profile with the analytical solutions given by Equation (5-22).

(c) Repeat part (a) for a Bingham plastic fluid by solving Equations (5-17) and (5-26), where $\tau_0 = 6.0$ kg/m·s^2, $\mu_0 = 0.20$ kg/m·s, and $\rho = 800$ kg/m^3. What is unique about the velocity profile for this fluid?

5.3.4   Solution (Suggestions)

(a) and (b) Similar differential equations are solved in Problem 3.6.

(c) Equation (5-26) along with the logic as discussed for a Bingham plastic fluid must be used instead of Equation (5-19).

5.4 LAMINAR FLOW OF NON-NEWTONIAN FLUIDS IN A HORIZONTAL ANNULUS

5.4.1   Concepts Demonstrated

Shear stress, velocity profiles, and average velocity calculations for laminar flow of Newtonian, dilatant, and pseudoplastic fluids in a horizontal annulus.

5.4.2   Numerical Methods Utilized

Solution of simultaneous first-order ordinary differential equations employing a shooting technique to converge on the desired boundary conditions.

5.4.3   Problem Statement

An incompressible fluid is flowing within an annulus between two concentric horizontal pipes, as shown in Figure 5–3 where the flow is fully developed. The shell momentum balance discussed in Section 5.1.3 applies.

$$\frac{d}{dr}(r\tau_{rx}) = \left(\frac{\Delta p}{L}\right)r \tag{5-27}$$

The particular fluid will determine the relationship between the shear rate and the shear stress. For a Newtonian fluid, this can be written as

$$\frac{dv_x}{dr} = -\frac{\tau_{rx}}{\mu} \tag{5-28}$$

These two equations describe the flow in the annulus, but the boundary

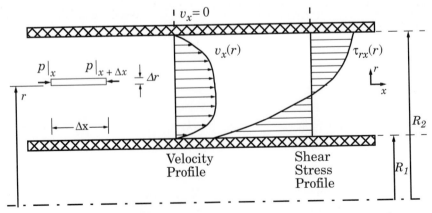

Figure 5–3 Differential Element for Laminar Flow in a Horizontal Annulus

conditions are that the velocity at each wall is zero. Thus

$$v_x = 0 \quad \text{at } r = R_1 \tag{5-29}$$

$$v_x = 0 \quad \text{at } r = R_2 \tag{5-30}$$

An analytical solution for the velocity profile in the annulus with a Newtonian fluid is

$$v_x = \frac{\Delta p}{4\mu L}\left[R_2^2 - r^2 + \frac{(R_2^2 - R_1^2)}{\ln(R_2/R_1)}\left(\ln\frac{r}{R_2} \right) \right] \tag{5-31}$$

where the pressure drop is given by $\Delta p = p_0 - p_L$. The average velocity $v_{x,av}$ is calculated from

$$v_{x,av} = \frac{1}{\pi(R_2^2 - R_1^2)} \int_{R_1}^{R_2} v_x 2\pi r \, dr \tag{5-32}$$

to give the analytical solution

$$v_{x,av} = \frac{\Delta p}{8\mu L}\left[R_1^2 + R_2^2 - \frac{(R_2^2 - R_1^2)}{\ln(R_2/R_1)} \right] \tag{5-33}$$

For a Newtonian fluid such as water, Equation (5-9) relates the derivative of the velocity to shear stress. A number of non-Newtonian fluids can be described by

$$\tau_{rx} = -K\left| \frac{dv_x}{dr} \right|^{n-1} \frac{dv_x}{dr} \tag{5-34}$$

which is discussed in Section 5.2.3 for both pseudoplastic and dilatant fluids. Relationships between the velocity derivative and the shear stress are given in Equations (5-14) and (5-15).

(a) Numerically solve Equations (5-27) and (5-28) with the boundary condi-
     tions given by Equations (5-29) and (5-30) for water, a Newtonian fluid,
     at 25 °C with $\mu = 8.937 \times 10^{-4}$ kg/m·s, $\Delta p = 100$ Pa, $L = 10$ m, $R_1 =$
     0.02223 m, and $R_2 = 0.03129$ m. The shear stress τ_{rx} and velocity v_x
     should be plotted versus the radius r in the annulus from R_1 to R_2. Cal-
     culate the average velocity in the annulus from the differential equa-
     tion form of Equation (5-32), as detailed in Problem 5.1.
(b) Compare the calculated velocity profile with the analytical solutions
     given by Equation (5-31). Compare the calculated average velocity with
     the analytical solution given by Equation (5-32).
(c) Repeat part (a) for a dilatant fluid with $K = 1.0 \times 10^{-6}$ and $n = 2$.
(d) Repeat part (a) for a pseudoplastic fluid with $K = 0.01$ and $n = 0.5$.

5.4.4   Solution (Suggestion)

(a)–(c) A good initial estimate for τ_{rx} at R_1 is –0.33 for all fluids.

5.5 TEMPERATURE DEPENDENCY OF DENSITY AND VISCOSITY OF VARIOUS LIQUIDS

5.5.1   Concepts Demonstrated

General correlation of density and viscosity for liquids as a function of temperature in both English and SI units.

5.5.2   Numerical Methods Utilized

Regression of experimental data to different functional forms.

5.5.3   Problem Statement

Fluid flow calculations typically require correlations of the density and viscosity of liquids at various temperatures. In this section, various equations representing these properties as a function of temperature will be determined for water as a representative case.

Tables D–1 to D–3 in Appendix D present density and absolute viscosity of various liquids as a function of temperature in both English and SI units.

(a) Determine appropriate correlations for the temperature dependency of density and viscosity of liquid water in both English and SI units using the data of Tables D–1 and D–2.

(b) Select one of the components in Table D–3 and provide similar correlations that give the results for density and viscosity in SI units.

5.5.4   Solution (Partial)

The techniques for finding the most appropriate correlations are discussed in Chapter 2. Using those methods and the data in Table D–1 and Table D–2 yields the following results for water.

(a) In English units, the data of Table D–1 for water yield

$$\rho = 62.122 + 0.0122T - 1.54 \times 10^{-4}T^2 + 2.65 \times 10^{-7}T^3 - 2.24 \times 10^{-10}T^4 \qquad \textbf{(5-35)}$$

$$\ln \mu = -11.0318 + \frac{1057.51}{T + 214.624} \qquad \textbf{(5-36)}$$

where T is in °F, ρ is in lb_m/ft^3, and μ is in $lb_m/ft \cdot s$.

In SI units, the data of Table D–2 for water yield

$$\rho = 46.048 + 9.418T - 0.0329T^2 + 4.882\times10^{-5}T^3 - 2.895\times10^{-8}T^4 \qquad \textbf{(5-37)}$$

$$\ln \mu = -10.547 + \frac{541.69}{T - 144.53} \qquad \textbf{(5-38)}$$

where T is in K, ρ is in kg/m^3, and μ is in Pa·s or kg/m·s.

(b) Table D–3 provides only data in English units. Use appropriate conversion factors to first convert the data to SI units and then regress equations similar to Equations (5-35) to (5-38) in SI units.

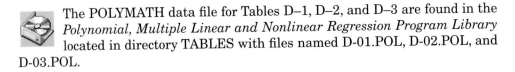 The POLYMATH data file for Tables D–1, D–2, and D–3 are found in the *Polynomial, Multiple Linear and Nonlinear Regression Program Library* located in directory **TABLES** with files named D-01.POL, D-02.POL, and D-03.POL.

5.6 TERMINAL VELOCITY OF FALLING PARTICLES

5.6.1 Concepts Demonstrated

Calculation of terminal velocity of solid particles falling in liquid or gas under gravity and additional forces.

5.6.2 Numerical Methods Utilized

Solution of single nonlinear equation.

5.6.3 Problem Statement

(a) Calculate the terminal velocity for 65 mesh particles of coal $\rho_p = 1800$ kg/m³ falling in one of the liquids listed in Tables D–1 and D–3 at 25 °C.

(b) Estimate the terminal velocity of the coal particles in a centrifugal separator where the acceleration is 30g.

Additional Information and Data

Assuming that the coal particles are spherical, a force balance on a particle yields

$$v_t = \sqrt{\frac{4g(\rho_p - \rho)D_p}{3C_D\rho}} \tag{5-39}$$

where v_t is the terminal velocity in m/s, g is the acceleration of gravity given by $g = 9.80665$ m/s², ρ_p is the particle's density in kg/m³, ρ is the fluid density in kg/m³, D_p is the diameter of the spherical particle in m, and C_D is a dimensionless drag coefficient.

The drag coefficient on a spherical particle at terminal velocity varies with the Reynolds number (Re) as follows (pp. 5-63 and 5-64 in Perry et al.[9]):

$$C_D = 24/Re \quad \text{for} \quad Re < 0.1 \tag{5-40}$$

$$C_D = (24/Re)(1 + 0.14Re^{0.7}) \quad \text{for} \quad 0.1 \le Re \le 1000 \tag{5-41}$$

$$C_D = 0.44 \quad \text{for} \quad 1000 < Re \le 350000 \tag{5-42}$$

$$C_D = 0.19 - 8 \times 10^4/Re \quad \text{for} \quad 350000 < Re \tag{5-43}$$

where $Re = D_p v_t \rho/\mu$ and μ is the viscosity in Pa·s or kg/m·s.

5.6.4 Solution

(a) For conditions similar to those of this problem, the Reynolds number will not exceed 1000 so that only Equations (5-40) and (5-41) need to be applied. The logic that selects the proper equation based on the value of Re can be employed using the "if … then … else … " capability within the POLYMATH *Simultaneous Algebraic Equation Solver.*

$$C_D = \text{if(} Re < 0.1) \quad \text{then} \quad (24/Re) \quad \text{else} \quad ((24/Re) \times (1 + 0.14 Re^{0.7}))$$

The numerical values of the additional variables are $D_p = 0.208$ mm (p. 941 in McCabe et al.[7]) with $\rho = 994.6$ kg/m^3 and $\mu = 8.931 \times 10^{-4}$ kg/m·s, as calculated using Equations (5-37) and (5-38) at $T = 298.15$ K.

Equation (5-39) should be rearranged in order to avoid possible division by zero and negative square roots as it is entered into the form of a nonlinear equation for POLYMATH.

$$f(v_t) = v_t^2(3C_D\rho) - 4g(\rho_p - \rho)D_p \tag{5-44}$$

The following equation set can be solved by POLYMATH:

```
Equations:
f(vt)=vt^2*(3*CD*rho)-4*g*(rhop-rho)*Dp
g=9.80665
rhop=1800
rho=994.6
Dp=0.208e-3
vis=8.931e-4
Re=Dp*vt*rho/vis
CD=if (Re<0.1) then (24/Re) else ((24/Re)*(1+0.14*Re^0.7))
vt(min)=0.0001, vt(max)=0.05
```

Specifying $v_{t,\,\min} = 0.0001$ and $v_{t,\,\max} = 0.05$ leads to the results summarized in Table 5–3.

Table 5–3 Terminal Velocity Solution

Variable	Value	$f()$
vt	0.0157816	–8.882e–16
Re	3.65564	
CD	8.84266	

(b) The terminal velocity in the centrifugal separator can be calculated by replacing the g in Equation (5-44) by $30g$. Introduction of this change to the equation set gives the results that $v_t = 0.2060$ m/s, $Re = 47.72$, and $C_D = 1.5566$. Only the viscosity and the density of the liquid should be changed to carry out the calculations for other fluids.

The POLYMATH problem solution files are found in the *Simultaneous Algebraic Equation Solver Library* located in directory CHAP5 with files named P5-06A.POL and P5-06B.POL. This problem is also solved with Excel, Maple, MathCAD, MATLAB, Mathematica, and POLYMATH as problem 5 in the Set of Ten Problems discussed in Appendix F.

5.7 COMPARISON OF FRICTION FACTOR CORRELATIONS FOR TURBULENT PIPE FLOW

5.7.1   Concepts Demonstrated

Various correlations for calculating friction factors of turbulent fluid flow in pipes.

5.7.2   Numerical Methods Utilized

Solution of nonlinear algebraic equations.

5.7.3   Problem Statement

The Fanning friction factor can be used to calculate the friction loss for iso-thermal liquid flow. This friction factor is dependent upon the Reynolds number and the surface roughness factor ε in m. The Reynolds number is $Re = Dv\rho/\mu$, where ρ is the density of the fluid in kg/m^3 and μ is the viscosity in $kg/m \cdot s$. A widely used chart gives experimental values of the friction factor as a function of Re and ε. (See, for example, Geankoplis[4] or Perry et al.[9])

There are also a number of implicit and explicit correlations for the friction factor in turbulent flow where $Re > 3000$. For hydraulically smooth pipes where $\varepsilon/D = 0$, an implicit equation is

$$\frac{1}{\sqrt{f_F}} = 4.0 \log(Re\sqrt{f_F}) - 0.4 \quad \text{(Nikuradse[8] equation)} \quad \textbf{(5-45)}$$

and an explicit equation is

$$f_F = 0.0791 Re^{-1/4} \quad \text{(Blasius equation)} \quad \textbf{(5-46)}$$

For rough pipes, where surface roughness characterized by ε/D ratios is important, a widely used implicit equation is

$$\frac{1}{\sqrt{f_F}} = -4.0 \log\left(\frac{\varepsilon}{D} + \frac{4.67}{Re\sqrt{f_F}}\right) + 2.28 \quad \text{(Colebrook and White[3] equation)} \quad \textbf{(5-47)}$$

and explicit equations are given by

$$f_F = \frac{1}{16\left\{\log\left[\frac{\varepsilon/D}{3.7} - \frac{5.02}{Re}\log\left(\frac{\varepsilon/D}{3.7} + \frac{14.5}{Re}\right)\right]\right\}^2} \quad \text{(Shacham[12] equation)} \quad \textbf{(5-48)}$$

and

$$f_F = \left\{ -3.6 \log\left[\frac{6.9}{Re} + \left(\frac{\varepsilon/D}{3.7} \right)^{10/9} \right] \right\}^{-2} \qquad \text{(Haaland}^5 \text{ equation)} \qquad \textbf{(5-49)}$$

(a)   Summarize calculated friction factors from Equations (5-45) through (5-49) at $Re = 10^4$ and 10^7 for smooth pipes where $\varepsilon/D = 0$. Comment on the differences between the implicit and explicit equations.

(b)   Summarize calculated friction factors from Equations (5-47) through (5-49) at $Re = 10^4$ and 10^7 for rough pipes where $\varepsilon/D = 0.0001$ and 0.01. Comment on the differences between the implicit and explicit equations.

(c)   Compare the results of (a) and (b) with the friction factors obtained from a general graphical presentation of the Fanning friction factor.

5.7.4   Solution (Suggestion)

All of the explicit and implicit equations can be solved simultaneously for a particular set of conditions by using the POLYMATH *Simultaneous Algebraic Equation Solver*. Results can easily be compared in a table.

5.8 CALCULATIONS INVOLVING FRICTION FACTORS FOR FLOW IN PIPES

5.8.1 Concepts Demonstrated

Calculation of friction factor and pressure drop for both laminar and turbulent liquid flow in pipes.

5.8.2 Numerical Methods Utilized

Solution of a system of simultaneous nonlinear algebraic equations.

5.8.3 Problem Statement

The Fanning friction factor can be used to calculate the friction loss for isothermal liquid flow in uniform circular pipes by the formula

$$F_f = 2f_F \frac{\Delta L v^2}{D} \tag{5-50}$$

where F_f is the friction loss in $N \cdot m/kg$ or J/kg, f_F is the Fanning friction factor (dimensionless), ΔL is the length of the pipe in m, v is the fluid velocity in m/s, and D is the pipe diameter in m.

The friction factor can be used to predict the pressure drop due to friction loss from

$$\Delta p_f = \rho F_f = 2f_F \rho \frac{\Delta L v^2}{D} \tag{5-51}$$

where Δp_f is the pressure drop in Pa and ρ is the density in kg/m^3. The average fluid velocity can be calculated from

$$v = \frac{q}{\pi D^2/4} \tag{5-52}$$

where q is the volumetric flow rate in m^3/s.

If the flow is in the laminar region when $Re < 2100$, the Fanning friction factor can be calculated from

$$f_F = \frac{16}{Re} \tag{5-53}$$

Otherwise for turbulent flow in a smooth tube when $Re > 2100$, the Nikuradse[8] correlation represented by Equation (5-45) can be utilized to calculate the Fanning friction factor.

(a)   A heat exchanger is required that will be able to handle $Q = 2.5$ liter/s
       of water through a smooth pipe with an equivalent length of $L = 100$ m.
       The maximum total pressure drop is to be $\Delta P = 103$ kPa at a tempera-
       ture of 25 °C. Calculate the diameter D of the pipe required for this
       application and select the appropriate tube from Table D–4.
(b)   What will be the pressure drop for the pipe selected in part (a) if the
       temperature drops to 5 °C?
(c)   Repeat parts (a) and (b) for another liquid from Table D–3.

5.8.4   Solution (Partial)

(a) The calculations in this problem require a friction factor correlation,
which in turn depends upon the Reynolds number Re. Both correlations can be
combined in the POLYMATH *Simultaneous Algebraic Equation Solver* by using
the "if ... then ... else ... " statement in the nonlinear equation, which represents
both laminar and turbulent flow.

For turbulent flow, the Nikuradse correlation of Equation (5-45) can be used
and rearranged as

$$f_F = \frac{1}{[4.0 \log(Re \sqrt{f_F}) - 0.4]^2} \qquad \text{(5-54)}$$

In POLYMATH, this nonlinear equation can be entered as

```
f(fF)=fF-1/(4*log(Re*sqrt(fF))-0.4)^2
```
                                                                                      (5-55)

For laminar flow, Equation (5-53), which is a linear equation, can also be
expressed as a nonlinear equation as

```
f(fF)=fF-16/R
```
                                                                                      (5-56)

Equations (5-55) and (5-56) can be used in a single POLYMATH "if ... then
... else ... " expression to achieve a solution for the friction factor which depends
upon the value of the Reynolds number Re.

Equation (5-51) is also a nonlinear equation that can be entered into POLY-
MATH as a function of the diameter D. Thus two nonlinear equations can be
solved simultaneously by entering the following equation set where the density
and viscosity of water at 25 °C are calculated from equations developed in Problem 5.5.

```
Equations:
f(D)=dp-2*fF*rho*v*v*L/D
f(fF)=if (Re<2100) then (fF-16/Re) else (fF-1
   /(4*log(Re*sqrt(fF))-0.4)^2)
dp=103000
L=100
T=25+273.15
```

```
pi=3.1416
q=0.0025
rho=46.048+T*(9.418+T*(-0.0329+T*(4.882e-5-T*2.895e-8)))
vis=exp(-10.547+541.69/(T-144.53))
v=q/(pi*D^2/4)
Re=D*v*rho/vis
Initial Estimates:
D(0)=0.01
fF(0)=0.001
```

This set of equations yields the results summarized in Table 5–4, where the required pipe diameter is determined to be $D = 0.039$ m or 39 mm. A comparison of this diameter with the data of available heat exchanger tube diameters in Table D–4 indicates that the minimal available tube is a 2-inch OD BWG No. 10 tube with an inside diameter of 43.99 mm.

Table 5–4 Solution for Problem 5.8

Variable	Value	f( )
D	0.0389653	−1.455e-11
fF	0.00459053	8.67e-19
dp	103000	
L	100	
T	298.15	
pi	3.1416	'
q	0.0025	
rho	994.572	
vis	0.000893083	
v	2.09649	
Re	90973.6	

(b) Changing the temperature to 5 °C and setting $D = 0.04399$ m results in a calculated pressure drop of $\Delta p = 64.82$ kPa.

(c) The calculations for other liquids can be carried out by modification of the correlation equations for density and viscosity of the particular liquid.

 The POLYMATH problem solution files are found in the *Simultaneous Algebraic Equation Solver Library* located in directory CHAP5 with files named P5-08A.POL and P5-08B.POL.

5.9 Average Velocity in Turbulent Smooth Pipe Flow from Maximum Velocity

5.9.1 Concepts Demonstrated

Conversion of maximum velocity into average velocity using the universal velocity distribution and comparison with experimental values.

5.9.2 Numerical Methods Utilized

Solution of a single nonlinear equation and regression of general expressions to experimental data.

5.9.3 Problem Statement

It is often necessary to convert maximum flow velocity in a pipe v_{max} to average velocity v_{av}. A common example is the use of a pitot tube, which measures v_{max}. The ratio of v_{av}/v_{max} can be calculated for a smooth tube from the following equation given by McCabe et al.[7]:

$$v_{av} = \frac{v_{max}}{1 + (3.45\sqrt{f_F/2})} \tag{5-57}$$

where f_F is the Fanning friction factor. For smooth tubes, the friction factor can be calculated from the Nikuradse[8] Equation (5-45) using the Reynolds number based on the average velocity.

Water is flowing at 25 °C inside a 1-inch OD BWG–10 tube. (See Table D–4 for tube specifications.) A technician needs a simple relationship that can be used to estimate the average velocity from a measured maximum velocity.

(a) Calculate the average velocities for the maximum velocities given in Table 5–5 and enter these values in the table.

(b) Obtain a simple but accurate correlation of the predicted velocity as a function of the maximum velocity that a technician could use.

(c) Enter the correlation values in the Table 5–5 for comparison purposes.

Table 5–5 Comparison of Measured and Calculated Water Velocities

No.	Measured v_{max} (m/s)	Calculated v_{av} (m/s)	Correlated v_{av} (m/s)
1	0.2		
2	0.4		
3	0.8		
4	1.6		
5	3.2		
6	6.4		
7	12.8		
8	25.6		
9	51.2		
10	102.4		

5.10 CALCULATION OF THE FLOW RATE IN A PIPELINE

5.10.1   Concepts Demonstrated

Application of the general mechanical energy balance for incompressible fluids, and calculation of flow rate in a pipeline for various pressure drops.

5.10.2   Numerical Methods Utilized

Solution of a single nonlinear algebraic equation.

5.10.3   Problem Statement

Figure 5–4 shows a pipeline which delivers liquid at constant temperature T

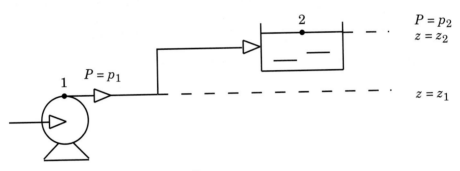

Figure 5–4 Liquid Flow in a Pipeline

from point 1, where the pressure is p_1 and the elevation is z_1 to point 2, where the pressure is p_2 and the elevation z_2. The effective length of the pipe line, including fittings and expansion losses, is L and its diameter is D.

(a) Calculate the flow rate q (in gal/min) in the pipeline for water at 60 °F. The pipeline is nominal 6-inch-diameter schedule 40 commercial steel pipe with L = 5000 ft, P_1= 150 psig, P_2 = 0 psig, z_1 = 0 ft, z_2 = 300 ft.

(b) Plot the calculated flow rate as a function of pressure difference (200 psig max). What is the minimum pressure difference needed to start flow?

(c) Add curves to the plot created in part (b) for both 4- and 8-inch-diameter schedule 40 commercial steel pipe and summarize any observed general trends.

Additional Information and Data

The general mechanical energy balance on an incompressible liquid in turbulent flow is given by Geankoplis[4] as

$$\frac{1}{2}(v_2^2 - v_1^2) + g(z_2 - z_1) + \frac{P_2 - P_1}{\rho} + \Sigma F + W_S = 0 \qquad \textbf{(5-58)}$$

where ΣF is the summation of frictional losses given by Equation (5-50) and W_S is the shaft work done by the system in J/kg. Other terms in Equation (5-58) are defined in Section 5.8.4. For English units, the conversion factor $g_c = 32.174$ ft·lbm/lbf·s^2 must be used to maintain consistent units on some terms of Equation (5-58).

5.10.4   Solution (Suggestions)

The inside diameter of the 6-inch steel pipe is shown in Table D–5, and its surface roughness can be found in Table D–6. The explicit Shacham Equation (5-48) is convenient for calculation of the Fanning friction factor in order to use Equation (5-50) to evaluate the ΣF term in Equation (5-58).   Equations (5-35) and (5-36) can be used to determine the viscosity and density of water in English units.

(a) The calculated flow rate is $q = 368.8$ gal/min.

5.11 FLOW DISTRIBUTION IN A PIPELINE NETWORK

5.11.1   Concepts Demonstrated

Calculation of flow rates and pressure drops in a pipeline network using Fanning friction factors.

5.11.2   Numerical Methods Utilized

Solution of systems of nonlinear algebraic equations.

5.11.3   Problem Statement

Water at 25 °C is flowing in the pipeline network given in Figure 5–5. The pressure at the exit of the pump is 15 bar (15×10^5 Pa) above atmospheric, and the water is discharged at atmospheric pressure at the end of the pipeline. All the pipes are 6-inch schedule 40 steel with an inside diameter of 0.154 m. The equivalent lengths of the pipes connecting different nodes are the following: $L_{01} = 100$ m, $L_{12} = L_{23} = L_{45} = 300$ m, and $L_{13} = L_{24} = L_{34} = 1200$ m.

(a)   Calculate all the flow rates and pressures at nodes 1, 2, 3, and 4 for the pipeline network shown in Figure 5–5. The Fanning friction factor can be assumed to be constant at $f_F = 0.005$ for all pipelines.

(b)   Investigate how the flow rates and pressure drops change when one of the pipelines becomes blocked. Select one of the flow rates (q_{12}, q_{23}, q_{24}, or q_{13}) equal to zero and repeat part (a).

(c)   Rework a more accurate solution for part (a) by using Equation (5-48) to calculate each Fanning friction factor where $\varepsilon = 4.6 \times 10^{-5}$ m. Which calculated flow rate from part (a) has the greatest error?

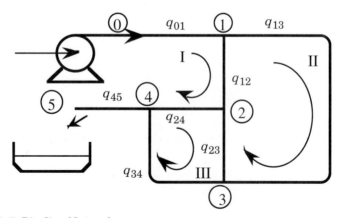

Figure 5–5 Pipeline Network

Additional Information and Data

For the solution of this problem it is convenient to express the pressure drop from node i to node j as

$$\Delta P_{ij} = k_{ij}(q_{ij})^2 \tag{5-59}$$

where ΔP_{ij} is the pressure drop and q_{ij} is the volumetric flow rate between nodes i and j. The k_{ij} terms in Equation (5-59) are related to the Fanning friction factors and average fluid velocities, as described by Equations (5-51) and (5-52). Thus

$$k_{ij} = \frac{2f_F \rho \dfrac{\Delta L_{ij}}{D}}{(\pi D^2/4)^2} = \frac{32 f_F \rho \Delta L_{ij}}{\pi^2 D^5} \tag{5-60}$$

There are two relationships that govern the steady-state flow rate in pipeline networks. First, *the algebraic sum of the flow rates at each node must be zero.* Second, *the algebraic sum of all pressure drops in a closed loop must be zero.*

The summation of flow rates at each node is just an application of the conservation of mass balance, which can be expressed as

$$\text{INPUT} - \text{OUTPUT} = \text{ZERO}$$

The balance equations are written based on assumed flow directions. If the solution yields some negative flow rates, then the actual flow direction is opposite to the assumed flow direction.

Balance on Node 1

$$q_{01} - q_{12} - q_{13} = 0 \tag{5-61}$$

Balance on Node 2

$$q_{12} - q_{24} - q_{23} = 0 \tag{5-62}$$

Balance on Node 3

$$q_{23} + q_{13} - q_{34} = 0 \tag{5-63}$$

Balance on Node 4

$$q_{24} + q_{34} - q_{45} = 0 \tag{5-64}$$

The summation of pressure drops around various loops utilizes ΔP_{ij} with a + sign if the loop direction is the same as the assumed flow direction; otherwise it is negative. The pressure drop across a pump is negative when in the direction of flow.

Summation on Loop I

$$\Delta P_{01} + \Delta P_{12} + \Delta P_{24} + \Delta P_{45} + \Delta P_{\text{PUMP}} = 0 \tag{5-65}$$

Summation on Loop II

$$\Delta P_{13} - \Delta P_{23} - \Delta P_{12} = 0 \qquad (5\text{-}66)$$

Summation on Loop III

$$\Delta P_{23} + \Delta P_{34} - \Delta P_{24} = 0 \qquad (5\text{-}67)$$

(a) The pressure drops can be expressed as functions of q_{ij} using Equation (5-59). This substitution leads to seven equations with seven unknown flow rates designated as $q_{01}, q_{12}, q_{13}, q_{24}, q_{23}, q_{34}$, and q_{45}. Introducing this system of equations (including the numerical values of the different k_{ij}'s) into the POLYMATH *Simultaneous Algebraic Equation Solver* yields

```
Equations:
f(q01)=q01-q12-q13
f(q12)=q12-q24-q23
f(q13)=q23+q13-q34
f(q24)=q24+q34-q45
f(q23)=k01*q01^2+k12*q12^2+k24*q24^2+k45*q45^2+deltaPUMP
f(q34)=k13*q13^2-k23*q23^2-k12*q12^2
f(q45)=k23*q23^2+k34*q34^2-k24*q24^2
fF=0.005
rho=997.08
D=0.154
pi=3.1416
deltaPUMP=-15.e5
k01=32*fF*rho*100/(pi^2*D^5)
k12=32*fF*rho*300/(pi^2*D^5)
k24=32*fF*rho*1200/(pi^2*D^5)
k45=32*fF*rho*300/(pi^2*D^5)
k13=32*fF*rho*1200/(pi^2*D^5)
k23=32*fF*rho*300/(pi^2*D^5)
k34=32*fF*rho*1200/(pi^2*D^5)
Initial Estimates:
q01(0)=0.1
q12(0)=0.1
q13(0)=0.1
q24(0)=0.1
q23(0)=0.1
q34(0)=0.1
q45(0)=0.1
```

The initial estimates for all the volumetric flow rates can be set at 0.1 m³/s.

 The POLYMATH data file for part (a) is found in the *Simultaneous Algebraic Equation Solver Library* located in directory CHAP5 with file named P5-11A.POL.

(b) Note that if one of the flow rates is set to zero, the number of unknowns is reduced by one. Consequently, the number of equations must be reduced by one. Remember, with POLYMATH, the variable within a nonlinear equation entry on the left side in the parentheses does not need to be a part of the expression on the right side. However, all nonlinear expression variables need to be within the parentheses on the left to identify all the problem variables to POLYMATH.

5.12 WATER DISTRIBUTION NETWORK

5.12.1 Concepts Demonstrated

Calculation of flow rates and pressure drops in a pipeline network using Fanning friction factors.

5.12.2 Numerical Methods Utilized

Solution of systems of nonlinear algebraic equations.

5.12.3 Problem Statement

Ingels and Powers[6] investigated the following water distribution network shown in Figure 5–6. There is one input source to node 1, which can deliver 1400 gpm. The demand at nodes 2, 3, and 5 is 420 gpm and at node 6 is 140.0 gpm.

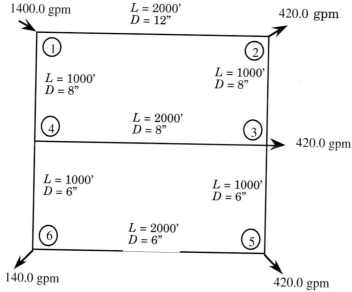

Figure 5–6 Pipeline Network for Problem 5.12

(a) Calculate the flow rates through the different pipes and the head differ-
     ences between the supply node and the different customers for water at
     60 °F. The pipe lengths and diameters are shown in Figure 5–6. A con-
     stant Fanning friction factor of $f_F = 0.005$ can be assumed for all calcu-
     lations.

(b) What pressure in ft of water head must be developed by the supply
     pump at node 1 to maintain all flow rates if the line between nodes 1
     and 2 must be shut down for repairs?

5.12.4 Solution (Suggestions)

(a) Background information can be found in Problems 5.8 through 5.11.
Only five of the possible six node balances for flow rates give independent relationships. Thus pressure drop considerations around the two loops will be
needed. Conversion factors including g_c may be needed to maintain consistent
units.

5.13 PIPE AND PUMP NETWORK

5.13.1 Concepts Demonstrated

Calculations of flow rates in a pipe and pump network using an overall mechanical energy balance and accounting for various frictional losses.

5.13.2 Numerical Methods Utilized

Solution of systems of nonlinear algebraic equations.

5.13.3 Problem Statement

Water at 60 °F and one atmosphere is being transferred from tank 1 to tank 2 with a 2-hp pump that is 75% efficient, as shown in Figure 5–7. All the piping is 4-inch schedule 40 steel pipe except for the last section, which is 2-inch schedule 40 steel pipe. All elbows are 4-inch diameter, and a reducer is used to connect to the 2-inch pipe. The change in elevation between points 1 and 2 is $z_2 - z_1 = 60$ ft.

(a) Calculate the expected flow rate in gal/min when all frictional losses are considered.
(b) Repeat part (a) but only consider the frictional losses in the straight pipes.
(c) What is the % error in flow rate for part (b) relative to part (a)?
(d) Repeat parts (a), (b), and (c) for one of the liquids given in Table D–3.

Additional Information and Data

The various frictional losses and the mechanical energy balance are discussed by Geankoplis[4] and Perry et al.[9]

5.13.4 Solution (Suggestions)

(a) Frictional losses for this problem include contraction loss at tank 1 exit, friction in three 4-inch elbows, contraction loss from 4-inch to 2-inch pipe, friction in both the 2-inch and 4-inch pipes, and expansion loss at the tank 2 entrance. The explicit equation given by Equation (5-48) can be used to calculate the friction factor for both sizes of pipe. Equations (5-35) and (5-36) can be used to determine the viscosity and density of water in English units. Additional data are found in Appendix D.

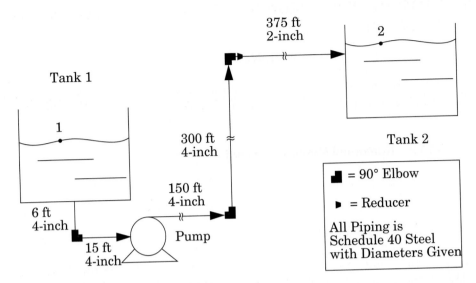

Figure 5–7 Pipe and Pump Network

5.14 OPTIMAL PIPE LENGTH FOR DRAINING A CYLINDRICAL TANK IN TURBULENT FLOW

5.14.1   Concepts Demonstrated

Use of the overall mechanical energy balance and unsteady-state material balance in fluid flow calculations.

5.14.2   Numerical Methods Utilized

Solution of nonlinear equations and differential algebraic equations.

5.14.3   Problem Statement

There is a need to design a piping system for the rapid draining of a tank in an emergency, and you are called upon to settle a disagreement. A simplified view of the tank is shown in Figure 5–8. Some engineers say that the shorter the length of the pipe, the less time it will take to drain the tank. Others argue that a longer pipe will lead to shorter drain times. Still others say that there is an optimal length. You will have to calculate the draining time for different lengths of the pipe and determine the length L that will minimize the draining time t_f for water, which is expected to remain in *turbulent flow* during the draining time.

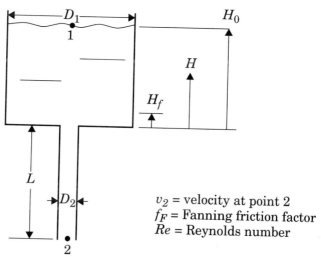

Figure 5–8 Tank with Draining Pipe

The drain pipe is a nominal 1/2-inch schedule 40 steel pipe with ε = 0.00015. The tank diameter D_1 is 3 ft, and the initial liquid level is always $H_0 = 6$ ft. The liquid in the tank is water at 60 °F. The tank is considered to be drained when the final level H_f reaches 1 inch. The variables are limited to the ranges 1 in $\leq H \leq 6$ ft and 1 in $\leq L \leq 10$ ft.

(a) Investigate the effects of various values of L and H on v_2, f_F, and Re by completing Table 5–6 by using the steady-state mechanical energy balance and assuming friction losses only in the pipe. Comment on the assumption of a constant Fanning friction factor for an approximate solution of this problem and verify that the flow is turbulent.

(b) Show that an approximate solution for the draining time assuming a constant value of the friction factor is given by

$$t_f = \frac{D_1^2}{D_2^2}\left(\sqrt{H_0 + L} - \sqrt{H_f + L}\right)\sqrt{\frac{2}{g}\left(1 + \frac{4f_F L}{D_2}\right)} \tag{5-68}$$

where t_f is the time (in seconds) that it takes to reach a final level where $H_f = 1$ inch. Here the f_F is to be the average of the friction factors calculated from Table 5–6 for the case where H_0 is 6 ft. Perform calculations for $L = 0.0833$ ft to 10 ft until you find an optimum or a clear trend in the change of the draining time that will support your recommended pipe length and estimated draining time.

(c) Solve part (b) numerically where the entrance and flow constriction effects can be neglected but the friction factor can vary. The steady-state mechanical energy equation can be used in this case. Are there meaningful differences between the results obtained in this part of the question and in part (b)? What is your recommended pipe length L?

(d) Repeat parts (b) and (c) for the case where H_0 is 3 ft. What is your recommended pipe length?

Table 5–6 Calculated Values for Tank Draining of Water at 60 °F

L ft	H ft	f_F	v_2 ft/s	Re
1/12	1/12			
1/12	6			
10	1/12			
10	6			
5	3			

Additional Information and Data

The unsteady-state mechanical energy balance can be applied between points 1 and 2 shown in Figure 5–8. The assumption of an incompressible fluid leads to the isothermal mechanical energy balance given in Equation (5-58). Note that this equation is only an approximation as the time-dependent terms have been neglected, which is reasonable in this case. Application to this problem yields an expression for the outlet velocity v_2

$$v_2^2 = \frac{2g(H+L)}{1 + 4f_F\left(\dfrac{L}{D_2}\right)} \qquad \text{(5-69)}$$

where the Fanning friction factor can be evaluated from explicit Equation (5-48). An unsteady-state material balance on the fluid in the tank yields

$$\frac{dH}{dt} = -\frac{D_2^2 v_2}{D_1^2} \qquad \text{(5-70)}$$

5.14.4   Solution (Suggestions)

(a) This problem becomes one of solving nonlinear Equation (5-69) for the exit velocity, where the friction factor is calculated from explicit Equation (5-48). Equations (5-35) and (5-36) can be used to determine the viscosity and density of water in English units.

(b) The approximate solution can be derived by substituting Equation (5-69) into differential Equation (5-70) and integrating from H_0 to H_f. The calculations involved only require repeated evaluations of Equation (5-68) with the average friction factor for various L's.

(c) This part involves the solution of a differential equation and a nonlinear algebraic equation, which requires a DAE problem solution (discussed in Problem 3.8).

5.15 OPTIMAL PIPE LENGTH FOR DRAINING A CYLINDRICAL TANK IN LAMINAR FLOW

5.15.1 Concepts Demonstrated

Use of the overall mechanical energy balance and unsteady-state material balance in fluid flow calculations.

5.15.2 Numerical Methods Utilized

Solution of nonlinear equations and differential algebraic equations.

5.15.3 Problem Statement

The liquid to be placed in the tank discussed in Problem 5.14 is hydraulic fluid MIL-M-5606 at 30 °F whose properties are given in Table D–3. As in Problem 5.14, you are to determine the length of the pipe that will minimize the draining time for this hydraulic fluid, which is expected to remain in *laminar flow* during the draining time. The variables are limited to the ranges 1 in $\leq H \leq 6$ ft and 1 in $\leq L \leq 10$ ft.

(a) Investigate the effects of various values of L and H on v_2, f_F and Re by completing Table 5–7 using the steady-state mechanical energy balance and assuming friction losses only in the pipe. Comment on the assumption of a constant Fanning friction factor for an approximate solution of this problem and verify that the flow is laminar.

(b) Solve this problem numerically where the entrance and flow constriction effects can be neglected but the friction factor can vary. The steady-state mechanical energy equation can be used in this case, and H_0 is 2.0 feet. What is your recommended pipe length L?

(c) Repeat part (b) for H_0 of 1.0 ft.

(d) Determine an analytical solution to part (b) and compare the results with the numerical solution of part (b).

Table 5–7 Calculated Values for Tank Draining of Hydraulic Fluid at 30 °F

L ft	H ft	f_F	v_2 ft/s	Re
1/12	1/12			
1/12	6			
10	1/12			

Table 5–7 Calculated Values for Tank Draining of Hydraulic Fluid at 30 °F

L ft	H ft	f_F	v_2 ft/s	Re
10	6			
5	3			

5.15.4 Solution (Suggestions)

The Fanning friction factor for laminar flow in pipes is given by Equation (5-53).

5.16 BASEBALL TRAJECTORIES AS A FUNCTION OF ELEVATION

5.16.1 Concepts Demonstrated

Effect of drag and fluid density on a solid object moving through a fluid.

5.16.2 Numerical Methods Utilized

Conversion of higher order differential equations to simultaneous first order differential equations and solution of a two point boundary value problem.

5.16.3 Problem Statement

Enthusiastic New York Mets baseball fans maintain that the Colorado Rockies baseball team has an unfair advantage because balls travel further in the thin air of Denver, Colorado. It is a good engineering assumption that the major difference between the conditions in New York and Denver is the decrease of the air density because of the much higher elevation in Denver.

(a) A good batter can impart an initial velocity of 120 ft/s to a baseball. What distance will the ball travel in Shea Stadium in New York when the initial angle is 30° with the horizontal?

(b) Use the data in Table 5–8 to calculate the distance that the ball will travel in Coors Stadium in Denver for the same initial velocity and angle as given in part (a)?

(c) What would be the best angle to optimize the resulting distance for the initial velocity of 120 ft/s in Shea Stadium? Plot the results for y versus x.

Additional Information and Data

Table 5–8 Data for Baseball Trajectories

Description	Variable	Value
Mass of the baseball	m	$0.313 \ \text{lb}_\text{m}$
Projected area	A	$0.046 \ \text{ft}^2$
Drag coefficient (assuming turbulent flow)	C_D	0.44
The acceleration of gravity (the conversion factor g_c has the same numerical value as g)	g	$32.174 \ \text{ft/s}^2$
Height of the ball with initial velocity imparted	y_0	5 ft
The angle with the horizontal in which the ball is being contacted	θ	30° for part (a)

Table 5–8 Data for Baseball Trajectories

Description	Variable	Value
The elevation of the field above sea level at Shea Stadium in New York	Z_1	100 ft
The elevation of the field above sea level in Coors Stadium in Denver	Z_2	5280 ft

It can be shown (Riggs[10]) that the trajectory of a baseball can be represented by the following pair of second-order differential equations:

$$m\frac{d^2x}{dt^2} = -k\frac{dx}{dt}\sqrt{\left(\frac{dx}{dt}\right)^2 + \left(\frac{dy}{dt}\right)^2}$$

$$m\frac{d^2y}{dt^2} = -k\frac{dy}{dt}\sqrt{\left(\frac{dx}{dt}\right)^2 + \left(\frac{dy}{dt}\right)^2} - mg$$

(5-71)

where x and y are the horizontal and vertical distances and k is the drag coefficient given by

$$k = \frac{C_D A\rho}{2g_c}$$

(5-72)

In the preceding equation, C_D is the drag coefficient, ρ is the density of the air, and g_c is a conversion factor. The air density changes with elevation above sea level given by Z according to

$$\rho = 0.07647 \exp\left(\frac{-Z}{3.33 \times 10^4}\right)$$

(5-73)

where Z is in ft and the units of ρ are lb_m/ft^3.

5.16.4   Solution (Suggestions)

The POLYMATH *Simultaneous Differential Equation Solver* program can be used to solve this problem. First the set of two second-order differential equations can be converted into a set of four first-order equations. The technique was discussed in Problem 3.6.

For this problem, it is convenient to define

$$v_x = \frac{dx}{dt}$$

$$v_y = \frac{dy}{dt}$$

(5-74)

so that the pair of Equation (5-71) becomes four first-order ordinary differential equations.

$$\frac{dx}{dt} = v_x$$

$$\frac{dy}{dt} = v_y$$

$$\frac{d(v_x)}{dt} = -\frac{k}{m}v_x\sqrt{v_x^2 + v_y^2}$$ (5-75)

$$\frac{d(v_y)}{dt} = -\frac{k}{m}v_y\sqrt{v_x^2 + v_y^2} - g$$

The corresponding initial values at $t = 0$ are given by

$$x = 0$$

$$y = y_0$$

$$v_x = V_0\cos\theta$$ (5-76)

$$v_y = V_0\sin\theta$$

where V_0 is the initial velocity of the ball.

(a) The problem is defined by the differential equation set (5-75) and the initial conditions (5-76). Equations (5-72) and (5-73) must be evaluated during the numerical solution. For Shea Stadium, $Z = 100 + y$. The boundary condition that indicates that the ball has impacted the ground is that $y = 0$, and the corresponding x is the distance.

Retaining a Value when a Condition Is Satisfied

When using POLYMATH, a convenient way to save the value of x when y = 0 for this problem is to use the "if ... then ... else ... " capability, as illustrated by

```
d(x)/d(t)=if (y>0) then (Vx) else (0)
```

The preceding statement conveniently holds the value of x when y becomes negative.

(b) The solution for Coors Stadium requires the modification that $Z = 5280 + y$. The equations must again be integrated to $y = 0$ to determine the x distance that the ball will travel.

(c) This solution uses the equations of part (a), but the angle θ can be varied so as to maximize the value of x when y becomes zero.

5.17 VELOCITY PROFILES FOR A WALL SUDDENLY SET IN MOTION—LAMINAR FLOW

5.17.1 Concepts Demonstrated

Unsteady-state one-dimensional boundary layer velocity profiles for a Newtonian fluid in laminar flow near a wall, which is suddenly set in motion at a constant velocity.

5.17.2 Numerical Methods Utilized

Application of the numerical method of lines to solve a partial differential equation and solution of simultaneous ordinary differential equations.

5.17.3 Problem Statement

A large volume of water at 25 °C with $\mu = 8.931 \times 10^{-4}$ kg/m·s and $\rho = 994.6$ kg/m³ is bounded on one side by a plane surface, which is initially at rest with $v_x = 0$. At time $t = 0$ the plane is suddenly set in motion with a constant velocity V. The flow is expected to be laminar, and there is no gravity force or pressure gradients. Water may be considered a Newtonian fluid. This situation is shown in Figure 5–9 where the velocity profile is shown at a particular time t.

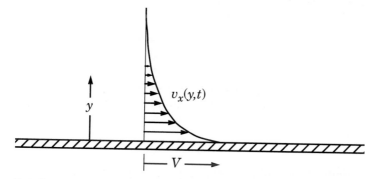

Figure 5–9 Velocity Profile of a Newtonian Fluid Near a Wall Suddenly Set in Motion.

For this situation, the equation of continuity and the equation of motion for one-dimensional flow of a Newtonian fluid (see Bird et al.[1]) yield

$$\frac{\partial v_x}{\partial t} = \frac{\mu}{\rho} \frac{\partial^2 v_x}{\partial y^2} \qquad\qquad (5\text{-}77)$$

where the initial and boundary conditions are given by

$$
\begin{aligned}
v_x &= 0 && \text{at any } y && t \le 0 \\
v_x &= V && \text{at } y = 0 && t > 0 \\
v_x &= 0 && \text{at } y = \infty && t > 0
\end{aligned}
\qquad\text{(5-78)}
$$

The analytical solution to this problem by Bird et al.[1] uses the error function, which is available in many mathematical tables and given in Table A–3 of Appendix A.

$$
\frac{v_x}{V} = 1 - \frac{2}{\sqrt{\pi}} \int_0^{\eta} e^{-\eta^2}\, d\eta = 1 - \mathrm{erf}(\eta)
\qquad\text{(5-79)}
$$

where

$$
\eta = \frac{y}{\sqrt{\dfrac{4\mu t}{\rho}}}
\qquad\text{(5-80)}
$$

(a) Use the numerical method of lines to solve Equation (5-77) with the boundary conditions given by Equation (5-78). The velocity of the wall is given as $V = 0.2$ m/s. The y distance to be considered is 0.1 m and the recommended Δy distance between nodes is 0.01 m. Generate the numerical solution to a time of 500 s. Plot the velocities with time for $y = 0.01, 0.02, 0.03,$ and 0.04 m as a function of time to 500 s.

(b) Repeat the calculations of part (a) with $\Delta y = 0.005$ m to verify a reasonable solution. Make a comparison of the results of parts (a) and (b) at 500 s for the y distances used in the plots for part (a).

(c) Make a separate plot similar to Figure 5–9 at $t = 500$ s using the results of the part (b) calculations.

(d) Compare the results with the error function solution given by Equations (5-79) and (5-80) with the results generated for the four locations investigated in parts (a) and (b) at 500 s.

5.17.4 Solution (Suggestions)

(a) The numerical method of lines is discussed in Problem 3.9. In this application, the number of nodes must be adequate to describe the fluid, which is actually in movement. Typically 10 to 20 nodes should be adequate if the Δy spacing is properly chosen to be within the boundary layer as it is being established. The boundary layer is the region near the wall where the fluid velocity v_x is greater than 1% of the velocity V.

The set of equations can be entered into the POLYMATH *Simultaneous Differential Equation Solver.* The capability to duplicate equations is convenient for entry of the ordinary differential equations for the various nodes.

(c) The error function is summarized in many textbooks and compilations of mathematical tables. It can also easily be generated by numerically solving the ordinary differential equation given by

$$\frac{d}{dz}[\mathrm{erf}(z)] = \frac{2e^{-z^2}}{\sqrt{\pi}} \tag{5-81}$$

with an initial condition of $\mathrm{erf}(z) = 0$ and the integration continued to z yielding $\mathrm{erf}(z)$. Thus a plot or table of $\mathrm{erf}(z)$ versus x can easily be calculated, as has been accomplished in Table A–3 of Appendix A.

The POLYMATH file for calculation of the error function is found in the *Simultaneous Differential Equation Solver Library* located in directory CHAP5 with file named P5-ERROR.POL.

5.18 BOUNDARY LAYER FLOW OF A NEWTONIAN FLUID ON A FLAT PLATE

5.18.1   Concepts Demonstrated

Numerical solution of the equations of motion and continuity for a laminar boundary layer flow of a Newtonian fluid over a flat plate. Prediction of boundary layer thickness and drag force.

5.18.2   Numerical Methods Utilized

Transformation of partial differential equations into an ordinary differential equation, reduction of a higher order differential equation to a system of first-order differential equations, and numerical solution of simultaneous ordinary differential equations employing a shooting technique.

5.18.3   Problem Statement

An important case in laminar boundary layer theory involves the flow of a Newtonian fluid on a very thin flat plate, as shown in Figure 5–10. For this situation,

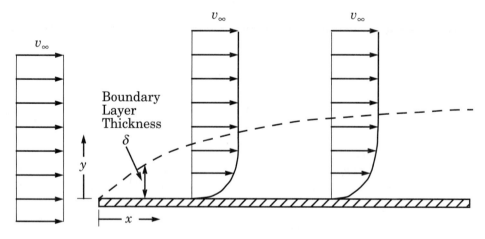

Figure 5–10 Laminar Boundary Layer for Flow Past a Flat Plate

the equation of motion and the continuity equation become (Geankoplis[4])

$$v_x\frac{\partial v_x}{\partial x} + v_y\frac{\partial v_x}{\partial y} = \frac{\mu}{\rho}\frac{\partial^2 v_x}{\partial y^2} \tag{5-82}$$

$$\frac{\partial v_x}{\partial x} + \frac{\partial v_y}{\partial y} = 0 \tag{5-83}$$

where the boundary conditions are $v_x = v_y = 0$ at $y = 0$ (y is the distance from the plate) and $v_x = v_\infty$ at $y = \infty$.

These partial differential equations can be transformed into an ordinary differential equation (see Schlichting[11] for details) by considering the boundary layer thickness δ to be proportional to $\sqrt{\dfrac{\mu x}{\rho v_\infty}}$.

A dimensionless coordinate can be defined by $\eta = y/\delta$ so that

$$\eta = y\sqrt{\frac{\rho v_\infty}{\mu x}} \tag{5-84}$$

Also, a stream function that satisfies the continuity equation, Equation (5-83), may be defined by

$$\psi = \sqrt{\frac{\mu x v_\infty}{\rho}} f(\eta) \tag{5-85}$$

where

$$v_x = \frac{\partial \psi}{\partial y} \text{ and } v_y = \frac{\partial \psi}{\partial x} \tag{5-86}$$

Thus the velocities become a function of the dimensionless stream function $f(\eta)$ given by

$$v_x = v_\infty f'(\eta) \tag{5-87}$$

$$v_y = \frac{1}{2}\sqrt{\frac{\mu v_\infty}{\rho x}}(\eta f' - f) \tag{5-88}$$

where the prime indicates differentiation with respect to η.

After introduction of Equations (5-87) and (5-88) into Equation (5-82), the partial differential equations are transformed into the ordinary differential equation in the dimensionless stream function given by

$$ff'' + 2f''' = 0 \tag{5-89}$$

with the initial conditions

$$f = 0 \text{ and } f' = 0 \text{ at } \eta = 0 \tag{5-90}$$

and the final condition

$$f' = 1 \text{ at } \eta = \infty \tag{5-91}$$

Boundary Layer Thickness

The boundary layer thickness is usually described as where the velocity is less than $0.99 v_\infty$, and this is where $f' \cong 0.99$. Thus Equation (5-84) can be used to determine the boundary layer thickness by

$$\delta = (\eta|_{f' = 0.99}) \sqrt{\frac{\mu x}{\rho v_\infty}} \tag{5-92}$$

Drag Force Due to Skin Friction

The total drag force acting upon the plate with length L and width W is given by (Geankoplis[4])

$$F_D = W \int_0^L \tau_0 \, dx \tag{5-93}$$

where the shear stress at the surface of the plate can be calculated from

$$\tau_0 = \mu \left(\frac{dv_x}{dy}\bigg|_{y = 0}\right) = \mu v_\infty \sqrt{\frac{\rho v_\infty}{\mu x}} (f''|_{\eta = 0}) \tag{5-94}$$

Thus evaluation of Equation (5-93) utilizing Equation (5-94) yields

$$F_D = 2(f''|_{\eta = 0}) \sqrt{\mu \rho v_\infty^2 L} \tag{5-95}$$

(a) Solve the differential equation given by Equation (5-89) with initial conditions and boundary conditions given by Equations (5-90) and (5-91) for $0 \le \eta \le 10$.

(b) Present the boundary layer thickness result of Equation (5-92) utilizing the result of part (a).

(c) Present the drag force result of Equation (5-95) utilizing the result of part (a).

(d) Compare the results of parts (b) and (c) with those presented by Geankoplis[4] or Schlichting[11] or a fluid dynamics textbook.

5.18.4   Solution (Partial)

(a) The third-order differential equation given by Equation (5-89) can be solved by transformation to a system of three simultaneous ordinary differential equations which can be solved by the *POLYMATH Simultaneous Ordinary Differential Equation Solver.* This can be accomplished by defining additional functions given by $g_1 = f$, $g_2 = f'$, $g_3 = f''$ (as per Carnahan et al.[2]), resulting in the differential equations

$$\frac{dg_1}{d\eta} = g_2 \tag{5-96}$$

$$\frac{dg_2}{d\eta} = g_3 \tag{5-97}$$

$$\frac{dg_3}{d\eta} = -\frac{1}{2}g_1g_3 \tag{5-98}$$

with the initial conditions

$$g_1 = 0 \text{ and } g_2 = 0 \text{ at } \eta = 0 \tag{5-99}$$

and final condition

$$g_2 = 1 \text{ at } \eta = \infty \tag{5-100}$$

A shooting technique can be used to determine (optimize) the initial condition for g_3 that will satisfy the final condition when η becomes large. Problem 3.6 discusses this solution technique and single-variable optimization in detail.

REFERENCES

1. Bird, R. B., Stewart, W. E., and Lightfoot, E. N. *Transport Phenomena*, New York: Wiley, 1960.
2. Carnahan, B., Luther, H. A., and Wilkes, J. O. *Applied Numerical Methods*, New York: Wiley, 1969.
3. Colebrook, C. F., and White, C. M. *J. Inst. Civil Eng.*, *10* (1) 99–118 (1937–38).
4. Geankoplis, C. J. *Transport Processes and Unit Operations*, 3rd ed., Englewood Cliffs, NJ: Prentice Hall, 1993.
5. Haaland, S. E. *Trans. ASME*, JFE, *105*, 89 (1983).
6. Ingels, D. M., and Powers, J. E. *Chem Eng. Progr.*, *60* (2), 65 (1964).
7. McCabe, W. L., Smith, J. C., and Harriot, P. *Unit Operations of Chemical Engineering*, 5th ed., New York: McGraw-Hill, 1993.
8. Nikuradse, J. *VDI-Forschungsheft*, 356 (1932).
9. Perry, R.H., Green, D. W., and Malorey, J. D., Eds. *Perry's Chemical Engineers Handbook*, 7th ed, New York: McGraw-Hill, 1997.
10. Riggs, J. B. *An Introduction to Numerical Methods for Chemical Engineers*, 2nd ed., Lubbock, TX: Texas Tech University Press, 1994.
11. Schlichting, H., *Boundary Layer Theory*, 6th ed., New York: McGraw-Hill, 1969.
12. Shacham, M. *Ind. Eng. Chem. Fund.*, *19*, 228–229 (1980).

Heat Transfer

6.1 ONE-DIMENSIONAL HEAT TRANSFER THROUGH A MULTILAYERED WALL

6.1.1 Concepts Demonstrated

Calculation of one-dimensional heat flux and temperature distributions for heat conduction including variation of thermal conductivity with temperature and convection as a boundary condition.

6.1.2 Numerical Methods Utilized

Solution of ordinary differential equations in sequence which require matching of boundary conditions.

6.1.3 Problem Statement

Heat transfer by conduction in solids and fluids follows Fourier's law. One-dimensional heat transfer by conduction is described by

$$\frac{q_x}{A} = -k\frac{dT}{dx} \tag{6-1}$$

where q_x is the heat transfer in the x direction in W or J/s, A is the cross-sectional area that is normal to the direction of heat conduction in m^2, k is the thermal conductivity of the medium in W/m·K, and x is the distance in m.

Convective heat transfer between solids and fluids is described by

$$\frac{q_x}{A} = h(T_w - T_f) \tag{6-2}$$

where h is the heat transfer coefficient in W/m^2·K, T_w is the temperature of the solid surface in K, and T_f is the temperature of the fluid in K.

A common boundary condition at the solid/fluid interface is continuity of heat flux given by

$$\left.\frac{q_x}{A}\right|_S = \left.-k\frac{dT}{dx}\right|_S = h(T_w - T_f) \tag{6-3}$$

where S represents the solid surface whose temperature is T_w. At interfaces between two different solids, the temperature is continuous:

$$T = \left.T_1\right|_{x=I} = \left.T_2\right|_{x=I} \tag{6-4}$$

where I represents the interface between the two different solids, and the heat flux continuity is given by

$$\frac{q_x}{A} = \left.-k_1\frac{dT_1}{dx}\right|_I = \left.-k_2\frac{dT_2}{dx}\right|_I \tag{6-5}$$

For steady-state heat transfer by conduction, an energy balance yields

$$\frac{dq_x}{dx} = \dot{q}A \tag{6-6}$$

where $\dot{q}$ is the rate of heat generated per unit volume in W/m^3. When the generation term in Equation (6-6) is zero, then the rate of heat transfer is constant at any value of x.

Consider a problem similar to that given by Geankoplis,[4] (adapted with permission) in which the wall of a cold-storage room is constructed of layers of pine, cork board, and concrete. The thicknesses of the layers are 15.0 mm, 100.0 mm, and 75.0 mm, respectively. The corresponding thermal conductivities are 0.151, 0.0433, and 0.762 in W/m·K. Figure 6–1 shows a diagram of the wall, where the pine, cork board, and concrete layers are labeled A, B, and C. The temperatures at each interface are shown in the diagram.

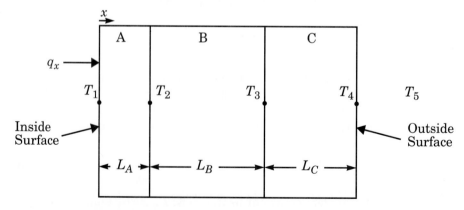

Figure 6–1 Heat Transfer through a Multilayered Wall

(a) Calculate the heat flux through the wall if the interior surface is at 255 K and the exterior surface is at 298 K.

(b) It is proposed to reduce the heat loss by 50% by increasing the thickness of the cork board. What total thickness of cork board would be required?

(c) A new material is to be used instead of the cork board in part (a). Its thermal conductivity is given by $k = 2.5 \times \exp(-1225/T)$ with T in K. Repeat part (a) for this material and plot the temperature profile within the wall.

(d) The natural convection heat transfer coefficient for a vertical plane representing the outside wall is given by $h = 1.37 \left| \dfrac{T_5 - T_4}{6} \right|^{1/4}$, where the T's are in K. Note that the absolute value is to be used in calculating h. Repeat part (c) using this information and consider that only the interior surface temperature is known to be 255 K. In this case the exterior surface temperature must be calculated when the ambient temperature T_5 is 298 K. Plot the temperature profile within the wall.

6.1.4   Solution (Partial)

(a) Since there is no heat generation within the multilayered wall, the heat transfer in the x direction is constant, as inferred from Equation (6-6). Equation (6-1) can therefore be rearranged and written in turn for each of the wall layers giving the differential equation set as

$$\frac{dT}{dx} = -\left(\frac{q_x}{A}\right)/k_A \qquad 0 \le x \le L_A$$

$$\frac{dT}{dx} = -\left(\frac{q_x}{A}\right)/k_B \qquad L_A < x \le (L_A + L_B) \tag{6-7}$$

$$\frac{dT}{dx} = -\left(\frac{q_x}{A}\right)/k_C \qquad (L_A + L_B) < x \le (L_A + L_B + L_C)$$

where the initial condition is given by

$$T = T_1 \qquad \text{when } x = 0 \tag{6-8}$$

and the interface temperatures are T_2, T_3, and T_4.

For constant thermal conductivities, the differential equations of Equation Set (6-7) can be integrated and the results incorporated into a single equation for

q_x/A and the overall temperature change (details given by Geankoplis[4]), yielding

$$q_x/A = \frac{T_1 - T_4}{L_A/k_A + L_B/k_B + L_C/k_C}$$

(6-9)

Thus q_x/A can be directly calculated to be -17.15 W/m^2. Note that the negative sign indicates that the heat flux is actually in the negative x direction as the energy is actually conducted into the cold-storage room. *A useful convention is to set up the heat transfer equations with the heat transfer assumed to be in the positive x direction and then use the sign of the result to indicate actual heat flow.*

(b) This part may be solved by setting the heat flux at the desired value of $0.5 \times (-17.15$ W/m$^2)$ and solving for L_B from Equation (6-9).

(c) When the thermal conductivity is a function of temperature, the differential equation for that material may be difficult to solve analytically. One approach for this problem is to solve the differential equation given in Equation Set (6-7), where the thermal conductivity k_B is expressed as a function of temperature T. This differential equation can be solved by the POLYMATH *Simultaneous Differential Equation Solver* with the use of the "if ... then ... else ... " statement to control the differential equation for temperature T depending upon the position within the wall. The shooting technique can be employed to determine a value of the heat flux so that the final boundary condition is satisfied. The initial value is $T = 255$ at $x = 0$, and the boundary condition to be satisfied is that $T = 298$ when $x = L_A + L_B + L_C = 0.19$. The POLYMATH equation set where variable Q_x is used for the heat flux (q_x/A) is given by

```
Equations:
d(T)/d(x)=if(x<=LA)then(-Qx/kA)else(if(x<=(LA+LB)&(x>LA))then(-
    Qx/kB)else(-Qx/kC))
err=T-298
LA=.015
Qx=-12.2
kA=.151
LB=.1
kB=2.5*exp(-1225/T)
kC=.762
Initial Conditions:
x(0)=0
T(0)=255
Final Value:
x(f)=0.19
```

A single-variable optimization technique as discussed in Problem 3.5 and a shooting technique as discussed in Problem 3.6 can be used to determine the heat flux Q_x.

This solution also allows convenient plotting of the temperature profile, as shown in Figure 6–2. Note that the T is not a linear function of distance in the new material because of the temperature variation of the thermal conductivity.

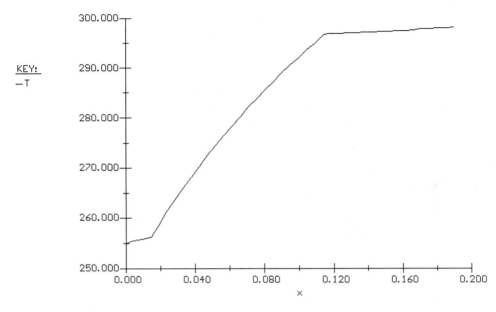

Figure 6–2 Temperature Profile through Multilayered Wall

 The POLYMATH problem solution file for part (c) is found in the *Simultaneous Differential Equation Solver Library* located in directory CHAP6 with file named P6-01C.POL.

   (d) This solution should utilize Equation (6-3) to calculate T_5 since T_4 and Q_x are known at the completion of an integration. Again, a technique must be used to converge upon a value for Q_x that gives $T_5 = 298$.

6.2 HEAT CONDUCTION IN A WIRE WITH ELECTRICAL HEAT SOURCE AND INSULATION

6.2.1   Concepts Demonstrated

One-dimensional heat conduction with generation in cylindrical coordinates with a free convection boundary condition, and conduction through several layers of materials with different thermal conductivities.

6.2.2   Numerical Methods Utilized

Solution of simultaneous ordinary differential equations with a shooting technique to optimize an initial condition for various final boundary conditions, and sequential solution of ordinary differential equations involving matching boundary conditions.

6.2.3   Problem Statement

An insulated wire is carrying an electrical current. The passage of the current generates heat (thermal energy) according to

$$\dot{q} = \frac{I^2}{k_e} \tag{6-10}$$

where $\dot{q}$ has units of W/m^3, I is the current density in amps/m^2, and the electrical conductivity is given by k_e in ohm^{-1}m^{-1}. The wire has a uniform radius R_1 in m. A reasonable assumption is that $\dot{q}$ is constant and does not vary with position.

An energy balance on a cylindrical shell of thickness Δr and length L, as shown in Figure 6–3, leads to the differential equation

$$\frac{d}{dr}(rQ_r) = \dot{q}r \tag{6-11}$$

where the heat flux Q_r can be calculated from Fourier's law given by

$$Q_r = -k\frac{dT}{dr} \tag{6-12}$$

Conduction and convection are discussed in Problem 6.1.

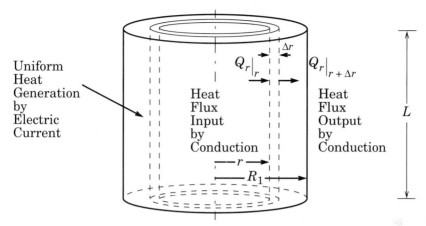

Figure 6–3 Differential Volume for Shell Balance

(a) Calculate and plot the temperature and the heat flux within the wire if the wire surface is maintained at T_1 = 15 °C = 288.15 K, and the electrical and thermal conductivities are given by $k_e = 1.4 \times 10^5 \exp(0.003T)$ ohm^{-1}m^{-1} and k = 5 W/m·K. The wire radius is R_1 = 0.004 m and the *total* current is maintained at I = 400 amps.

(b) The wire surface has a heat transfer coefficient due to natural convection given by $h = 1.32(|\Delta T|/D)^{1/4}$, where h is in W/m^2·K, ΔT is in K, and D is in m. Calculate the heat flux at the wire surface and plot the temperature distribution within the wire. The temperature of the surroundings is T_b =15 °C = 288.15 K. The wire radius is R_1 = 0.004 m and the *total* current is maintained at I = 50 amps.

(c) The wire is covered by an insulating layer whose outer radius is R_2 = 0.015 m and whose constant thermal conductivity is k_I = 0.2 W/m·K. Calculate the heat flux at the exterior surface of the insulation and plot the temperature distribution within the wire and the insulation layer. Comment on the use of insulation in this application.

6.2.4   Solution (Partial)

(a) The energy generation expression of Equation (6-10) can be substituted into Equation (6-11) to give

$$\frac{d}{dr}(rQ_r) = \frac{I^2 r}{k_e} \tag{6-13}$$

where the initial condition is that the combined variable (rQ_r) = 0 at the centerline for the wire where both $r = 0$ and $Q_r = 0$. Fourier's law represented by Equation (6-12) can be rearranged to yield

$$\frac{dT}{dr} = -\frac{Q_r}{k} \tag{6-14}$$

where the final condition is known that $T = T_1$ at $r = R_1$. Thus the problem involves split boundary conditions and a shooting technique with optimization of the initial condition for T at $r = 0$, T_0, so that the final condition for T at $r = R_1$ is satisfied. Using the notation and logic developed in Problem 3.5, the objective function representing the error at the final condition can be written as

$$\varepsilon(T_0) = T\big|_{r = R_1} - 288.15 \tag{6-15}$$

Thus the initial condition on temperature T must be optimized to yield a final value of T, which is the known temperature of 288.15 K. When the optimized initial condition is determined, the value of the right-hand side of Equation (6-15) will be approximately zero at the final condition.

This problem formulation also requires use of the algebraic equation

$$Q_r = \frac{(rQ_r)}{r} \tag{6-16}$$

which is necessary to determine Q_r for Equation (6-14) from the combined variable (rQ_r). Division by zero at $r = 0$ in Equation (6-16) is avoided by setting $Q_r = 0$ at $r = 0$ through use of the "if ... then ... else ... " statement in the POLYMATH *Simultaneous Ordinary Differential Equation Solver*.

Thus this problem requires the simultaneous solution of Equations (6-13) through (6-16), as follows. This equation set shows the final result of a single-variable search that determines the initial condition for T at $r = 0$.

```
Equations:
d(T)/d(r)=-Qr/k
d(rQr)/d(r)=qdot*r
Qr=if(r>0)then(rQr/r)else(0)
k=5.
ke=1.4e5*exp(0.0035*T)
R1=.004
err=T-288.15
I=400/(3.1416*R1^2)
qdot=I^2/ke
Initial Conditions:
r(0)=0
T(0)=389.25
rQr(0)=0
Final Value:
r(f)=0.004
```

A plot of the temperature profile is shown in Figure 6–4. The calculated temperature profile exhibits the zero slope condition at $r = 0$ and matches the

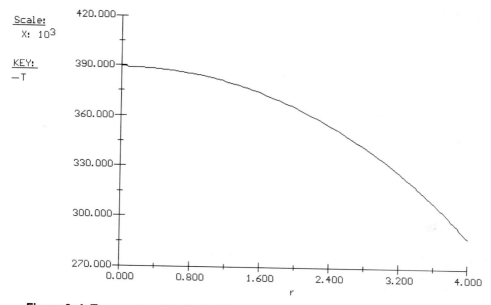

Figure 6–4 Temperature Profile for Wire with Surface Held at 288.15 K (15 °C)

desired boundary condition $T = 288.15$ K (15 °C) at $r = 0.004$ m. The heat flux Q_r is nonlinearly related to the radius, as indicated in Figure 6–5. Note that this

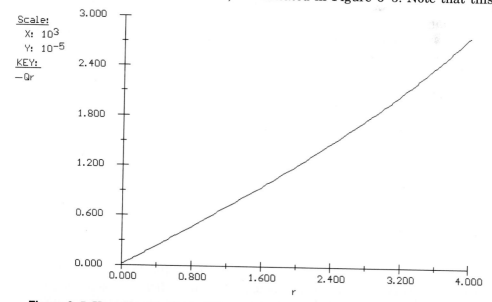

Figure 6–5 Heat Flux Profile for Wire with Surface Held at 288.15 K (15 °C)

numerical solution is similar to that of Problem 5.1, demonstrating the analogy between viscous flow in a pipe and this heated wire problem, as emphasized in the approach by Bird et al.[1]

The POLYMATH data file for part (a) is found in the *Simultaneous Differential Equation Solver Library* located in directory CHAP6 with file named P6-02A.POL.

(b) The same differential equations apply as used in part (a), but the final boundary condition must account for the convection from the wire surface to the ambient temperature at $r = R_1$. The energy balance at the wire surface can be expressed, as developed in Equation (6-3), by setting the conduction (flux) to the wire surface equal to the convection (flux) from the wire surface to the surroundings. Thus

$$Q_r\big|_{r = R_1} = -k\frac{dT}{dr}\bigg|_{r = R_1} = h(T\big|_{r = R_1} - T_b) \qquad (6\text{-}17)$$

where the temperature of the surroundings is represented by T_b. An objective function based on the error in the energy balance of Equation (6-17) can be expressed as

$$\varepsilon(T_0) = Q_r\big|_{r = R_1} - h(T\big|_{r = R_1} - T_b) \qquad (6\text{-}18)$$

which will be zero when the variables Q_r and T are evaluated at the final condition. Thus this requires a single-variable search to optimize the initial condition for T.

The resulting temperature profile is plotted in Figure 6–6, which indicates a rather flat temperature profile with a large temperature driving force for the natural convection from the wire surface to the surroundings.

(c) The equation set for conduction without the generation terms must be used for the insulation layer, and a different thermal conductivity must be used. Thus the equation set within the insulation is given by

$$\frac{d}{dr}(rQ_r) = 0 \qquad R_1 < r \le R_2$$

$$\frac{dT}{dr} = -\frac{Q_r}{k_I} \qquad R_1 < r \le R_2 \qquad (6\text{-}19)$$

$$Q_r = \frac{(Q_r r)}{r} \qquad R_1 < r \le R_2$$

The preceding changes can be accomplished within the POLYMATH *Simultaneous Differential Equation Solver* by using the "if ... then ... else ... " capability, as discussed in Problem 6.1. The boundary condition of part (b) is then applied to the preceding equations at R_2.

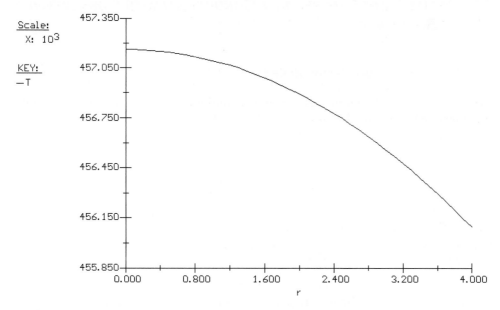

Figure 6–6 Temperature Profile in Wire with Natural Convection Boundary
Condition

6.3 RADIAL HEAT TRANSFER BY CONDUCTION WITH CONVECTION AT BOUNDARIES

6.3.1   Concepts Demonstrated

Heat transfer in the radial direction with constant thermal conductivities with convection at both boundaries.

6.3.2   Numerical Methods Utilized

Solution of a nonlinear algebraic equation with some additional explicit algebraic equations.

6.3.3   Problem Statement

A low-pressure steam system at 60 psia (292.73 °F) has saturated steam that is contained in a 2-inch schedule 40 steel pipe with a thermal conductivity of $k = 26$ btu/h·ft·°F. The pipe insulation has a thermal conductivity of $k = 0.05$ btu/h·ft·°F. The steam side heat transfer coefficient is $h_i = 2000$ btu/h·ft^2·°F and the air side heat transfer coefficient is $h_0 = 4$ btu/h·ft^2·°F. The ambient air temperature is 70 °F.

(a)   Calculate the heat loss per foot if the insulation is 1 inch thick.
(b)   What is the outer radius of insulation in inches that is required to keep the heat loss at 50 btu/h per foot of pipe?

Additional Information and Data

The overall radial heat transfer through a cylindrical geometry with various resistances and two materials in series can be expressed (see Geankoplis[4]) as

$$q = \frac{T_1 - T_5}{1/(h_i A_i) + (r_1 - r_i)/k_A A_{A \text{ lm}} + (r_o - r_1)/k_B A_{B \text{ lm}} + 1/h_o A_o} \qquad \text{(6-20)}$$

for the case that is illustrated in Figure 6–7. Here the inside surface area is $A_i = 2\pi L r_i$, the outside surface area is $A_o = 2\pi L r_o$, and the area between the two materials is $A_1 = 2\pi L r_1$. The analytical solution when the thermal conductivities are constant requires the use of log mean areas for material A (the steel pipe) and material B (the insulation), given by

$$A_{A \text{ lm}} = \frac{A_1 - A_i}{\ln(A_1/A_i)} \qquad A_{B \text{ lm}} = \frac{A_o - A_1}{\ln(A_o/A_1)} \qquad \text{(6-21)}$$

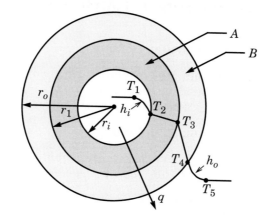

L = length

Figure 6–7 General Radial Temperature Profiles for Conduction through Two
Materials with Convection at the Boundaries

6.3.4   Solution (Suggestions)

(a) The heat loss can be directly calculated from Equation (6-20) and the
appropriate areas. This can conveniently be accomplished in the POLYMATH
Simultaneous Algebraic Equation Solver since all of the equations are explicit.

(b) The value of r_o can be determined using Equation (6-20) as a nonlinear
equation with the additional calculations as explicit algebraic equations. Simple
modifications to the POLYMATH equation for the solution to part (a) are
required.

6.4 ENERGY LOSS FROM AN INSULATED PIPE

6.4.1   Concepts Demonstrated

Radial heat conduction within pipe insulation with variable thermal conductivity and a free convection boundary condition.

6.4.2   Numerical Methods Utilized

Solution of simultaneous ordinary differential equations with a shooting technique to optimize a boundary condition.

6.4.3   Problem Statement

A horizontal steel pipe carrying steam at 450 K is to be insulated with a newly developed material whose thermal conductivity varies with temperature, as summarized in Table 6–1. The pipe can be assumed to be at the steam temperature due to an effective internal heat transfer coefficient and the large thermal conductivity of steel. The external diameter of the steel pipe is 0.033 m, and the ambient temperature is 300 K.

The external heat transfer coefficient from the surface of a cylinder due to natural convection in air can be approximated by (see Geankoplis[4])

$$h = 1.32(|\Delta T|/D)^{1/4} \tag{6-22}$$

where h is the heat transfer coefficient in $W/m^2 \cdot K$, ΔT is the temperature difference between the surface and the air in K, and D is the cylinder diameter in m.

(a) Calculate the heat loss per meter from the uninsulated pipe.
(b) Calculate the heat loss per meter from the pipe when the insulation thickness is 0.04 m and plot the temperature distribution in the insulation.

Table 6–1 Thermal Conductivity Data

T in K	k in W/m · K
250	0.042
300	0.061
350	0.077
400	0.087
450	0.094
500	0.098

6.5 HEAT LOSS THROUGH PIPE FLANGES

6.5.1   Concepts Demonstrated

Radial heat conduction in one dimension with simultaneous heat convection. Effects of various ambient temperatures on convective losses, and calculation of heat loss through pipe flanges made of different metals.

6.5.2   Numerical Methods Utilized

Solution of two point boundary value problem involving ordinary differential equations using a shooting technique and single-variable optimization.

6.5.3   Problem Statement

Metal pipes may be connected by the bolting together of two flanges on the pipe ends, as shown in Figure 6–8, without the bolts. Often these unions are not insu-

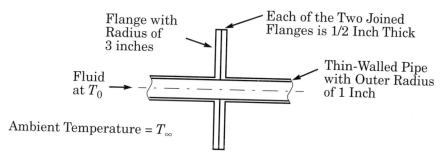

Figure 6–8 Cross Section of Union Consisting of Two Pipe Flanges

lated so that the pipes can be quickly disconnected. For heat transfer purposes, this union can be considered as a cylinder of solid metal made from two pipe flanges. Consider the uninsulated union of Figure 6–8 where the fluid in the pipe is at 260 °F and the heat transfer coefficient to the surroundings is constant at $h = 3$ btu/h $\cdot$ ft^2 $\cdot$ °F.

(a)   Select one of the metals listed in Table 6–2 and calculate the total heat loss from a single flange for an average day when the average ambient temperature $T_\infty = 60$ °F. Assume that the thermal conductivity is constant at the value reported for 212 °F in Table 6–2.

(b)   Plot the heat transfer rate and temperature versus radius for part (a). What percentage of the total heat loss is from the end of the flange?

(c)   Repeat the total heat loss calculations of part (a) for the same flange material for a cold winter day ($T_\infty = 10$ °F) and a very hot summer day ($T_\infty = 100$ °F).

Table 6–2 Thermal Conductivity of Selected Metals (Welty[6] with permission.)

Metal	Thermal Conductivity (btu/h · ft · °F)		
	68 °F	212 °F	572 °F
Aluminum	132	133	133
Copper	223	219	213
Iron	42.3	39	31.6
Nickel	53.7	47.7	36.9
Stainless Steel	9.4	10.0	13
Steel (1% C)	24.8	24.8	22.9

6.5.4   Solution (Partial)

The union formed by two flanges is symmetrical, and thus only one flange has to be considered with heat loss from one exposed circular face and an exposed rim as shown in Figure 6–9. The other side of the circular face, which is the line of symmetry, is treated as insulated due to symmetry with no heat transfer across this surface. The problem becomes one dimensional, with the heat transfer being a function of the radius. The boundary condition at the inner surface where $r = R_1$ is the temperature of the fluid. The thickness of the pipe is neglected as it is very thin. The boundary condition at the outer surface where $r = R_2$ is determined by heat convection from the surface to the ambient temperature T_∞.

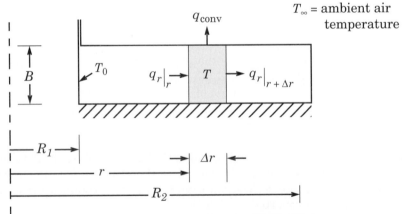

Figure 6–9 Cross Section of Pipe Flange Forming Half of Union

An energy balance can be performed on the differential volume within the flange shown in Figure 6–9, where q_r is the total heat rate in the r direction due to convection in btu/min.

$$\text{INPUT} + \text{GENERATION} = \text{OUTPUT} + \text{ACCUMULATION}$$

$$q_r\Big|_r + 0 = q_r\Big|_{r+\Delta r} + h(2\pi r)\Delta r(T - T_\infty) + 0 \tag{6-23}$$

Rearrangement of the preceding equation and utilization of the definition of the derivative for q_r by taking the limit as $\Delta r \to 0$ yields

$$\frac{dq_r}{dr} = -h(2\pi r)(T - T_\infty) \tag{6-24}$$

The relationship between heat rate q_r and heat flux Q_r at radius r requires the cross-sectional area for heat conduction at radius r, giving

$$q_r = Q_r(2\pi r)B \tag{6-25}$$

Substitution of Equation (6-25) into Equation (6-24) yields

$$\frac{d}{dr}(rQ_r) = \frac{-hr(T - T_\infty)}{B} \tag{6-26}$$

where the heat flux Q_r in btu/ft$^2 \cdot$h can be calculated from Fourier's law, given by

$$Q_r = -k\frac{dT}{dr} \tag{6-27}$$

which can be rearranged to

$$\frac{dT}{dr} = (-Q_r)/k \tag{6-28}$$

In the preceding equations, h is the heat transfer coefficient in btu/h $\cdot$ ft$^2 \cdot$ °F, k is the thermal conductivity of the metal in btu/h $\cdot$ ft $\cdot$ °F, and B is the thickness of the flange in ft. The algebraic equation is given by

$$Q_r = \frac{rQ_r}{r} \tag{6-29}$$

Thus the differential equations that describe the heat transfer processes consist of Equations (6-26) and (6-28) along with the algebraic relationship given Equation (6-29), which is needed to calculate Q_r since the product of two-variable rQ_r is used as a problem variable.

The boundary (final) condition on Equation (6-26) is obtained from an energy balance on the rim, which equates the heat flux by conduction to the rim to the heat that is transported from the rim by convection. Thus

$$Q_r\Big|_{r=R_2} = h(T - T_\infty)\Big|_{r=R_2} \tag{6-30}$$

and thus the variable rQ_r is calculated from

$$(rQ_r)\big|_{r = R_2} = R_2 h(T - T_\infty)\big|_{r = R_2} \tag{6-31}$$

The initial condition for Equation (6-28) is that the temperature is the temperature of the fluid in the pipe designated by T_0 at radius $r = R_1$.

$$T\big|_{r = R_1} = T_0 \tag{6-32}$$

(a) For the case of the flange and pipe being made of aluminum, the parameter values are given by $k = 133$ btu/h·ft·°F, $T_0 = 260$ °F, $T_\infty = 60$ °F, and $h = 3$ btu/h·ft^2·°F. The solution can utilize a shooting technique to solve the differential Equations (6-26) and (6-28), which is discussed in Problems 3.5, 6.2, and 6.4. The initial condition for Equation (6-26) must be determined for the variable rQ_r so that the boundary condition (final condition) of Equation (6-31) is met. This can be accomplished by defining an objective function as

$$\varepsilon(rQ_r) = (rQ_r)\big|_{r = R_2} - R_2 h(T - T_\infty)\big|_{r = R_2} \tag{6-33}$$

Thus the initial value of (rQ_r) must be optimized using a single-variable search to minimize the objective function calculated from Equation (6-33).

An equation set with the converged solution to this problem using the POLYMATH *Ordinary Differential Equation Solver* is as follows:

```
Equations:
d(rQ)/d(r)=-h*r*(T-Tinf)/B
d(T)/d(r)=-Q/k
Q=rQ/r
h=3
Tinf=60
B=.5/12
k=133
err=rQ-r*h*(T-Tinf)
q=Q*2*3.1416*r*B
Initial Conditions:
r(0)=0.08333
rQ(0)=542.4
T(0)=260
Final Value:
r(f)=0.25
```

 The POLYMATH problem solution file for part (a) is found in the *Simultaneous Differential Equation Solver Library* located in directory CHAP6 with file named P6-05A.POL.

(b) The actual values of the heat rate q at various values of r can be calculated from the converged solution and Equation (6-25), which is included in the preceding equation set. The results are shown in Figure 6–10. The heat rate at

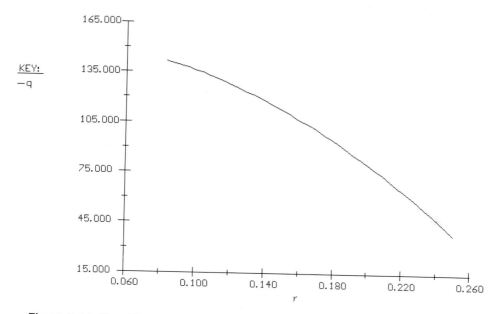

Figure 6–10 Heat Flux in Flange as a Function of Radius

the end of the flange is 38.6 btu/h compared to the total heat rate at R_1 of 142. btu/h. Thus the percentage of the total lost from the end is 27.2%.

6.6 HEAT TRANSFER FROM A HORIZONTAL CYLINDER ATTACHED TO A HEATED WALL

6.6.1   Concepts Demonstrated

One-dimensional heat transfer due to conduction in a rod subject to convection and radiation from the surface.

6.6.2   Numerical Methods Utilized

Solution of simultaneous ordinary differential equations with a shooting technique in order to converge upon a variety of boundary conditions.

6.6.3   Problem Statement

A horizontal circular rod (cylinder) is used to transfer heat from a vertical wall that is held at a constant temperature $T_0 = 275$ °F to the surrounding air at temperature T_∞, as illustrated in Figure 6–11. The natural convection from the side

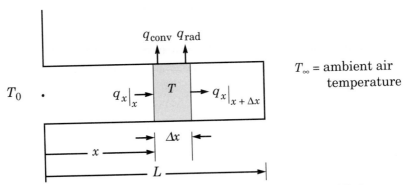

Figure 6–11 Differential Element for Heat Transfer for a Horizontal Rod

surface of the rod is given by

$$h_s = 0.27|\Delta T/D|^{1/4} \tag{6-34}$$

where h_s is the heat transfer coefficient in btu/h·ft^2·°F, ΔT is the temperature difference between the rod surface temperature T and the ambient air temperature T_∞ in °F, and D is the diameter of the rod in ft.

At the end of the rod that is exposed to the air, the heat transfer coefficient is given by

$$h_e = 0.18|\Delta T|^{1/3} \tag{6-35}$$

where the units are the same as for Equation (6-34).

The radiative heat transfer from the rod surfaces is described by the Stefan-Boltzmann equation with an effective emissivity of ε to the surroundings at temperature T_∞:

$$q_{rad} = \varepsilon \sigma A (T - T_\infty)^4 \tag{6-36}$$

where $\sigma = 0.1714 \times 10^{-8}$ btu/h$\cdot$ft$^2\cdot$°R^4 and A is the surface area in ft^2.

For the usual case where the diameter of the rod is small relative to the length L of the rod, there will be little temperature change in the radial direction but there will be a significant temperature change in the x direction. Thus a mathematical description involves one-dimensional conduction within the rod subject to convection and radiation from the rod surface to the ambient air and surroundings.

(a) Calculate the temperature profile down the length of the horizontal rod and the total rate of heat removed from the heated wall when *only* convection from *all* rod surfaces is considered. The wall temperature is $T_0 = 275$ °F, the surroundings are at $T_\infty = 70$ °F, the rod diameter $D = 0.1$ ft, the rod length $L = 0.5$ ft, and the rod thermal conductivity is $k = 30$ btu/h$\cdot$ft$\cdot$°F.

(b) Repeat part (a) with the assumption that there is no convection from the exposed end of the rod.

(c) Repeat part (a) if radiation from all rod surfaces with an effective emissivity of $\varepsilon = 0.85$ is included with the convection.

(d) What conclusions can be reached about the boundary condition assumption made in part (b) and the inclusion of radiation in part (c)?

(e) Repeat parts (a) through (d) for $T_0 = 575$ °F.

Additional Information and Data

Modeling Equations for Heat Transfer The differential equation describing the temperature in the rod can be obtained by an energy balance on the differential element depicted in Figure 6–11.

INPUT + GENERATION = OUTPUT + ACCUMULATION

$$q_x \big|_x + 0 = q_{conv} + q_{rad} + q_x \big|_{x + \Delta x} + 0 \tag{6-37}$$

Fourier's law [Equation (6-1)] applies to the conduction within the rod and can be expressed as

$$\frac{dT}{dx} = (-Q_x)/k \tag{6-38}$$

where the heat flux is represented by $Q_x = q_x/A$. The initial condition for Equation (6-38) is that $T = T_0$ at $x = 0$.

The expressions for heat flux, convection, and radiation with the corresponding areas can be introduced into Equation (6-37):

$$Q_x\left(\frac{\pi D^2}{4}\right)\Bigg|_x = h(\pi D \Delta x)(T - T_\infty) + \varepsilon\sigma(\pi D \Delta x)(T-T_\infty)^4 + Q_x\left(\frac{\pi D^2}{4}\right)\Bigg|_{x+\Delta x} \quad \text{(6-39)}$$

Taking the limit of the preceding equation as $\Delta x \to 0$ yields

$$\frac{dQ_x}{dx} = -\left(\frac{4}{D}\right)[h_s(T - T_\infty) + \varepsilon\sigma(T-T_\infty)^4] \quad \text{(6-40)}$$

The boundary condition at the end of the rod is determined from a steady-state energy balance at the end surface where the area is $A = \pi D^2/4$.

$$\text{INPUT} + \text{GENERATION} = \text{OUTPUT} + \text{ACCUMULATION}$$

$$q_x\Big|_{x=L} + 0 = q_{conv} + q_{rad} + 0 \quad \text{(6-41)}$$

Insertion of the various terms into Equation (6-41) and solving for the heat flux yields

$$Q_x\Big|_{x=L} = [h(T - T_\infty) + \varepsilon\sigma(T-T_\infty)^4]\Big|_{x=L} \quad \text{(6-42)}$$

Equation for Heat Rate into Rod The total heat lost from the rod due to convection and radiation is equal to the heat conducted into the rod at the base where $x = 0$. This can be expressed by

$$q_x\Big|_{x=0} = -k\left(\frac{\pi D^2}{4}\right)\frac{dT}{dx}\Big|_{x=0} \quad \text{(6-43)}$$

and evaluated once the numerical solution to Equation (6-38) has been achieved.

6.6.4   Solution (Suggestions)

The numerical solution requires the simultaneous integration of Equations (6-38) and (6-40). A shooting technique, as outlined in Problem 3.6, can be used to converge upon the initial condition for $(dQ_x)/dx\big|_{x=0}$, which results in the desired final condition for Q_x. The corresponding objective function for the final condition, which should converge to zero, can be written from Equation (6-42) as

$$\text{err} = Q_x\Big|_{x=L} - [h_e(T - T_\infty) + \varepsilon\sigma(T-T_\infty)^4]\Big|_{x=L} \quad \text{(6-44)}$$

The heat transfer rate into the rod can then be obtained from Equation (6-43). The various modes of heat transfer can be modeled by adjusting the terms in the equations, and the boundary conditions can be similarly modified.

6.7 HEAT TRANSFER FROM A TRIANGULAR FIN

6.7.1   Concepts Demonstrated

One dimensional heat transfer due to conduction in a triangular fin subject to convection from the surface.

6.7.2   Numerical Methods Utilized

Solution of simultaneous ordinary differential equations with a shooting technique in order to converge upon split boundary conditions.

6.7.3   Problem Statement

Triangular fins are widely used to increase heat transfer with a minimal amount of fin material. Such a fin is shown in Figure 6–12. A simplified treatment of this

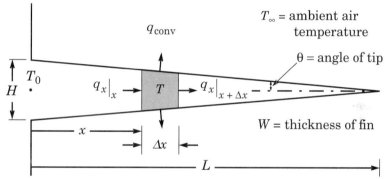

Figure 6–12 Differential Element for Heat Transfer in a Triangular Fin

fin can utilize the assumptions that the temperature T varies only in the x direction, no energy is lost from either the sides or the end of the fin, and simple convection with a constant heat transfer coefficient describes the convection from the fin surfaces.

Thus an energy balance on the differential volume of the fin as shown in Figure 6–12 yields

INPUT + GENERATION = OUTPUT + ACCUMULATION

$$q_x\big|_x + 0 = q_{conv} + q_x\big|_{x + \Delta x} + 0 \qquad (6\text{-}45)$$

Utilization of the convective heat transfer from the upper and lower fine surfaces gives

$$q_{conv} = 2W\Delta x \sec\theta\, h_s(T - T_\infty) \qquad (6\text{-}46)$$

where H is the height of the fin and θ is angle at the tip of the fin whose tangent is $W/2L$. Definitions and units are the same as in Problem 6.6.

Equation (6-45) can be divided by Δx and rearranged to the definition of a derivative. As the limit is taken where $\Delta x \to 0$, the following differential equation results:

$$\frac{dq}{dx} = -2W \sec\theta\, h_s(T - T_\infty) \tag{6-47}$$

Fourier's law for heat conduction in the fin can be written as

$$q = -kA\frac{dT}{dx} \tag{6-48}$$

where the cross-sectional area given by A can be expressed as a function of x by

$$A = \left(1 - \frac{x}{L}\right)H \tag{6-49}$$

Thus Equation (6-48) can be rewritten using Equation (6-49) for A as

$$\frac{dT}{dx} = \frac{q}{-k\left(1 - \dfrac{x}{L}\right)H} \tag{6-50}$$

The final condition for Equation (6-47) is $q|_{x = L} = 0$ and the initial condition for Equation (6-50) is $T|_{x = 0} = T_0$.

(a) Select one of the metals in Table 6–2 for use in a triangular cross-sectional fin. Calculate the temperature profile down the length of the fin and the total rate of heat removed from the heated wall. The wall temperature is $T_0 = 212$ °F, the fin length is $L = 1.0$ ft, and the rod length is $L = 0.5$ ft. The surroundings are at temperature $T_\infty = 60$ °F and $h_s = 2.5$ btu/h·ft²·°F. Assume that the metal thermal conductivity remains constant at the value given in Table 6–2 for 212 °F.

(b) What is the efficiency factor η_f for the fin in part (a) that is defined as the ratio of the heat rate actually transferred from the base to the heat rate if the entire fin surface was at the base temperature T_0?

(c) How much error is introduced into the result for part (b) by assuming a constant thermal conductivity at 212 °F for the metal used in the fin?

(d) What second-order ordinary differential equation and boundary conditions can also be used to describe the combined energy balance and Fourier's law for the constant thermal conductivity case?

6.8 SINGLE-PASS HEAT EXCHANGER WITH CONVECTIVE HEAT TRANSFER ON TUBE SIDE

6.8.1   Concepts Demonstrated

Simple heat exchanger with average, log mean, and point temperature driving forces, and convective heat transfer involving turbulent flow where physical properties vary with temperature, and overall and differential energy balances applied to simple heat exchangers.

6.8.2   Numerical Methods Utilized

Solutions of nonlinear equations and ordinary differential equations.

6.8.3   Problem Statement

A simple heat exchanger consists of a single tube with an inside diameter of $D = 0.01033$ m and an equivalent length $L = 8$ m in which $m = 1$ kg/s of a light hydrocarbon oil is heated by condensing steam at $T_S = 170$ °C. Because of the high heat transfer coefficient from the steam to the tube surface and the high thermal conductivity of the tube wall, the inside temperature of the tube wall can be assumed to be the steam temperature. The oil has a viscosity which varies significantly with temperature and is given by

$$\mu = \exp[-12.86 + 1436/(T-153)] \tag{6-51}$$

in kg/m·s with temperature T in K. Other physical properties are approximately constant with temperature and are given by $\rho = 850$ kg/m^3, $C_p = 2000$ J/kg·K, and $k = 0.140$ W/m·K. The inlet oil temperature is $T_1 = 40$ °C.

(a)   Calculate the outlet temperature of the oil stream in °C using the bulk mean oil temperature for physical properties and average ΔT for the temperature driving force for the heat transfer. What is the total amount of heat transferred? The Sieder-Tate expression given by Equation (2-21) should be used to determine h_i for the oil side in turbulent flow.

(b)   Repeat part (a) utilizing the log mean temperature difference for the heat transfer.

(c)   Repeat part (a) with a differential equation approach in which the local energy balance is solved down the length of the heat exchanger, with the Sieder-Tate correlation of Equation (2-21) providing the local heat transfer coefficient from local physical properties.

Additional Information and Data

The local or point energy balance utilizing the overall heat transfer coefficient can be expressed (see Geankoplis[4]) as

$$q = h_i A \Delta T \qquad\qquad (6\text{-}52)$$

where h_i is the convective heat transfer coefficient based on the inside area of the tube in $W/m^2 \cdot K$ and the inside surface area is A in m^2.

An overall energy balance on the tube side fluid yields

$$q_T = m C_p (T_2 - T_1) \qquad\qquad (6\text{-}53)$$

where q_T is the total heat transfer rate and $(T_2 - T_1)$ represents the temperature increase in the fluid as it passes through the heat exchanger.

Average Temperature Driving Force When the driving force does not vary greatly down the heat exchanger, the mean temperature difference at the inlet and outlet can be used in Equation (6-52):

$$\Delta T = \Delta T_m = [(T_1' - T_1) + (T_2' - T_2)]/2 \qquad\qquad (6\text{-}54)$$

where the subscript 1 represents the tube entrance location and subscript 2 represents the tube outlet location as diagrammed for the general cocurrent case in Figure 6–13. The prime (′) represents the hotter stream, which for this problem is the constant steam temperature T_S.

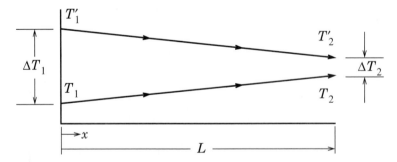

Figure 6–13 General Temperature Distribution for One-Pass Cocurrent Heat Exchanger

Log Mean Temperature Driving Force When the driving force varies significantly down the heat exchanger, the log mean temperature difference at the inlet and outlet should be used in Equation (6-52):

$$\Delta T = \Delta T_{lm} = \frac{[(T_1' - T_1) - (T_2' - T_2)]}{\ln[(T_1' - T_1)/(T_2' - T_2)]} \qquad\qquad (6\text{-}55)$$

Local Temperature Driving Force A differential energy balance can be made at a point that is located at distance x from the entrance, as shown in Figure 6–14. The steady-state energy balance on the differential element of the tube

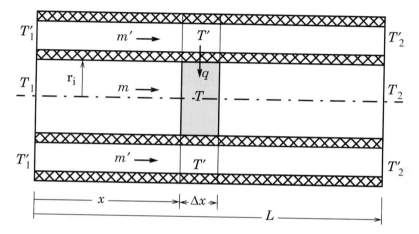

Figure 6–14 Differential Element for Energy Balance in One-Pass Cocurrent Heat Exchanger

yields

$$\text{INPUT} + \text{GENERATION} = \text{OUTPUT} + \text{ACCUMULATION}$$

$$mC_pT\big|_x + q + 0 = mC_pT\big|_{x+\Delta x} + 0 \qquad \text{(6-56)}$$

where m is the mass flow rate in the tube in kg/s. The q term, which represents the input from the hot stream to the tube stream, can be expressed using Equation (6-52) and the differential area for heat transfer as

$$q = h_i(2\pi r_i \Delta x)(T' - T) \qquad \text{(6-57)}$$

Rearranging and taking the limit as Δx goes to zero yields the differential equation

$$\frac{d}{dx}(mC_pT) = h_i(2\pi r_i)(T' - T) \qquad \text{(6-58)}$$

For constant m and C_p, the differential energy balance simplifies to

$$\frac{dT}{dx} = \frac{h_i(2\pi r_i)(T' - T)}{mC_p} \qquad \text{(6-59)}$$

This ordinary differential equation can be used to predict the temperature in the tube side of the heat exchanger as a function of distance down the exchanger. The initial condition is that $T = T_1$ at $x = 0$. Integration of this equation to $x = L$ yields the exit tube fluid temperature.

6.8.4   Solution (Partial)

(a) & (b) During steady-state operation of this heat exchanger, the total heat transferred from the steam (shell side) as calculated from Equation (6-52), where $q = q_T$, must equal the heat transferred to the light oil (tube side) given by Equation (6-53). Thus elimination of q_T from these equations yields a nonlinear equation which can be solved for the outlet temperature T_2.

$$h_i A \Delta T = m C_p (T_2 - T_1) \qquad\qquad (6\text{-}60)$$

The area term in the preceding equation is the inside surface area of the tube given by $A = 2\pi r_i L$. The total heat transferred can then be calculated from Equation (6-53).

Note that the log mean temperature driving force is used. It is helpful to rearrange the nonlinear equation by multiplying by the ln term to give

$$h_i[(T_1' - T_1) - (T_2' - T_2)] = m C_p (T_2 - T_1) \ln[(T_1' - T_1)/(T_2' - T_2)] \qquad (6\text{-}61)$$

6.9 DOUBLE-PIPE HEAT EXCHANGER

6.9.1 Concepts Demonstrated

Calculations of cocurrent and countercurrent heat exchangers using differential energy balances that allow for local variations in physical properties and heat transfer coefficients.

6.9.2 Numerical Methods Utilized

Solution of coupled first-order ordinary differential equations with known initial conditions (cocurrent flow) and split boundary conditions (countercurrent flow).

6.9.3 Problem Statement

One mode of operation of a double-pipe heat exchanger, which is diagramed in Figure 6–15 for cocurrent operation, is to cool a steady stream of m' lb_m/h of fluid in the tube from inlet temperature T_1 to an exit temperature T_2. The fluid in the shell side with a flow rate of m lb_m/h is correspondingly heated from inlet temperature T'_1 to exit temperature T'_2.

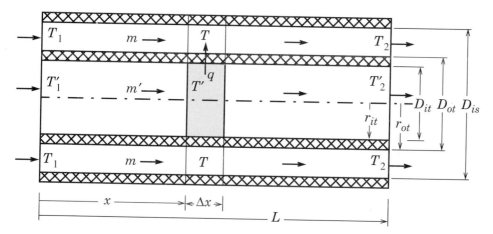

Figure 6–15 Double-Pipe Heat Exchanger for Cocurrent (Parallel) Flow

Select one of the fluids listed in Table 6–3 to be cooled by water in a double-pipe heat exchanger. The mass flow rate of cooling water is always to be 300% or three times that of the fluid mass flow rate for liquids and thirty times that of the fluid mass flow rate for gases. The cooling water is continuously available at 65 °F. The local convective heat transfer coefficients on both the tube and shell sides should be calculated from the Dittus-Boelter correlation,[3] given by

$$Nu = 0.023 Re^{0.8} Pr^n \tag{6-62}$$

or

$$\frac{hD}{k} = 0.023 \left(\frac{Dv\rho}{\mu}\right)^{0.8} \left(\frac{C_p\mu}{k}\right)^n \tag{6-63}$$

where $n = 0.4$ for heating and $n = 0.3$ for cooling. The various terms are discussed in Problem 2.5. Note that the $v\rho$ term in the Re can also be conveniently expressed as the mass flux or mass velocity where $v\rho = m/A_c$, with m representing the mass flow rate and A_c representing the cross-sectional area for flow.

The center heat exchanger tube is made of copper with a constant thermal conductivity of 220 btu/ft·h·°F, and the exterior of the steel pipe shell is very well insulated. Scale formation is not a problem. The physical properties of the different fluids at various temperature are given in Tables E–2 and E–3 of Appendix E.

Select a fluid from Table 6–3 for all calculations.
(a) Calculate the length required to reach the desired exit temperature when the selected fluid is in the heat exchanger tube and the cooling water is in cocurrent flow in the shell.
(b) Calculate the length when the selected fluid is in the heat exchanger tube and the cooling water is in countercurrent flow in the tube.
(c) Calculate the length when the cooling water is in the heat exchanger tube and the selected fluid is in cocurrent flow in the shell.
(d) Calculate the length when the cooling water is in the heat exchanger tube and the selected fluid is in countercurrent flow in the shell.

Table 6–3 Data on Heat Exchanger Operation for Various Fluids

Fluid	Flow	Inlet	Outlet	Inside Tube		Outside Pipe	
	m	T	T	OD		Nominal	
	lb_m/h	°F	°F	in.	BWG	Pipe, in.	Schedule
Benzene (liquid)	3000	150	80	7/8	16	$1\frac{1}{2}$	40
Kerosene (liquid)	4000	140	85	1	14	2	80
Ammonia (liquid)	2000	120	82	3/4	12	$1\frac{1}{4}$	40
Carbon Dioxide (gas)	25	250	85	5/8	16	1	40
Sulfur Dioxide (gas)	35	300	90	3/4	14	$1\frac{1}{4}$	80

Additional Information and Data

The heat exchanger can be operated with cocurrent flow in both the tube and shell, as shown in Figure 6–15, or with countercurrent flow, where the flow in the shell is opposite to that in the tube.

Cocurrent or Parallel Flow A differential energy balance on the tube fluid in concurrent flow in a manner similar to that detailed in Problem 6.8, with the prime indicating the fluid temperature or property of the higher temperature fluid, leads to

$$\frac{d}{dx}(T') = -\frac{U_i(\pi D_i)(T' - T)}{m'C_p'} \tag{6-64}$$

where the overall heat transfer coefficient U_i in btu/h$\cdot$ft$^2\cdot$°F is based on the inside area given by πD_i with the inside tube diameter D_i in ft. The initial condition for Equation (6-64) is just the inlet temperature T_1' to the tube, and the final condition is the outlet temperature T_2' from the tube.

Similarly, for the well insulated shell, a differential energy balance on the shell fluid in concurrent flow and the inside heat transfer area yields

$$\frac{dT}{dx} = \frac{U_i(\pi D_i)(T' - T)}{mC_p} \tag{6-65}$$

where the convective heat transfer coefficient for the shell is also based on the outside diameter D_i of the inner pipe in ft. The inlet and outlet shell temperatures are T_1 and T_2, respectively. Figure 6–13 illustrates the general temperature distributions for cocurrent flow using this notation.

Overall Heat Transfer Coefficients The local overall heat transfer coefficient based on the inside area is calculated from (Geankoplis[4])

$$U_i = \frac{1}{1/h_i + ((r_{ot} - r_{it})D_{it})/(k_t D_{tlm}) + D_{it}/(D_{ot}h_o)} \tag{6-66}$$

where t represents the tube wall material and dimensions. The D_{tlm} term is the log mean diameter (or area) for the tube wall.

Heat Transfer within the Shell The heat transfer coefficient for the fluid in the shell annulus is calculated by using an equivalent diameter that is determined by taking the outer diameter of the annulus and subtracting the inner diameter of the annulus. This equivalent diameter is then used in calculating the Reynolds and Nusselt numbers in the heat transfer correlations.

Countercurrent Flow The differential energy balance on the tube fluid in countercurrent flow remains the same as Equation (6-64); however, the reversal of the flow direction in the shell yields a differential energy balance given by

$$\frac{dT}{dx} = -\frac{U_i(\pi D_i)(T' - T)}{mC_p} \tag{6-67}$$

which is just the negative of the concurrent equation. The initial condition, however, is now the outlet shell temperature T_1, and the final condition is the inlet shell temperature T_2. Thus in this situation, the simultaneous solution of Equations (6-64) and (6-67) is a split boundary problem, which typically involves assuming an outlet temperature T_1 for the shell followed by integration to determine the inlet temperature T_2 of at the desired heat exchanger length. The solution of two-point boundary value problems is discussed in Problem 3.6.

6.9.4   Solution (Suggestions)

The properties of liquid water and the other fluids are found in Appendix E. These properties can be represented as a function of the temperature in °F by using the techniques and suggested correlation equations introduced in Problems 2.3 and 5.5.

The POLYMATH table data files for the different fluids are found in the *Polynomial, Multiple Linear and Nonlinear Regression Program Library* located in directory TABLES with files named E-02A.POL, E-02B.POL, and E-03A.POL through E-03C.POL.

6.10 HEAT LOSSES FROM AN UNINSULATED TANK DUE TO CONVECTION

6.10.1 Concepts Demonstrated

Calculation of temperature in a well-mixed and uninsulated tank with convective heat losses to the surroundings.

6.10.2 Numerical Methods Utilized

Solution of a nonlinear algebraic equation and explicit algebraic equations.

6.10.3 Problem Statement

Cooling water exits a heat exchanger at a rate of $Q = 0.5$ m³/h and temperature of $T = 80$ °C. The water is collected in a cylindrical tank with a diameter $D = 1$ m and a height of $H = 2$ m. It can be assumed that the contents of the tank are well mixed and that all the heat loss is due to natural convection to the surrounding air. Water is being withdrawn from the tank at the same rate as it enters. The external surface of the tank can be assumed to be at the tank temperature, and the bottom of the tank can be considered to be very well insulated.

The properties of air at atmospheric pressure and for various temperatures are given in Table E–1 of Appendix E.

(a) Calculate the energy lost from the tank due to heat transfer in W and the temperature of the water in °C within the tank for different surrounding air temperatures of –10, 15, and 40 °C in calm air.
(b) Repeat part (a) for windy conditions with a constant velocity of 30 mph.

Additional Information and Data

Natural Convection For natural convection from the top surface under calm conditions, the following correlations (Geankoplis[4]) can be used:

$$Nu = 0.54Ra^{1/4} \quad \text{for} \quad 10^5 < Ra < 2 \times 10^7$$
$$Nu = 0.14Ra^{1/3} \quad \text{for} \quad 2\times10^7 < Ra < 3 \times 10^{10} \tag{6-68}$$

where

$$Nu = \frac{hL}{k} \text{ is the Nusselt number}$$

$$Ra = GrPr \text{ is the Rayleigh number}$$

$$Gr = \frac{\beta g \rho^2 L^3 \Delta T}{\mu^2} \text{ is the Grashof number}$$

$$Pr = \frac{\mu C_p}{k} \text{ is the Prandtl number}$$

In the preceding equations, h is the heat transfer coefficient in W/m$^2 \cdot$K, L is a characteristic length in m, k is the air thermal conductivity in W/m$\cdot$K, β is the air coefficient of thermal expansion in 1/K, g is the gravitational acceleration given by 9.80665 m/s^2, ρ is air density in kg/m^3, ΔT is the positive temperature difference between the tank surface and the air in K, μ is the air viscosity in kg/m$\cdot$s, and C_p is the heat capacity of air in J/kg$\cdot$K. In this problem, $L = 0.8862D$, as this is the length of a square that has the same area as the top of the tank with diameter D.

For natural convection over a wide range of Ra, the Churchill and Chu correlation[2] can be used for the vertical side of the cylindrical tank.

$$Nu = \left(0.825 + \frac{0.387 Ra^{1/6}}{[1 + (0.492/Pr)^{9/16}]^{8/27}} \right)^2 \tag{6-69}$$

Equations (6-68) and (6-69) should be evaluated with the properties of air at the film temperature T_f, which is given by

$$T_f = \frac{T + T_{\text{air}}}{2} \tag{6-70}$$

Forced Convection For forced convection the heat transfer coefficients can be approximated for both the top and the side of the cylindrical tank by using

$$Nu = 0.0366 Re_L^{0.8} Pr^{1/3} \tag{6-71}$$

where the Reynolds number is defined as $Re_L = (Lv\rho)/\mu$. Note that the characteristic length for the circular flat plate, which is the top of the cylindrical tank, is estimated by $L = 0.8862D$, while the length for convection from the side of the cylindrical tank is given by $L = D$. The velocity v is defined as the wind velocity in m/s. The air properties in Equations (6-71) are evaluated at the film temperature given by Equation (6-70).

6.10.4   Solution (Suggestions)

An enthalpy balance on the tank contents yields

$$Q\rho_w C_{pw}(80 - T) = h_1 A_1(T - T_{air}) + h_2 A_2(T - T_{air}) \tag{6-72}$$

where ρ_w is the density of water, C_{pw} is the heat capacity of the water, T is the outlet temperature from the tank, h_1 is the horizontal heat transfer coefficient, A_1 is the horizontal surface area for heat transfer, h_2 is the vertical heat transfer coefficient, and A_2 is the vertical surface area for heat transfer.

The properties for air can be represented as functions of temperature by applying the expressions and techniques suggested in Chapter 2 to the data of Table E–1.

(a) Calm Air The problem solution involves solving nonlinear Equation (6-72) along with algebraic Equation (6-70), while the appropriate values of h_1 and h_2 can be calculated from Equations (6-68) and (6-69).

(b) Windy Conditions Both horizontal and vertical heat transfer coefficients, h_1 and h_2, can be calculated from Equation (6-71) with care to define the Reynolds number properly for each case.

 The POLYMATH table data file for air is found in the *Polynomial, Multiple Linear and Nonlinear Regression Program Library* located in directory TABLES with file named E-01.POL.

6.11   UNSTEADY-STATE RADIATION TO A THIN PLATE

6.11.1   Concepts Demonstrated

Unsteady-state heat transfer via radiation to an object with high thermal conductivity whose temperature can be considered uniform throughout, and determination of transient and steady state behavior of temperature and heat flux.

6.11.2   Numerical Methods Utilized

Solution of an ordinary differential equation.

6.11.3   Problem Statement

A thin metal plate is to be heat treated for a period of time in a high-temperature vacuum furnace. The metal plate is 0.5 m by 0.5 m with a thickness of 0.0015 m. The initial temperature of the plate is 20 °C, and the furnace temperature is 1000 °C. The plate is to be suspended in the furnace so that radiation from the furnace walls provides rapid heating to both plate surfaces. The interior of the furnace and the surfaces of the plate can be considered to radiate as black bodies.

Select one of the metals whose properties are summarized in Table 6–4 and can be assumed to be constant with temperature.
(a)   Plot the temperature of the plate as a function of time to steady state.
(b)   Plot the heat flux to the plate as a function of time to steady state.
(c)   What is the time required to reach steady state defined as 99% of the possible change in plate temperature?

Table 6–4   Properties of Selected Metals (from Thomas[5])

Metal	Density ρ (kg/m^3)	Heat Capacity C_p (kJ/kg $\cdot$ K)
Copper	8950	0.383
Iron	7870	0.452
Nickel	8900	0.446
Silver	10500	0.234
Stainless Steel	8238	0.468
Steel (1% C)	7801	0.473
Zirconium	6750	0.272

Additional Information and Data

This is an unsteady-state problem in which the radiative heat transfer between the furnace wall and the plate causes the thin metal plate to heat. The plate itself might be subjected to temperature variations within its thickness if it were thick and/or had a low thermal conductivity. In this case for a thin plate with high thermal conductivity, it is a good assumption that the conduction is rapid throughout the plate and that the temperature is uniform throughout.

A simple unsteady-state energy balance on the plate can be made over a time increment Δt.

$$\text{INPUT} + \text{GENERATION} = \text{OUTPUT} + \text{ACCUMULATION}$$

$$\sigma A_P F_{12}(T_F^4 - T^4)\Delta t + 0 = 0 + V_P \rho C_P(T|_{t+\Delta t} - T|_t) \tag{6-73}$$

Here σ is the Stefan-Boltzmann constant with a value of 5.676×10^{-8} W/m$^2 \cdot$K^4, A_P is the surface area of both sides of the plate in m^2, V_P is the volume of the plate in m^3, F_{12} is the view factor, which can be assumed to be unity, and T_F is the temperature of the furnace in K. All other variables refer to the properties of the metal plate.

Equation (6-73) can be rearranged and the limit taken as Δt goes to zero to give the ordinary differential equation

$$\frac{dT}{dt} = \frac{\sigma A_P F_{12}(T_F^4 - T^4)}{V_P \rho C_P} \tag{6-74}$$

in which the properties of the metal are assumed constant.

The heat flux Q to the plate in W/m^2 is given by

$$Q = \sigma F_{12}(T_F^4 - T^4) \tag{6-75}$$

6.12 UNSTEADY-STATE CONDUCTION WITHIN A SEMI-INFINITE SLAB

6.12.1   Concepts Demonstrated

Unsteady-state heat conduction in a one-dimensional semi-infinite slab with constant properties and subjected to time-dependent boundary conditions.

6.12.2   Numerical Methods Utilized

Application of the numerical method of lines to solve a partial differential equation, and solution of simultaneous ordinary differential equations and explicit algebraic equations.

6.12.3   Problem Statement

Unsteady-state heat transfer in one dimension, when the physical properties are constant, follows the partial differential equation

$$\frac{\partial T}{\partial t} = \alpha \frac{\partial^2 T}{\partial x^2} \qquad\qquad (6\text{-}76)$$

which has been discussed in Problem 3.9.

An interesting application of this type of heat transfer involves the variation of the temperature of soil at various depths, which has been considered by Thomas.[5] Let us consider the annual variations in the surface temperature of the soil and how this will affect the temperature at various soil depths due to heat conduction. The variation of the surface soil temperature in °F during a typical year is reported by Thomas[5] to be represented by

$$T_s(t) = T_M - \Delta T_s \cos\left[\frac{2\pi}{\tau}(t - t_0)\right] \qquad\qquad (6\text{-}77)$$

where t is time in days of the year, T_M is the annual mean earth temperature in °F, ΔT_s is the amplitude of the annual variation in surface soil temperature, τ is the period that is 365 days, and t_0 is the phase constant in days. Data for the parameters are available for various locations, and typical values are given in Table 6–5.

Table 6–5 Soil Temperature Parameters for Selected U. S. Cities (from Thomas[5])

City	T_M (°F)	ΔT_s (°F)	t_0 (days)
Bismarck, ND	44	31	33
Burlington, VT	46	26	37
Chicago, IL	51	25	37
Las Vegas, NV	69	23	32
Phoenix, AZ	73	23	33

An example of a plot of annual soil surface temperature variation for Chicago is given in Figure 6–16 by utilizing the data of Table 6–5 in Equation (6-73) for $(0 \leq t \leq 365)$.

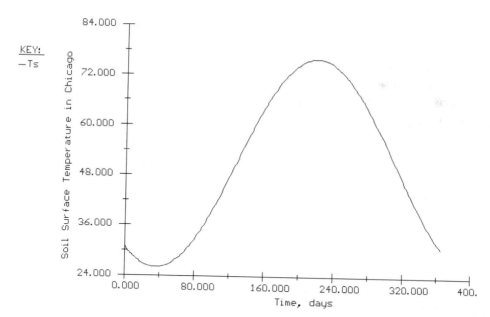

Figure 6–16 Annual Surface Temperature Variation for Chicago

Select one of the cities whose surface temperature data are given in Table 6–5. Assume that the soil's thermal diffusivity $\alpha = 0.9$ ft^2/day.

(a) Use the numerical method of lines to solve Equation (6-76) for the temperatures at various levels beneath the soil surface by applying the surface temperature of Equation (6-77) as a boundary condition. Plot the temperatures at values of x corresponding to 8, 16, 24, and 72 ft over a four-year period. It is suggested that an initial simulation with 11 nodes and 10 evenly spaced intervals of 8 ft should be adequate. The initial condition is suggested to be the annual mean earth temperature, T_M.

(b) Verify that the intervals you used in part (a) are adequate by doubling the number of intervals and halving the interval spacing. Compare results with part (a).

(c) Why is there a phase change in the temperatures between the soil surface and the soils levels below? Construct some plots that clearly show this effect, and indicate why a number of years of solution time is desirable for this problem.

6.12.4   Solution (Suggestions)

(a) & (b) The numerical method of lines has been discussed in Problem 3.9 and applied to a semi-infinite fluid in Problem 5.17. In achieving a solution to this type of problem, it is desirable to conduct trial solutions with various number of nodes and intervals between nodes. The objective of these trial solutions for the semi-infinite slab is to have enough node intervals so that the variables at the greatest depth (large x in this case) do not change appreciably with time. The numbers of nodes can then be adjusted so that the solution is not greatly dependent upon the number of nodes as the nodes are increased over the same total interval.

The POLYMATH *Simultaneous Differential Equation Solver* can be used in the process of adjusting the number of nodes and their spacing. The capability to duplicate an equation is very useful in writing the finite differences for each of the nodes. Some large problems may require very large numbers of differential equations to be solved simultaneously for accurate solutions.

6.13 COOLING OF A SOLID SPHERE IN A FINITE WATER BATH

6.13.1 Concepts Demonstrated

Analysis of heat transfer via both lumped and distributed systems—formulation of the lumped and distributed models for unsteady-state conduction within a sphere and lumped model for a water bath.

6.13.2 Numerical Methods Utilized

Solution of partial differential equations with the numerical method of lines and integration of simultaneous ordinary differential equations.

6.13.3 Problem Statement

A metal fabrication company has a need for rapid cooling of solid spheres of uniform diameters made from various metals. This cooling is to be accomplished by a well-mixed water bath into which individual spheres are dropped after manufacture from an elevated temperature. A sketch of this operation is shown in the upper left of Figure 6–17.

(a) Consider a lumped analysis in which the conduction within the sphere is so rapid that the sphere maintains a uniform temperature as it cools. The sphere has a diameter $D = 0.1$ m with a constant thermal conductivity given by $k = 10$ W/m·°C, a density of $\rho = 8200$ kg/m³, and a heat capacity $C_p = 0.41$ kJ/kg·K. The heat transfer coefficient between the sphere and the bath is $h = 220$ W/m²·°C. The initial temperature of the sphere is 300 °C, and the water bath is initially at 20 °C. Assume that the water bath maintains a constant temperature, and neglect heat transfer from the sphere until it enters the water bath. Calculate and make separate plots of the sphere's temperature T and heat rate q as a function of time to steady state.

(b) Repeat part (a) but take into account the heat conduction within the sphere. Plot the temperatures within the sphere at radii of 0 m, 0.05 m, and 0.1 m to steady state.

(c) Repeat part (b) but additionally consider the dynamic heating of the water bath. The volume of the bath is $V = 0.1$ m³, and the properties of water can be considered constant with density $\rho_W = 965$ kg/m³ and heat capacity $C_{pW} = 4.199$ kJ/kg·K. Plot the temperatures T_W and the temperatures at radii of 0 m, 0.05 m, and 0.1 m to steady state.

6.13.4 Solution (Partial)

This is an unsteady-state problem in which the solid sphere will cool down upon entry into the water bath. The conduction within the sphere will be rapid if the

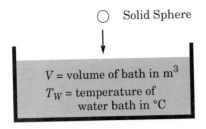

○ Solid Sphere

V = volume of bath in m³

T_W = temperature of
water bath in °C

T_S = temperature of sphere's surface

T = temperature in differential element

$q|_r$ = heat rate at radius r

$q|_{r+\Delta r}$ = heat rate at radius $r+\Delta r$

h = convective heat transfer coefficient at

sphere's surface

Enlarged   View   of   Solid
Sphere in Water Bath Showing Differential Element

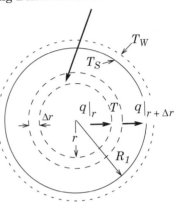

Figure 6–17 Cooling of a Spherical Solid in a Water Bath

thermal conductivity is high and/or the sphere diameter is small. The water bath
will be heated by the energy transferred from the cooling sphere.

(a) The "lumped" treatment will allow a general unsteady-state energy balance to be performed on the entire sphere, where the temperature within the
sphere is assumed to be uniform throughout at any time. In this case, there is no
input to the sphere nor any generation within the sphere. The output is via heat
transfer to the water bath, and the accumulation is related to the energy change
within the sphere. Thus an unsteady-state balance over a Δt time interval yields

$$INPUT + GENERATION = OUTPUT + ACCUMULATION$$

$$0 + 0 = h(4\pi R_1^2)(T - T_W)\Delta t + \left(\frac{4\pi R_1^3}{3}\right)\rho C_p(T|_{t+\Delta t} - T|_t) \tag{6-78}$$

Taking the limit as Δt goes to zero followed by rearrangement yields

$$\frac{dT}{dt} = \frac{-h(4\pi R_1^2)(T - T_W)}{\left(\frac{4\pi R_1^3}{3}\right)\rho C_p} = \frac{-3h(T - T_W)}{R_1 \rho C_p} \tag{6-79}$$

where the initial condition is $T = 300$ °C at time $t = 0$.
The rate of heat lost from the sphere can be calculated from

$$q = h(4\pi R_1^2)(T - T_W) \tag{6-80}$$

which is just the output term of Equation (6-78).

(b) The heat conduction within the sphere can be described by making an unsteady-state energy balance on the differential element with incremental radius Δr shown in Figure 6–17 over an incremental time interval Δt.

$$\text{INPUT} + \text{GENERATION} = \text{OUTPUT} + \text{ACCUMULATION}$$

$$q|_r \Delta t + 0 = q|_{r+\Delta r} \Delta t + 4\pi r^2 \Delta r \rho C_p (T|_{t+\Delta t} - T|_t) \tag{6-81}$$

Rearrangement of the equation followed by take the limits as both Δr and Δt go to zero gives

$$\frac{\partial T}{\partial t} = -\left(\frac{1}{4\pi r^2 \rho C_p}\right)\frac{\partial q}{\partial r} \tag{6-82}$$

Fourier's law, expressed as

$$q = -k4\pi r^2 \frac{\partial T}{\partial r} \tag{6-83}$$

can be substituted into Equation (6-82) to yield the second-order partial differential equation given by

$$\frac{\partial T}{\partial t} = \frac{1}{r^2 \rho C_p}\frac{\partial}{\partial r}\left(kr^2 \frac{\partial T}{\partial r}\right) \tag{6-84}$$

If the thermal conductivity k is constant, it can be removed from the partial derivative on the right-hand side of the preceding equation.

A convenient numerical solution to Equation (6-84) utilizes the numerical method of lines, which is discussed in Problem 3.9. Instead of directly solving the second-order Equation (6-84), it is convenient numerically to solve the two equations that lead to it—namely, Equations (6-82) and (6-83). This can be accomplished by using a series of ordinary derivatives for $\partial T/\partial t$ and finite difference formulas for both $\partial q/\partial r$ and $\partial T/\partial r$. This solution will be illustrated by setting up 11 nodes for 10 sections of the radius, as shown in Figure 6–18. Thus, using

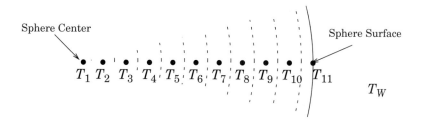

Figure 6–18 Nodes within the Sphere for Numerical Method of Lines Solution

the second-order central difference formula given by Equation (A-6) of Appendix A for the first derivatives with respect to r yields

$$\frac{\partial q_n}{\partial r} = \frac{(q_{n+1} - q_{n-1})}{2\Delta r} \quad \text{for } (2 \leq n \leq 10) \tag{6-85}$$

$$\frac{\partial T_n}{\partial r} = \frac{(T_{n+1} - T_{n-1})}{2\Delta r} \quad \text{for } (2 \leq n \leq 10) \tag{6-86}$$

Thus application of Equation (6-82) at each interior node point with Equation (6-85) results in

$$\frac{dT_n}{dt} = -\frac{1}{4\pi r_n^2 \rho C_p}\left(\frac{q_{n+1} - q_{n-1}}{2\Delta r}\right) \quad \text{for } (2 \leq n \leq 10) \tag{6-87}$$

Fourier's law of Equation (6-83) can be written for each interior node utilizing Equation (6-86) to give

$$q_n = -k 4\pi r_n^2 \left(\frac{T_{n+1} - T_{n-1}}{2\Delta r}\right) \quad \text{for } (2 \leq n \leq 10) \tag{6-88}$$

In the preceding equations, the arbitrary use of 11 nodes requires that $\Delta r = R_1/10$, and the radius to a particular node is given by $r_n = 0.1R_1(n-1)$.

Boundary Condition at Surface of the Sphere An energy balance at the surface of the sphere in the r direction relates that the rate of heat transferred to the surface by conduction within the sphere is equal to the convection from the sphere surface to the water bath. Thus

$$-k\frac{\partial T}{\partial r}\bigg|_{r = R_1} = h(T_{11} - T_W) \tag{6-89}$$

A second-order backward finite difference formula given by Equation (A-7) of Appendix A can be used for the partial derivative $\dfrac{\partial T}{\partial r}\bigg|_{r = R_1}$ to yield

$$-k\left(\frac{3T_{11} - 4T_{10} + T_9}{2\Delta r}\right) = h(T_{11} - T_W) \tag{6-90}$$

Equation (6-90) can be explicitly solved for T_{11}:

$$T_{11} = \frac{2\Delta r h T_W - 4kT_{10} + kT_9}{2\Delta r h - 3k} \tag{6-91}$$

and the corresponding heat transfer rate at node 11 is given by

$$q_{11} = h(4\pi R_1^2)(T_{11} - T_W) \tag{6-92}$$

Boundary Condition at Center of the Sphere At the center of the sphere, the heat transfer rate is zero; therefore,

$$q|_{r=0} = q_1 = 0 \tag{6-93}$$

which from Fourier's law also requires

$$\frac{\partial T}{\partial r}\bigg|_{r=0} = 0 \tag{6-94}$$

A second-order forward difference formula, Equation (A-5) of Appendix A, can be used in the preceding equation:

$$\frac{\partial T_1}{\partial r} = \frac{(-T_3 + 4T_2 - 3T_1)}{2\Delta r} \tag{6-95}$$

and T_1 can be determined explicitly at any time to be given by

$$T_1 = \frac{4T_2 - T_3}{3} \tag{6-96}$$

Numerical Solution The numerical solution requires the simultaneous solution of Equations (6-87) and (6-88) for the interior nodes along with Equations (6-91) and (6-92) for the boundary condition at the sphere surface and Equations (6-93) and (6-96) for the boundary condition at the center of the sphere. The initial temperature throughout the sphere is 300 °C, and the output heat rate from the sphere is given by q_{11}.

(c) The description of the water bath heating will require an unsteady-state energy balance on the well-mixed water within the tank. The tank will be assumed to be adiabatic, with the only heat transfer coming from the cooling sphere. Thus an unsteady-state balance over a Δt time interval yields

INPUT + GENERATION = OUTPUT + ACCUMULATION

$$q_{11} + 0 = 0 + V\rho_w C_{pW}(T_W|_{t+\Delta t} - T_W|_t) \tag{6-97}$$

where q_{11} as calculated in Equation (6-92) is the rate of heat transfer from the sphere at the outer surface to the water bath. Taking the limit as Δt goes to zero followed by rearrangement yields

$$\frac{dT_W}{dt} = \frac{q_{11}}{V\rho_W C_{pW}} \tag{6-98}$$

where the initial condition is $T_W = 20$ °C at time $t = 0$.

Thus the numerical solution only requires the addition of Equation (6-98) and the initial condition to the equation set used in part (b).

6.14 UNSTEADY-STATE CONDUCTION IN TWO DIMENSIONS

6.14.1   Concepts Demonstrated

Unsteady-state conduction in two dimensions with faces at known temperatures and constant thermal diffusivity.

6.14.2   Numerical Methods Utilized

Application of the numerical method of lines to solve a two-dimensional partial differential equation.

6.14.3   Problem Statement

Unsteady-state heat transfer in the x and y directions is described by the partial differential equation

$$\frac{\partial T}{\partial t} = \alpha\left(\frac{\partial^2 T}{\partial x^2} + \frac{\partial^2 T}{\partial y^2}\right) \tag{6-99}$$

where T is the temperature in K, t is the time in s, and α is the thermal diffusivity in m^2/s given by $k/\rho c_p$. In this treatment, the thermal conductivity k in W/m, the density ρ in kg/m^3, and the heat capacity c_p in J/kg are considered to be constant.

A hollow square chamber has the inside walls held at a temperature of 700 K while the outside walls are maintained at 300 K. The inside dimensions of the chamber are 1 m × 1 m and the outside dimensions are 2 m × 2 m, as shown in Figure 6–19. Grid spacing for a finite difference treatment of a symmetrical section of this chamber is also presented in Figure 6–19, where $\Delta x = \Delta y = 0.125$ m. Note that eight sections are required to describe the cross section of the chamber.

(a)   Use the numerical method of lines to solve Equation (6-99) to determine the temperatures at the note points of the chamber shown in Figure 6–19 as a function of time until steady state is reached. Plot temperatures $T_{1,2}$, $T_{2,2}$, $T_{3,2}$, and $T_{4,2}$ versus time. The thermal diffusivity of the chamber wall is $\alpha = 5 \times 10^{-5}$ m^2/s, the thermal conductivity is $k = 1.2$ W/m, and the initial temperature of all chamber material is 300 K.

(b)   Calculate and plot the heat flux q per meter of chamber length through the interior wall of the chamber as a function of time until steady state is reached.

(c)   Discuss how you could increase the accuracy of your calculations for parts (a) and (b).

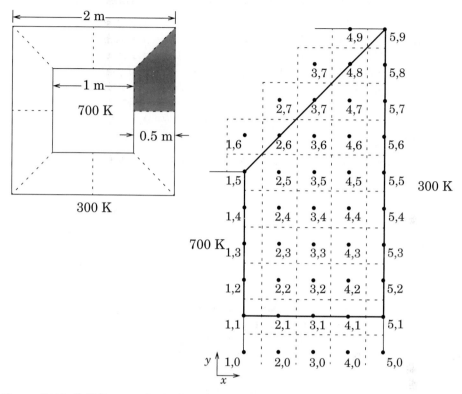

Figure 6–19 Grid Pattern for Hollow Square Chamber

The Numerical Method of Lines

The solution of a one-dimensional partial differential equation involving heat transfer utilizing the method of lines has been discussed in Problems 3.9, 6.12, and 6.13. For this problem, the time derivative can be replaced by ordinary derivatives and the two spacial derivatives can be written using finite difference approximations. Equation (6-99) can be rewritten using central difference formulas [Equation (A-3) of Appendix A] for each of the interior nodes in the grid spacing shown in Figure 6–19 as

$$\frac{dT_{n,m}}{dt} = \alpha \left[\frac{(T_{n+1,m} - 2T_{n,m} + T_{n-1,m})}{(\Delta x)^2} + \frac{(T_{n,m+1} - 2T_{n,m} + T_{n,m-1})}{(\Delta y)^2} \right] \quad \text{(6-100)}$$

where n represents a general node in the x direction and m represents a general node in the y direction (see Geankoplis[4]).

Due to the symmetry of this problem, the following temperature relations are true at any time

$$T_{1,6} = T_{2,5} \qquad T_{2,7} = T_{3,6} \qquad T_{3,7} = T_{4,7}$$
$$T_{2,2} = T_{2,0} \qquad T_{3,2} = T_{3,0} \qquad T_{4,2} = T_{4,0} \quad \text{(6-101)}$$

and can be used in Equation (6-100) to perform calculations limited to only points within the primary grid boundary.

6.14.4 Solution (Partial)

(a) The equations for the particular interior nodes with the boundary conditions and the symmetry relationships given by Equation (6-101) result in 10 simultaneous ordinary differential equations, which can be entered into the POLYMATH *Simultaneous Differential Solver.* There is an option in POLYMATH for duplicating equations that is useful in entering the equation set for this problem. The equation set is

```
Equations:
d(T21)/d(t)=alpha*((T31-2*T21+700)/deltax^2+(T22-2*T21+T22)/deltay^2)
d(T31)/d(t)=alpha*((T41-2*T31+T21)/deltax^2+(T32-2*T31+T32)/deltay^2)
d(T41)/d(t)=alpha*((300-2*T41+T31)/deltax^2+(T42-2*T41+T42)/deltay^2)
d(T22)/d(t)=alpha*((T32-2*T22+700)/deltax^2+(T23-2*T22+T21)/deltay^2)
d(T32)/d(t)=alpha*((T42-2*T32+T22)/deltax^2+(T33-2*T32+T31)/deltay^2)
d(T42)/d(t)=alpha*((300-2*T42+T32)/deltax^2+(T43-2*T42+T41)/deltay^2)
d(T23)/d(t)=alpha*((T33-2*T23+700)/deltax^2+(T24-2*T23+T22)/deltay^2)
d(T33)/d(t)=alpha*((T43-2*T33+T23)/deltax^2+(T34-2*T33+T32)/deltay^2)
d(T43)/d(t)=alpha*((300-2*T43+T33)/deltax^2+(T44-2*T43+T42)/deltay^2)
d(T24)/d(t)=alpha*((T34-2*T24+700)/deltax^2+(T25-2*T24+T23)/deltay^2)
d(T34)/d(t)=alpha*((T44-2*T34+T24)/deltax^2+(T35-2*T34+T33)/deltay^2)
d(T44)/d(t)=alpha*((300-2*T44+T34)/deltax^2+(T45-2*T44+T43)/deltay^2)
d(T25)/d(t)=alpha*((T35-2*T25+700)/deltax^2+(T26-2*T25+T24)/deltay^2)
d(T35)/d(t)=alpha*((T45-2*T35+T25)/deltax^2+(T36-2*T35+T34)/deltay^2)
d(T45)/d(t)=alpha*((300-2*T45+T35)/deltax^2+(T46-2*T45+T44)/deltay^2)
d(T26)/d(t)=alpha*((T36-2*T26+T25)/deltax^2+(T36-2*T26+T25)/deltay^2)
d(T36)/d(t)=alpha*((T46-2*T36+T26)/deltax^2+(T37-2*T36+T35)/deltay^2)
d(T46)/d(t)=alpha*((300-2*T46+T36)/deltax^2+(T47-2*T46+T45)/deltay^2)
d(T37)/d(t)=alpha*((T47-2*T37+T36)/deltax^2+(T47-2*T37+T36)/deltay^2)
d(T47)/d(t)=alpha*((300-2*T47+T37)/deltax^2+(T48-2*T47+T46)/deltay^2)
d(T48)/d(t)=alpha*((300-2*T48+T47)/deltax^2+(300-2*T48+T47)/deltay^2)
alpha=5e-5
deltax=0.25
deltay=0.25
Initial Conditions:
t(0)=0
T21(0)=300
T31(0)=300
T41(0)=300
T22(0)=300
T32(0)=300
T42(0)=300
T23(0)=300
T33(0)=300
T43(0)=300
T24(0)=300
T34(0)=300
T44(0)=300
T25(0)=300
T35(0)=300
```

```
T45(0)=300
T26(0)=300
T36(0)=300
T46(0)=300
T37(0)=300
T47(0)=300
T48(0)=300
Final Value:
t(f)=3000
```

(b) Once the temperature distribution has been calculated, the total heat loss at the interior wall of the chamber can be obtained by summing the heat fluxes at the various nodes due to the local temperature gradient. Fourier's law, given by

$$q_x = -kA\frac{dT}{dx}\bigg|_{x\,=\,\text{inner surface}} \tag{6-102}$$

can be applied with due consideration for the appropriate area A and the overall symmetry of the problem. The derivatives in Equation (6-102) can be obtained using the second-order forward finite difference expression whose general formula is given by [see Equation (A-5) of Appendix A]

$$\frac{dT_n}{dx} = \frac{-3T_n + 4T_{n+1} - T_{n+2}}{2\Delta x} \tag{6-103}$$

for the x direction. Thus the total heat flux through the interior wall surface (in positive x direction) for the inner chamber is given by

$$q = -8k\Delta x\left[\left(\frac{1}{2}\right)\frac{dT_{1,5}}{dx} + \frac{dT_{1,4}}{dx} + \frac{dT_{1,3}}{dx} + \frac{dT_{1,2}}{dx} + \left(\frac{1}{2}\right)\frac{dT_{1,1}}{dx}\right] \tag{6-104}$$

where the factor of 8 is for the eight sections with similar symmetry, and the factors of (1/2) are necessary because of the appropriate areas. This expression, with the derivatives calculated from Equation (6-103) for x, can be evaluated *at any time* to estimate the total heat flux.

The additions to the equation set for the POLYMATH solution of part (b) are

```
dT11dx=(-T31+4*T21-3*700)/(2*deltax)
dT12dx=(-T32+4*T22-3*700)/(2*deltax)
dT13dx=(-T33+4*T23-3*700)/(2*deltax)
dT14dx=(-T34+4*T24-3*700)/(2*deltax)
dT15dx=(-T35+4*T25-3*700)/(2*deltax)
q=-8*1.2*((1/2)*dT15dx+dT14dx+dT13dx+dT12dx+(1/2)*dT11dx)
```

The POLYMATH problem solution file for both parts (a) and (b) is found in the *Simultaneous Differential Equation Solver Library* located in directory CHAP6 with file named P6-14AB.POL.

REFERENCES

1. Bird, R. B., Stewart, W. E., and Lightfoot, E. N., *Transport Phenomena*, New York: Wiley, 1960.
2. Churchill, S. W., and Chu, H. H. S., *Int. J Heat & Mass Transfer*, 18, 1323 (1975).
3. Dittus, F. W., and L. M. K. Boelter, University of California-Berkeley, *Pub. Engr.*, 2, 443 (1930).
4. Geankoplis, C. J., *Transport Processes and Unit Operations*, 3rd ed., Englewood Cliffs, NJ: Prentice-Hall, 1993.
5. Thomas, L. C., *Heat Transfer*, Englewood Cliffs, NJ: Prentice Hall, 1992.
6. Welty J. R., Wicks, C. E., and Wilson, R.E., *Fundamentals of Momentum, Heat and Mass Transfer*, 3rd ed., New York: Wiley, 1984.

Mass Transfer

7.1 ONE-DIMENSIONAL BINARY MASS TRANSFER IN A STEFAN TUBE

7.1.1   Concepts Demonstrated

Binary gas phase diffusion of A through stagnant B during evaporation of a pure liquid in a simple diffusion tube.

7.1.2   Numerical Methods Utilized

Numerical integration of simultaneous ordinary differential equations by a shooting technique which must satisfy split boundary conditions.

7.1.3   Problem Statement

Liquid A is evaporating into a gas mixture of A and B from a liquid layer of pure A near the bottom of a cylindrical Stefan tube, as shown in Figure 7–1. The rate of evaporation is relatively slow, so it is a good assumption that the level of the liquid surface is constant. A gas mixture is passing over the upper surface of the Stefan tube. Thus the partial pressure of A, p_{A2}, and the mole fraction of A at point 2, x_{A2}, are both known at z_2. The surface of the liquid A contains no dissolved B since B is insoluble in liquid A; therefore, liquid A exerts its vapor pressure, p_{A1}, at location z_1. The mole fraction of A at the liquid surface is given by

$$x_{A1} = \frac{P_{A0}}{P} \tag{7-1}$$

where P_{A0} is the vapor pressure of component A and P is the total pressure.

The simplest assumptions for this system would be that the temperature and pressure are constant and that the gases A and B are ideal. Thus gas A is diffusing from the surface into the bulk stream above the surface of the Stefan tube though gas B which is stationary within the tube. This is the case of single component diffusion through a stagnant gas film.

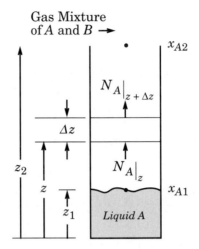

Figure 7–1 Gas Phase Diffusion of A through Stagnant B

Mass Balance on Component A within Diffusion Path

Consider a differential element between points z_1 and z_2 with a differential length of Δz. Since there is no reaction in this case, a steady-state mass balance in the positive z direction yields

$$\frac{dN_A}{dz} = 0 \qquad (7\text{-}2)$$

where N_A is the flux of A relative to stationary coordinates in kg-mol/m$^2 \cdot$s.

Fick's Law for Binary Diffusion

Also for stationary coordinates, the general expression of Fick's law for the flux of A can be written as

$$N_A = -D_{AB}C\frac{dx_A}{dz} + \frac{C_A}{C}(N_A + N_B) \qquad (7\text{-}3)$$

flux = diffusion   +   bulk flow (convection)

where C is the total concentration in kg-mol/m^3, D_{AB} is the molecular diffusivity of A in B in m^2/s, C_A is the concentration of A in kg-mol/m^3, and N_B is the flux of B in kg-mol/m$^2 \cdot$s. For this problem, the total gas concentration C is constant, and component B is stagnant. Thus N_B is zero. The mole fraction of A in the gas mixture, x_A, can be used to replace C_A/C. The modified expression for Fick's law can be written as

$$N_A = -D_{AB}\frac{dC_A}{dz} + x_A N_A \qquad (7\text{-}4)$$

Solving for N_A and rearranging Equation (7-3) yields

$$\frac{dx_A}{dz} = \frac{-(1 - x_A)N_A}{D_{AB}C} \tag{7-5}$$

Final Equations and Boundary Conditions

The diffusion with convection for this problem can be described by the simultaneous solution of Equations (7-2) and (7-5) with the initial condition for x_A at z_1, as given by Equation (7-1). The final value of x_{A2} at z_2 is the mole fraction of A in the gas mixture that is flowing across the top of the Stefan tube.

Analytical Solution

At constant temperature and pressure for ideal gases, the total concentration C and binary diffusivity D_{AB} may be considered constant. Thus Equation (7-5) can be solved for N_A and entered into Equation (7-2) and integrated twice using the boundary conditions given previously to give the analytical solution (see Bird et al.[2]) for the concentration profile as

$$\left(\frac{1 - x_A}{1 - x_{A1}}\right) = \left(\frac{1 - x_{A2}}{1 - x_{A1}}\right)^{\frac{(z - z_1)}{(z_2 - z_1)}} \tag{7-6}$$

The analytical solution for the flux at the liquid-gas interface gives

$$N_{Az}\big|_{z = z_1} = \frac{D_{AB}C}{(z_2 - z_1)(x_B)_{lm}}(x_{A1} - x_{A2}) \tag{7-7}$$

where

$$(x_B)_{lm} = \frac{(x_{B2} - x_{B1})}{\ln(x_{B2}/x_{B1})} = \frac{(x_{A1} - x_{A2})}{\ln[(1 - x_{A2})/(1 - x_{A1})]} \tag{7-8}$$

The evaporation of methanol into a stream of dry air is being studied in a Stefan tube apparatus with a diffusion path of 0.238 m. The measurements are made at a temperature of 328.5 K, where the vapor pressure of methanol is 68.4 kPa. The total pressure is 99.4 kPa. The binary molecular diffusion coefficient of methanol in air under these conditions is $D_{AB} = 1.991 \times 10^{-5}$ m^2/s, as estimated by Taylor and Krishna.[7]

(a) Calculate the constant molar flux of methanol within the Stefan tube at steady state using a numerical technique.

(b) Plot the mole fraction of methanol from the liquid methanol surface to the flowing air stream.

(c) Compare the result of part (a) with the result calculated from Equation (7-7).

(d) Verify several points on the numerical solution for the mole fraction profile with calculations from the analytical profile of Equation (7-6).

(e) Repeat parts (a) through (d) for a temperature of 298.15 K, where the vapor pressure of methanol is 16.0 kPa.

(f) Assume that the temperature within the Stefan tube varies linearly from 328.5 K at the methanol surface to 295 K in the air stream at the tube surface. Complete parts (a) and (b) for this condition.

Additional Information and Data

Geankoplis,[6] p. 396, recommends that the effect of temperature on binary diffusivities for gases vary according to the absolute temperature to the 1.75 power given by

$$D_{AB} = D_{AB}\Big|_{T_1}\left(\frac{T}{T_1}\right)^{1.75} \tag{7-9}$$

where the diffusivity is known at temperature T_1.

7.1.4   Solution (Partial with Suggestions)

(a), (b), & (c) The methanol flux and mole fraction profile require the numerical integration of Equations (7-2) and (7-5). For this case, the solution of Equation (7-2) is simply that N_A is a constant. Thus the initial and final conditions for Equation (7-2) are known for the methanol mole fraction x_A at both ends of the diffusion path. At $z = 0$, $x_A = 0.688$ [using Equation (7-1)] and at $z = 0.238$, $x_A = 0$.

The shooting technique as described in Problem 3.6 can be used to determine the constant value of N_A. Thus the initial value for x_A is known to be 0.688 at $z = 0$, and N_A can be optimized using the techniques of Problem 3.5 to satisfy the boundary condition, where $x_A = 0$ at $z = 0.238$. The value of C can be calculated from the perfect gas law.

The POLYMATH Simultaneous Differential Equation Solver can be used to solve Equation (7-5), and the POLYMATH equation set is given by

```
Equations:
d(xA)/d(z)=-(1-xA)*NA/(DAB*C)
xBlm=(xA1-0)/ln(1/(1-xA1))
```

```
NA=3.5461e-6
DAB=1.991e-5
P=99.4
R=8.31434
T=328.5
xA1=0.688
C=P/(R*T)
NACALC=DAB*C*(xA1-0)/((0.238-0)*xBlm)
Initial Conditions:
z(0)=0
xA(0)=0.688
Final Value:
z(f)=0.238
```

In the POLYMATH solution, the numerical solution for $N_A = 3.5461 \times 10^{-6}$ agrees with the analytical solution exactly to the five significant figures used. This is demonstrated by the results of various solutions with slightly different values of N_A. In solutions 1 and 3, the final value for x_A is further away from zero than for the best solution 2.

Table 7–1 Comparison of Final Values for Different Initial Conditions

Solution	N_A	$x_A\big|_{z = 0.238}$
#1	3.5460×10^{-6}	3.29891×10^{-5}
#2	3.5461×10^{-6}	1.43643×10^{-7}
#3	3.5462×10^{-6}	-3.27029×10^{-5}

The numerical solution allows convenient plotting of the methanol mole fraction x_A along the diffusion path. This is shown in Figure 7–2.

The POLYMATH problem solution file for parts (a), (b), and (c) is found in the *Simultaneous Differential Equation Solver Library* located in directory CHAP7 with file named P7-01ABC.POL.

(d) The analytical solution given by Equation (7-6) is a nonlinear equation that can be solved for the distance z if the mole fraction x_A is known or the equation can be solved for the mole fraction x_A if the distance z is known. Verification of the numerical solution will be carried out at three arbitrary values of z as given in Table 7–2, where the correspondence is exact to at least five significant figures.

(f) The temperature profile can be imposed on the numerical solution by adding an algebraic equation for T as a function of z. The effect of temperature on the diffusivity of methanol in air must be included by utilizing Equation (7-9).

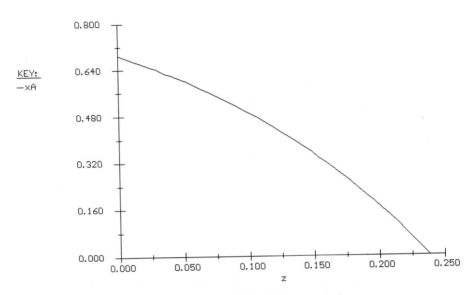

Figure 7–2 Mole Fraction of Methanol along Diffusion Path

Table 7–2 Comparison of Analytical and Numerical Methanol Mole Fraction

z	x_A Analytical	x_A Numerical
0.0714	0.557505	0.55750517
0.1428	0.37243	0.37243054
0.2142	0.109948	0.10994795

7.2 MASS TRANSFER IN A PACKED BED WITH KNOWN MASS TRANSFER COEFFICIENT

7.2.1   Concepts Demonstrated

Simple convective mass transfer from a surface to the bulk stream for a system involving transport of A through inert B with the fluid diffusivity dependent on concentration.

7.2.2   Numerical Methods Utilized

Numerical integration of an ordinary differential equation with a known initial condition and simultaneous explicit algebraic equations.

7.2.3   Problem Statement*

Pure water at 26.1 °C is slowly passing through a bed of benzoic acid spheres at a rate of 0.0701 ft^3/h. The spheres have a diameter of 0.251 inches, and the total surface area within the bed is 0.129 ft^2. The mass transfer coefficient for equimolar counterdiffusion varies with composition and is given by

$$k'_L = K_1 + K_2 x_A \qquad (7\text{-}10)$$

where k'_L is the mass transfer coefficient for equimolar counterdiffusion (see Geankoplis[6], p. 435) in m/s, $K_1 = 0.0551$ ft/h, $K_2 = 185.5$ ft/h, and x_A is the mole fraction of benzoic acid in the liquid phase. The saturation solubility of benzoic acid in water is 0.00184 lb-mol/ft^3 of solution at 26.1 °C. The total concentration of the liquid water phase is $C = 3.461$ lb-mol/ft^3, and the volumetric flow rate is $V = 0.0701$ ft^3/h.

(a)   Calculate and plot the concentration of benzoic acid within the bed as a function of the bed surface area to 0.129 ft^2.
(b)   Calculate the surface area necessary to achieve a liquid phase concentration of benzoic acid at the exit of the bed that is 50% of the saturation solubility.

Additional Information and Data

A general material balance on the differential surface area of the packed bed shown in Figure 7–3 yields

$$\frac{dC_A}{d(\text{Area})} = \frac{N_A|_i}{V} \qquad (7\text{-}11)$$

*Adapted from Geankoplis[6] with permission.

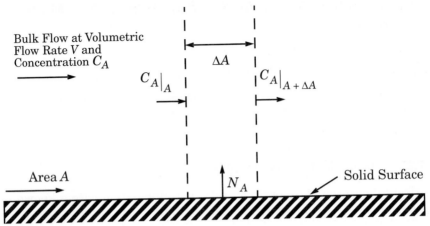

Figure 7–3 Differential Volume for Mass Transfer in Packed Bed Based on External Surface Area

The mass transfer flux for the diffusion of A (benzoic acid) through stagnant B (water) at the interface between the benzoic acid and the flowing stream is given by (Geankoplis[6], p. 435)

$$N_A\big|_i = \frac{k_L' \, C}{x_{BM}}(x_{Ai} - x_{Ab}) \tag{7-12}$$

where N_A has units of lb-mol/h $\cdot$ ft^2 and x_{BM} is given by

$$x_{BM} = \frac{x_{B2} - x_{B1}}{\ln(x_{B2}/x_{B1})} \tag{7-13}$$

between points 1 and 2, which in this case are the surface i and the bulk stream b respectively.

7.2.4   Solution (Suggestions)

This problem can be described by inserting Equation (7-12) into Equation (7-11) and using $C_A = x_A C$ to obtain an ordinary differential equation given by

$$\frac{dx_A}{d(\text{Area})} = \frac{k_L'}{V x_{BM}}(x_{Ai} - x_{Ab}) \tag{7-14}$$

with the initial condition that at the entrance to the packed bed, the mole fraction of benzoic acid is zero. Thus $x_A = 0$ when Area = 0. The POLYMATH *Simultaneous Ordinary Differential Equation Solver* can be used to integrate Equation (7-14) with the algebraic equations for k_L' and x_{BM} given by Equations (7-10) and (7-13), respectively.

7.3 SLOW SUBLIMATION OF A SOLID SPHERE

7.3.1   Concepts Demonstrated

Sublimation of a solid sphere by diffusion in still gas and by a mass transfer coefficient in a moving gas.

7.3.2   Numerical Methods Utilized

Solution of simultaneous ordinary differential equations while optimizing a single parameter to achieve split boundary conditions.

7.3.3   Problem Statement

Consider the sublimation of solid dichlorobenzene, designated by A, which is suspended in still air, designated by B, at 25 °C and atmospheric pressure. The particle is spherical with a radius of 3×10^{-3} m. The vapor pressure of A at this temperature is 1 mm Hg, and the diffusivity in air is 7.39×10^{-6} m^2/s. The density of A is 1458 kg/m^3, and the molecular weight is 147.

(a)   Estimate the initial rate of sublimation (flux) from the particle surface by using an approximate analytical solution to this diffusion problem. (See the following discussion for more information.)

(b)   Calculate the rate of sublimation (flux) from the surface of a sphere of solid dichlorobenzene in still air with a radius of 3×10^{-3} m with a numerical technique employing a shooting technique, with ordinary differential equations that describe the problem. Compare the result with part (a).

(c)   Show that expression for the rate of sublimation (flux) from the particle as predicted in part (a) is the same as that predicted by the external mass transfer coefficient for a still gas.

(d)   Calculate the time necessary for the complete sublimation of a single particle of dichlorobenzene if the particle is enclosed in a volume of 0.05 m^3.

Additional Information and Data

Diffusion The diffusion of A through stagnant B from the surface of a sphere is shown in Figure 7–4. A material balance on A in a differential volume between radius r and $r + \Delta r$ in a Δt time interval yields

$$\text{INPUT} + \text{GENERATION} = \text{OUTPUT} + \text{ACCUMULATION}$$

$$\left. (N_A 4\pi r^2) \right|_r \Delta t + 0 = \left. (N_A 4\pi r^2) \right|_{r + \Delta r} \Delta t + 0 \qquad \textbf{(7-15)}$$

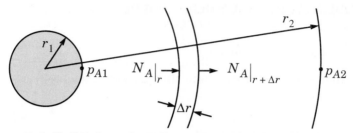

Figure 7–4 Shell Balance for Diffusion from the Surface of a Sphere

where N_A is the flux in kg-mol/m$^2 \cdot$s at radius r in m. The $4\pi r^2$ is the surface area of the sphere with radius r. Division by Δt and rearrangement of this equation while taking the limit as $\Delta r \to 0$ yields

$$\frac{d(N_A r^2)}{dr} = 0 \tag{7-16}$$

Fick's law for the diffusion of A through stagnant B in terms of partial pressures is expressed as

$$N_A = -\frac{D_{AB}}{RT}\frac{dp_A}{dr} + \frac{p_A}{P}N_A \tag{7-17}$$

where D_{AB} is the molecular diffusivity of A in B in m^2/s, R is the gas constant with a value of 8314.34 m$^3 \cdot$Pa/kg-mol$\cdot$K, T is the absolute temperature in K, and P is the total pressure in Pa.

Rearrangement of Equation (7-17) yields

$$\frac{dp_A}{dr} = -\frac{RTN_A\left(1-\dfrac{p_A}{P}\right)}{D_{AB}} \tag{7-18}$$

where the initial condition is that p_A = (1/760)1.01325$\times 10^5$ Pa = 133.32 Pa, which is the vapor pressure of the solid A at $r = 3\times 10^{-3}$ m. The final value is that $p_A = 0$ at some relatively large radius r.

The analytical solution to this problem can be obtained by integrating Equation (7-16) and introducing the result for N_A into Equation (7-18). The final solution as given by Geankoplis[6], p. 391, is

$$N_{A1} = \frac{D_{AB}P(p_{A1}-p_{A2})}{RTr_1 \quad p_{BM}} \tag{7-19}$$

where subscripts 1 and 2 indicate locations and p_{BM} is given by

$$p_{BM} = \frac{p_{B2}-p_{B1}}{\ln(p_{B2}/p_{B1})} = \frac{p_{A1}-p_{A2}}{\ln((P-p_{A2})/(P-p_{A1}))} \tag{7-20}$$

Mass Transfer Coefficient The transfer of A from the surface of the spherical particle to the surrounding gas can also be described by a mass transfer coefficient for transport of A through stagnant B.

A general relationship for gases that can be used to calculate the mass transfer coefficient for gases is presented by Geankoplis[6], p. 446, as

$$N_{Sh} = 2 + 0.552 N_{Re}^{0.53} N_{Sc}^{1/3} \tag{7-21}$$

Note that for a quiescent gas, the limiting value of N_{Sh} is 2 because the N_{Re} is zero.

The Sherwood number is defined as

$$N_{Sh} = k_c' \frac{D_p}{D_{AB}}$$

where k_c' is the mass transfer coefficient in m/s based on concentration and equimolar counterdiffusion, and D_p is the particle diameter in m. The particle Reynolds number is defined as

$$N_{Re} = \frac{D_p v \rho}{\mu}$$

where D_p is the particle diameter in m, v is the gas velocity in m/s, and μ is the gas viscosity in Pa·s. The Schmidt number is given by

$$N_{Sc} = \frac{\mu}{\rho D_{AB}}$$

with ρ representing the density of the gas in kg/m³.

The mass transfer coefficient (see Geankoplis[6], p. 435) can be used to describe the flux N_A from the surface of the sphere for transport through stagnant B by utilizing

$$N_A = \frac{k_c' P (p_{A1} - p_{A2})}{RT} \frac{1}{p_{BM}} \tag{7-22}$$

7.3.4   Solution (Partial with Suggestions)

(a) The analytical solution is given by Equations (7-19) and (7-20) which can be easily evaluated. Since comparisons will be made with the numbered solutions, these explicit equations can be entered into the particular POLYMATH Program that will be used subsequently.

(b) The numerical solution involves the simultaneous solution of Equations (7-16) and (7-18) along with the following algebraic equation, which is needed to calculate the flux N_A from the quantity $(N_A r^2)$:

$$N_A = \frac{(N_A r^2)}{r^2} \tag{7-23}$$

The POLYMATH *Simultaneous Differential Equation Solver* can be used to integrate these differential equations for the boundary conditions. The initial radius is the known radius of the sphere (initial condition), and the final value of the radius is much greater than the initial radius, so that the value of the partial pressure of A is effectively zero. The shooting technique for accomplishing this type of a solution is discussed in Problems 3.6 and 7.1. A POLYMATH equation set that provides the solution is given by

```
Equations:
d(NAr2)/d(r)=0
d(pA)/d(r)=-R*T*NA*(1-pA/P)/DAB
R=8314.34
T=298.15
NA=NAr2/(r^2)
P=1.01325e5
DAB=7.39e-6
PBM=(133.32-0)/ln((P-0)/(P-133.32))
NACALCatr1=DAB*P*(133.32-0)/(R*T*0.003*PBM)
NACALCr12=NACALCatr1*0.003^2
Initial Conditions:
r(0)=0.003
NAr2(0)=1.19335e-12
pA(0)=133.32
Final Value:
r(f)=16
```

The final value in this numerical solution is such that the desired value of the partial pressure of A is very nearly zero (−7.86e–4 Pa for the preceding equation set) at the very "large" final radius of $r = 16$ m. This "large" final radius is really somewhat arbitrary as long as the boundary condition that $p_A = 0$ is achieved, which is independent of the "large" final radius value.

A comparison of the calculated N_A with the analytical N_A at the surface of the sphere is quite satisfactory in that there is agreement to four significant places for the result of 1.326×10^{-7} kg-mol/m$^2 \cdot$ s.

The POLYMATH problem solution file for parts (a) and (b) is found in the *Simultaneous Differential Equation Solver Library* located in directory CHAP7 with file named P7-03AB.POL.

(c) The resulting analytical solutions should be identical with each other.

(d) This is an unsteady-state problem that can be solved utilizing the mass transfer coefficient by making the pseudo-steady-state assumption that the mass transfer can be described by Equation (7-22) at any time. The mass transfer coefficient increases as the particle diameter decreases because for a still gas

$$N_{Sh} = k'_c \frac{D_p}{D_{AB}} = 2 \tag{7-24}$$

as indicated by Equation (7-22); thus, in terms of the particle radius,

$$k'_c = \frac{D_{AB}}{r} \tag{7-25}$$

A material balance on component A in the well-mixed gas phase volume V with the only input due to the sublimation of A yields

INPUT + GENERATION = OUTPUT + ACCUMULATION

$$(N_A 4\pi r^2)\Big|_r \Delta t + 0 = 0 + \left(\frac{V p_A}{RT}\right)\Big|_{t+\Delta t} - \left(\frac{V p_A}{RT}\right)\Big|_t \tag{7-26}$$

The limit as $\Delta t \to 0$ and the use of Equation (7-22) for N_A give

$$\frac{dp_A}{dt} = \frac{4\pi r^2 k'_c}{V} \frac{P(p_{A1} - p_{A2})}{p_{BM}} \tag{7-27}$$

with p_{A1} being the vapor pressure of A at the solid surface and p_{A2} being the partial pressure of A in the gas volume.

A material balance on A within the solid sphere of radius r gives

INPUT + GENERATION = OUTPUT + ACCUMULATION

$$0 + 0 = (N_A 4\pi r^2)\Big|_r \Delta t + \left(\frac{4}{3}\pi r^3 \frac{\rho_A}{M_A}\right)\Big|_{t+\Delta t} - \left(\frac{4}{3}\pi r^3 \frac{\rho_A}{M_A}\right)\Big|_t \tag{7-28}$$

where ρ_A is the density of the solid and M_A is the molecular weight of the solid. Rearranging Equation (7-28) and taking the limit as $\Delta t \to 0$ yields

$$\frac{d(r^3)}{dt} = \frac{3r^2 dr}{dt} = -\frac{3N_A M_A r^2}{\rho_A} \tag{7-29}$$

Simplifying and introducing Equation (7-22) for N_A gives

$$\frac{dr}{dt} = -\frac{M_A k'_c}{\rho_A RT} \frac{P(p_{A1} - p_{A2})}{p_{BM}} \tag{7-30}$$

The complete sublimation of A is described by the simultaneous solution of differential Equations (7-27) and (7-30) along with Equation (7-25).

7.4 CONTROLLED DRUG DELIVERY BY DISSOLUTION OF PILL COATING

7.4.1   Concepts Demonstrated

Unsteady-state dissolution of a solid into a liquid with transport described by a mass transfer coefficient with subsequent reaction.

7.4.2   Numerical Methods Utilized

Solution of simultaneous ordinary differential equations with known initial conditions that have conditional alterations during the numerical solution.

7.4.3   Problem Statement (Adapted from Fogler[4], p. 600)

The pill to deliver a particular drug has a solid spherical inner core of pure drug D and is surrounded by a spherical outer coating of A that makes the pill palatable and helps to control the drug release. The outer coating and the drug dissolve at different rates in the stomach due to their difference in solubilities. Let C_{AS} = concentration of coating in the stomach in mg/cm^3, C_{DS} = concentration of drug in the stomach in mg/cm^3, and C_{DB} = concentration of drug in the body in mg/kg.

Three different pill formulations are available

Pill 1 — Diameter of A = 5 mm, Diameter of D = 3 mm

Pill 2 — Diameter of A = 4 mm, Diameter of D = 3 mm

Pill 3 — Diameter of A = 3.5 mm, Diameter of D = 3 mm

Additional Information and Data

Amount of drug in inner core of each pill = 20 mg

Density of inner and outer layers = 1414.7 mg/cm^3

Solubility of outer pill layer at stomach conditions = S_A = 1.0 mg/cm^3

Solubility of inner drug core at stomach conditions = S_D = 0.4 mg/cm^3

Volume of fluid in stomach = V = 1.2 liters

Residence time in stomach = $\tau = V/v_0$ = 4 hours, where v_0 is the volumetric flow rate

Typical body weight W = 75 kg

Sherwood number = $N_{Sh} = k_L \dfrac{D_p}{D_{AB}} = 2$ (see Problem 7.3 for details)

Effective Diffusivities of A and D in stomach $D_A = D_D = 0.6$ cm^2/min

A person takes all three pills at the same time. Assume that the stomach is well mixed and that the pills remain in the stomach while they are dissolving.

(a) Plot C_{AS} and C_{DS} as a function of time for up to 12 hours after the pills are taken.

(b) If the drug is absorbed into the body (from solution in the stomach into the blood stream) by an effective mass transfer rate of $10 \times C_{DS}$ in mg/min, plot C_{AS} and C_{DS} under these conditions.

(c) If the body metabolizes the drug with an effective first-order reaction rate of $1.0 \times C_{DB}$ in mg/kg.min, plot C_{DB} in the body under the conditions of (b).

(d) Under the conditions of (c), what is the estimated time that C_{DB} is greater than 2.0×10^{-3} mg/kg, which is the minimum effective concentration level for the drug?

(e) Please suggest a more effective layering of the three pills with the drug placed only in the center to increase the time for the effective level concentration of 2.0×10^{-3} mg/kg under the same conditions.

7.4.4 Solution (Suggestions)

(a) Differential equations will need to be solved simultaneously for the drug resulting from each of the three pills. No drug will be released until the outer coating is dissolved. See Problem 7.3 for help with the drug dissolution which is analogous to sublimation for gases.

A material balance on the volume of pill 1 yields

$$\frac{dD_1}{dt} = -\frac{2k_{L1}}{\rho}(S_A - C_{AS}) \quad \text{I. C.} \quad D_1 = 0.5 \text{ cm at } t = 0 \text{ min} \tag{7-31}$$

which becomes

$$\frac{dD_1}{dt} = -\frac{2k_{L1}}{\rho}(S_D - C_{DS}) \quad \text{for } 10^{-5} \le D_1 \le 0.3 \text{ cm} \tag{7-32}$$

and

$$\frac{dD_1}{dt} = 0 \quad \text{for } D_1 \le 10^{-5} \text{ cm} \tag{7-33}$$

where the mass transfer coefficient for pill 1, k_{L1}, depends upon diameter D_1.

$$k_{L1} = \frac{2(0.6)}{D_1} \tag{7-34}$$

Similar equations can be derived for pills 2 and 3 that describe pill diameters D_2 and D_3. A single differential equation can be constructed from Equations (7-31) through (7-33) using the logic capability of nested "if ... then ... else ... " statements in the POLYMATH *Simultaneous Ordinary Differential Equation Solver*.

The differential equation describing the material balance on A within the stomach is given by

$$\frac{dC_{AS}}{dt} = \frac{1}{V}[S_{W1}k_{L1}(S_A - C_{AS})\pi D_1^2 + S_{W2}k_{L2}(S_A - C_{AS})\pi D_2^2$$
$$+ S_{W3}k_{L3}(S_A - C_{AS})\pi D_3^2] - \frac{C_{AS}}{\tau} \tag{7-35}$$

each "switch" is unity when each diameter D is greater that 0.3 and zero at other times. This "switch" then provides the appropriate input of the pill coating to the stomach.

A similar material balance on the drug D yields

$$\frac{dC_{DS}}{dt} = \frac{1}{V}[(1 - S_{W1})k_{L1}(S_D - C_{DS})\pi D_1^2 + (1 - S_{W2})k_{L2}(S_D - C_{DS})\pi D_2^2$$
$$+ (1 - S_{W3})k_{L3}(S_D - C_{DS})\pi D_3^2] - \frac{C_{DS}}{\tau} \tag{7-36}$$

in which the "switch" defined previously provides the input of drug from each pill.

A POLYMATH equation set that describes the dissolution processes for the three pills is as follows:

```
Equations:
d(D1)/d(t)=if(D1>0.3)then(-2*kL1*(SA-CAS)/rho)else(if(D1>1e-
    6)then(-2*kL1*(SD-CDS)/rho)else(0.0))
d(D2)/d(t)=if(D2>0.3)then(-2*kL2*(SA-CAS)/rho)else(if(D2>1e-
    6)then(-2*kL2*(SD-CDS)/rho)else(0.0))
d(D3)/d(t)=if(D3>0.3)then(-2*kL3*(SA-CAS)/rho)else(if(D3>1e-
    6)then(-2*kL3*(SD-CDS)/rho)else(0.0))
d(CAS)/d(t)=(1/V)*(SW1*kL1*(SA-CAS)*3.1416*D1^2+SW2*kL2*(SA-
    CAS)*3.1416*D2^2+SW3*kL3*(SA-CAS)*3.1416*D3^2)-CAS/tau
d(CDS)/d(t)=(1/V)*((1-SW1)*kL1*(SD-CDS)*3.1416*D1^2+(1-
    SW2)*kL2*(SD-CDS)*3.1416*D2^2+(1-SW3)*kL3*(SD-
    CDS)*3.1416*D3^2)-CDS/tau
kL1=2*0.6/D1
SA=1.0
rho=1414.7
SD=0.4
V=1200
SW1=if(D1>0.3)then(1.0)else(0.0)
SW2=if(D2>0.3)then(1.0)else(0.0)
kL2=2*0.6/D2
SW3=if(D3>0.3)then(1.0)else(0.0)
kL3=2*0.6/D3
tau=240
Initial Conditions:
t(0)=0
```

```
D1(0)=0.5
D2(0)=0.4
D3(0)=0.35
CAS(0)=0
CDS(0)=0
Final Value:
t(f)=150
```

The partial results of this solution are shown for the three pill diameters in Figure 7–5.

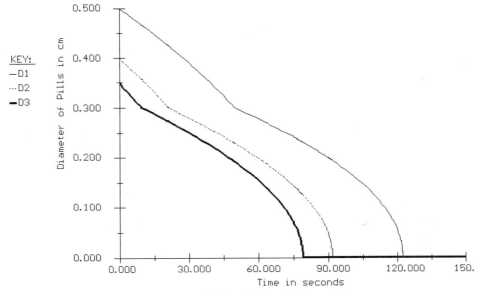

KEY:
—D1
···D2
▬D3

Figure 7–5 Diameter Variations for Dissolution of Three Pills

 The POLYMATH problem solution file for part (a) is found in the *Simultaneous Differential Equation Solver Library* located in directory CHAP7 with file named P7-04A.POL.

(b) The rate of absorption of the drug in terms of the mass transfer rate can be included in Equation (7-36), which was derived for part (a). Note that this absorption will tend to reduce the buildup of *D* within the stomach.

(c) A differential balance on the drug within the body utilizing C_{DB} needs to be made on the volume of the body, where the input is the rate of absorption is calculated as in part (b). The first-order reaction must also be included in the differential equation development. Assume the body volume to be completely mixed at all times.

(d) One technique to determine the dose time is to set up a differential equation for this time that has a derivative of unity when above the minimum effective concentration and that has a derivative of zero at all other times.

7.5 DIFFUSION WITH SIMULTANEOUS REACTION IN ISOTHERMAL CATALYST PARTICLES

7.5.1   Concepts Demonstrated

Determination of effectiveness factors for porous catalyst particles with cylindrical, and spherical geometries and various reaction orders under isothermal conditions.

7.5.2   Numerical Methods Utilized

Solution of simultaneous ordinary differential equations with split boundary values.

7.5.3   Problem Statement

The mathematical solution of simultaneous diffusion and reaction inside porous catalytic particles at constant temperature is typically formulated as an isothermal internal effectiveness factor problem. The differential equations that describe the diffusion with reaction are solved with the result expressed as

$$\eta = \frac{\text{average reaction rate within the particle}}{\text{reaction rate at the concentrations of the particle surface}} \qquad \textbf{(7-37)}$$

where η is the isothermal internal effectiveness factor.

The general solution involves the derivation of the ordinary differential equations that describe the material balance within the particle geometry. This treatment also requires the use of Fick's law for diffusion, which is usually assumed to involve only the diffusion of the reactant with an effective diffusivity for the catalyst particle. Details are found in Fogler, pp. 610–20.

The numerical solution for a spherical particle involves the material balance on a differential volume within the catalyst sphere, which yields

$$\frac{d}{dr}(N_A r^2) = -k''a C_A r^2 \qquad \textbf{(7-38)}$$

where N_A is the flux of reactant A, r is the radius of the spherical particle, k'' is the first-order rate constant based on particle volume, a is the surface area per unit volume of particle, and C_A is the concentration of reactant. The initial condition for Equation (7-38) is that there is no flux at the particle center; therefore, N_A or the combined variable $N_A r$ are both zero at $r = 0$.

Fick's law for the diffusion of reactant A can be written as

$$\frac{dC_A}{dr} = \frac{N_A}{-D_e} \qquad \textbf{(7-39)}$$

where D_e is the effective diffusivity for the diffusion of reactant A in the porous particle. The boundary condition for this equation is that the concentration of A at the particle surface is given by $C_A = C_{As}$ when $r = R$ (the radius of the spherical particle).

Since the combined variable $(N_A r^2)$ is used in Equation (7-38), an algebraic equation must be included with this problem formulation to provide N_A for Equation (7-39). Thus

$$N_A = \frac{N_A r^2}{r^2} \tag{7-40}$$

The effectiveness factor can be calculated from

$$\eta = \frac{\int_0^R k''a C_A (4\pi r^2) dr}{k''a C_{As}\left(\frac{4}{3}\pi R^3\right)} = \frac{3}{C_{As} R^3}\int_0^R C_A r^2 dr \tag{7-41}$$

which is the mathematical equivalent to Equation (7-37).

For convenience in calculation of the effectiveness factor during the solution of Equations (7-38) to (7-40), the effectiveness factor of Equation (7-41) can be differentiated with respect to r to obtain

$$\frac{d\eta}{dr} = \frac{3 C_A r^2}{C_{As} R^3} \tag{7-42}$$

whose initial condition is $\eta = 0$ and $r = 0$. This differential equation can be solved simultaneously with Equations (7-38) and (7-39). The final value of the effectiveness factor is given by the value of η at $r = R$ provided that the boundary conditions of Equations (7-38) and (7-39) are satisfied.

Similar problem solutions can be obtained for the slab and cylindrical geometries for a variety of reaction rate expressions. The numerical solution of these problems can provide the concentration profiles and the effectiveness factor when the boundary conditions are satisfied.

Analytical Solution

The analytical solution to the diffusion with first-order reaction is given by (Fogler[4], p. 617, or Bird et al.[2], p. 545)

$$\eta = \frac{3}{\phi^2}\{\phi[\coth(\phi)] - 1\} \tag{7-43}$$

where coth( ) is the hyperbolic cotangent and ϕ is the Thiele modulus defined by

$$\phi = R\sqrt{\frac{k''a C_{As}^{n-1}}{D_e}} \tag{7-44}$$

with n representing the order of the reaction.

Consider the simultaneous diffusion and reaction inside porous catalyst particle at constant temperature.

(a) Calculate the concentration profile numerically for C_A and simultaneously determine the effectiveness factor η for a first-order irreversible reaction in a spherical particle, where $R = 0.5$ cm, $D_e = 0.1$ cm^2/s, $C_{As} = 0.2$ g-mol/cm^3, and $k''a = 6.4$ s^{-1}.

(b) Compare the result in part (a) for the effectiveness factor η with the analytical solution for η given by Equations (7-43) and (7-44).

(c) Repeat part (a) for a catalyst particle that is a long cylinder with radius $R = 0.5$ cm, $D_e = 0.1$ cm^2/s, $C_{As} = 0.2$ g-mol/cm^3, and $k''a = 6.4$ s^{-1}.

(d) Repeat part (a) for a second-order irreversible reaction with $C_{As} = 0.2$ g-mol/cm^3, $D_e = 0.1$ cm^2/s, and $k''a = 32$ cm^3/g-mol·s. [Note that the Thiele modulus is the same as for part(a).]

(e) Repeat part (a) for a catalyst particle that is a long cylinder with radius $R = 0.5$ cm. The reaction is second order and irreversible with $C_{As} = 0.2$ g-mol/cm^3 and $k''a = 32$ cm^3/g-mol·s. [Note that the Thiele modulus is the same as for part(a).]

7.5.4   Solution (Partial)

(a) and (b) Sphere This problem requires the numerical integration of ordinary differential equations given by Equations (7-38), (7-39), and (7-41) along with algebraic equations (7-40), (7-43), and (7-44). A suitable equation set for the POLYMATH *Simultaneous Ordinary Differential Equation Solver* is given as follows:

```
Equations:
d(CA)/d(r)=NA/(-De)
d(eta)/d(r)=3*CA*r^2/(CAs*R^3)
d(NArr)/d(r)=-kppa*CA*r^2
NA=if(r==0)then(0.0)else(NArr/r^2)
kppa=6.4
De=0.1
CAs=0.2
R=0.5
err=CA-CAs
phi=R*sqrt(kppa/De)
etacalc=(3/phi^2)*(phi*(1/tanh(phi))-1)
Initial Conditions:
r(0)=0
CA(0)=0.029315
eta(0)=0
NArr(0)=0
Final Value:
r(f)=0.5
```

Division by Zero Note that the possible division by zero in the POLY-MATH equation set for Equation (7-40) is handled with "if ... then ... else ... " logic, as follows:

```
NA=if(r==0)then(0.0)else(NArr/r^2)
```

Boundary Condition Convergence The shooting technique can be used to determine the value of the initial concentration C_A at $r = 0$ that gives $C_A = C_{As}$ = 0.2 g-mol/cm³. The POLYMATH equation set statement that gives the error is

```
err=CA-CAs
```

Convergence to a low value of err can be accomplished by simple trial and error iterations or by more sophisticated methods, as discussed in Problem 3.7.

The POLYMATH problem solution file for parts (a) and (b) is found in the *Simultaneous Differential Equation Solver Library* located in directory CHAP7 with file named P7-05AB.POL.

(c) Cylinder The corresponding equations for the cylinder are as follows:

Material Balance Differential Equation

$$\frac{d}{dr}(N_A r) = -k''aC_A r \tag{7-45}$$

Fick's Law Differential Equation

$$\frac{dC_A}{dr} = \frac{N_A}{-D_e} \tag{7-46}$$

Algebraic Equation

$$N_A = \frac{N_A r}{r} \tag{7-47}$$

Effectiveness Factor Integral Equation

$$\eta = \frac{\int_0^R k''aC_A(2\pi r)dr}{k''aC_{As}(\pi R^2)} = \frac{2}{C_{As}R^2}\int_0^R C_A r\, dr \tag{7-48}$$

Effectiveness Factor in Differential Equation Form

$$\frac{d\eta}{dr} = \frac{2C_A r}{C_{As}R^2} \tag{7-49}$$

7.6 GENERAL EFFECTIVENESS FACTOR CALCULATIONS FOR FIRST-ORDER REACTIONS

7.6.1   Concepts Demonstrated

Demonstration of the similarity of effectiveness factor solutions for diffusion with first-order reaction in a slab, cylinder, and sphere using a modified Thiele modulus with the same volume to surface area ratios.

7.6.2   Numerical Methods Utilized

Numerical solution of simultaneous ordinary differential equations with split boundary conditions.

7.6.3   Problem Statement

The cases of diffusion with first-order reaction in isothermal catalyst particles in several geometries have been shown by Aris[1] to yield similar effectiveness factors when a modified Thiele modulus is utilized. (Effectiveness factors and Thiele modulus are discussed in Problem 7.5.)

The modified Thiele modulus is defined by

$$\phi_m = \left(\frac{V}{S}\right)\sqrt{\frac{k''a}{D_e}}$$
(7-50)

where $\frac{V}{S}$ represents the volume to surface ratio for the particle. The ratios are summarized in Table 7–3. A plot of this modified Thiele modulus is given by Fogler[4], p. 618.

Table 7–3 Volume to Surface Ratios for Particles

Particle Geometry	$\dfrac{V}{S}$
Slab of thickness L with reaction on both surfaces	$\dfrac{L}{2}$
Cylinder of radius R_c	$\dfrac{R_c}{2}$
Sphere of radius R_s	$\dfrac{R_s}{3}$

Consider the simultaneous diffusion and first-order reaction inside porous catalyst particle at constant temperature. This problem will help to verify that the calculated effectiveness factors are very similar when the volume to surface ratio of the particles given in Table 7–3 are the same. A complete numerical solution will be required for each particle to enable this comparison.

(a) Determine the effectiveness factor for the base case, a spherical particle where $R = 0.3$ cm, $D_e = 0.08$ cm^2/s, $C_{As} = 0.4$ g-mol/cm^3, and $k''a = 8$ s^{-1}.

(b) Determine the effectiveness factor for a slab whose thickness gives the same modified Thiele modules as the sphere in part (a). (This is equivalent to having the same volume to surface ratio as the base case sphere.)

(c) Determine the effectiveness factor for a long cylinder whose radius gives the same modified Thiele modules as the sphere in part (a). (This is equivalent to having the same volume to surface ratio as the base case sphere.)

(d) Repeat the effectiveness factor calculations for a sphere, slab, and cylinder whose dimensions are increased to give two times the volume to surface ratio (V/S) of the original spherical catalyst particle in part (a). Note that this doubles the modified Thiele modulus.

(e) Summarize the various effectiveness factor calculations made in parts (a) through (d) for the two values of the modified Thiele modulus. Do your results confirm that the effectiveness factor values are approximately equivalent at the same modified Thiele modulus?

7.7 SIMULTANEOUS DIFFUSION AND REVERSIBLE REACTION IN A CATALYTIC LAYER

7.7.1   Concepts Demonstrated

Modeling of binary gaseous diffusion with simultaneous, isothermal, reversible reaction in a porous catalyst layer.

7.7.2   Numerical Methods Utilized

Integration of simultaneous ordinary differential equations with split boundary conditions.

7.7.3   Problem Statement

The catalytic gas phase reaction between components A and B is occurring reversibly in a catalyst layer at a particular point in a monolithic reactor.

$$2A \leftrightarrow B \tag{7-51}$$

The catalytic reaction rate expression for reactant A is given by

$$r'_A = -k\left(C_A^2 - \frac{C_B}{K_C}\right) \tag{7-52}$$

where the rate is in g-mol/cm^3·s, the rate constant k has been determined to be 8×10^4 in cm^3/s·g-mol, and the equilibrium constant is known to be $K_C = 6\times10^5$ cm^3/g-mol. The catalytic layer has a thickness of $L = 0.2$ cm, and the effective diffusivity of A in B for this layer is $D_e = 0.01$ cm^2/s. The reactant mixture contains only gases A and B, so only binary gas diffusion need be considered. The total concentration of A and B is $C_T = 4\times10^{-5}$ g-mol/cm^3.

Calculation of the effectiveness factor (see Problem 7.5 for definition) requires consideration of the simultaneous diffusion of both components along with a material balance involving the reversible reaction rate expression.

Material Balances on A and B within the Porous Layer

The material balance on component A in the z direction from the top of the porous layer results in an ordinary differential equation involving the flux N_A:

$$\frac{dN_A}{dz} = -k\left(C_A^2 - \frac{C_B}{K_C}\right) \tag{7-53}$$

with the boundary condition that there is no mass transfer flux at the bottom of the porous layer, expressed as

$$N_A\big|_{z=L} = 0 \tag{7-54}$$

A similar differential equation could be derived for component B, but it is simpler to use the algebraic relationship between the molar fluxes of A and B due to the reaction stoichiometry.

$$N_B = -\frac{1}{2}N_A \qquad (7\text{-}55)$$

Note that this equation also satisfied the boundary condition at the bottom of the porous layer, where

$$N_B\big|_{z=L} = 0 \qquad (7\text{-}56)$$

Fick's Law for Binary Diffusion

The diffusion of A in this binary system is described by

$$N_A = -D_e\frac{dC_A}{dz} + x_A(N_A + N_B) \qquad (7\text{-}57)$$

where the effective diffusivity D_e of the catalyst layer is given by

$$D_e = \frac{D_{AB}\varepsilon_p\sigma}{\tilde{\tau}} \qquad (7\text{-}58)$$

where D_{AB} is the binary diffusivity, ε_p is the catalyst porosity, σ is the constriction factor, and $\tilde{\tau}$ is the tortuosity (see Fogler[4], p. 608, for details)

Equation (7-57) can be rearranged by incorporating Equation (7-55) and the definition of x_A in terms of concentrations to yield

$$\frac{dC_A}{dz} = \frac{\frac{C_A}{C_T}\left(\frac{N_A}{2}\right) - N_A}{D_e} \qquad (7\text{-}59)$$

with the initial condition

$$C_A\big|_{z=0} = C_{As} \qquad (7\text{-}60)$$

A similar differential equation can be derived for component B; however, it is much easier to use the overall mass balance for this gas phase system to calculate C_B. Thus

$$C_B = C_T - C_A \qquad (7\text{-}61)$$

and Equation (7-53) can therefore be expressed in terms of only C_A as

$$\frac{dN_A}{dz} = -k\left[C_A^2 - \frac{(C_T - C_A)}{K_C}\right] \qquad (7\text{-}62)$$

(a)   Use an implicit finite difference technique to calculate the effective-
      ness factor for the given reaction and summarize C_A and N_A in a table
      at 10 equally spaced intervals within the catalytic layer. The reactant
      concentrations at the surface of the layer are known to be $C_{As} =$
      3×10^{-5} g-mol/cm^3 and $C_{Bs} = 1 \times 10^{-5}$ g-mol/cm^3.
(b)   Repeat part(a) using the shooting technique for solving the problem,
      and enter the calculated results in the same table.
(c)   Compare the two solutions of parts (a) and (b). Please comment on
      which solution technique you prefer.
(d)   Repeat part (a) for $C_{As} = 1 \times 10^{-5}$ g-mol/cm^3 and $C_{Bs} = 3 \times 10^{-5}$ g-mol/
      cm^3 using the solution technique that you prefer.

7.7.4   Solution (Suggestions)

This problem requires the simultaneous solution of differential Equation (7-59)
with the initial condition given by Equation (7-60) and differential Equation (7-
62) with the boundary condition given by Equation (7-56). Thus the boundary
conditions for the two differential equations are split, with one being an initial
condition and the other being a final condition.

(a) Implicit Finite Difference (IFD) Solution

This solution involves utilizing finite difference formulas for the differential
equations and boundary conditions. The resulting system of algebraic equations
can be solved for the variables at the various node points utilized in the problem.
The 10 intervals with 11 node points for this problem are illustrated in Figure 7–
6. Equation (7-59) can be rewritten using a second-order central difference formula for the first derivative [see Equation (A-6) of Appendix A] as

$$\frac{dC_{A_n}}{dz} = \frac{C_{A_{n+1}} - C_{A_{n-1}}}{2\Delta z} = \frac{(C_{A_n}/C_T)(N_{A_n}/2) - N_{A_n}}{D_e} \quad \text{for } (2 \leq n \leq 10) \quad \textbf{(7-63)}$$

The known initial condition of Equation (7-60) gives $C_{A_1} = C_{As}$.

      For node 11 at the bottom of the catalyst layer, the second-order backward
difference approximation [see Equation (A-7) of Appendix A] for the first derivative can be used, giving

$$\left.\frac{dC_A}{dz}\right|_{n=11} = \frac{3C_{A_{11}} - 4C_{A_{10}} + C_{A_9}}{2\Delta z} = \frac{(C_{A_{11}}/C_T)(N_{A_{11}}/2) - N_{A_{11}}}{D_e} \quad \textbf{(7-64)}$$

The boundary condition for N_A, Equation (7-54), results in the right-hand side of
Equation (7-64) becoming zero; therefore, the resulting simplification yields the

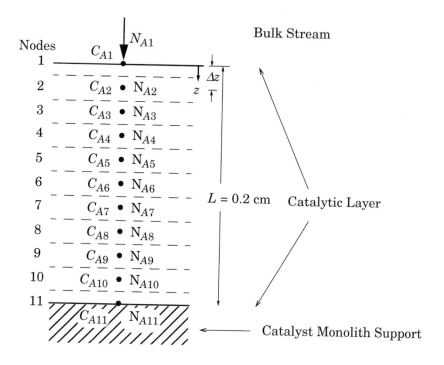

Figure 7–6 Diffusion with Reversible Chemical Reaction in a Catalytic Layer

following algebraic equation at node 11:

$$C_{A_{11}} = (4C_{A_{10}} - C_{A_9})/3 \tag{7-65}$$

Similarly, for the flux N_A, differential Equation (7-62) becomes

$$\frac{dN_{A_n}}{dz} = \frac{N_{A_{n+1}} - N_{A_{n-1}}}{2\Delta z} = -k\left[C_{A_n}^2 - \frac{(C_T - C_{A_n})}{K_C} \right] \quad \text{for } (2 \le n \le 10) \tag{7-66}$$

The known final condition of Equation (7-54) gives zero flux at the bottom of the catalyst layer.

$$N_{A_{11}} = 0 \tag{7-67}$$

For node 1 at the top of the catalyst layer, the second-order forward difference approximation for the first derivative [Equation (A-5) of Appendix A] can be used, giving

$$\frac{dN_A}{dz}\bigg|_{n=1} = \frac{-3N_{A_1} + 4N_{A_2} - N_{A_3}}{2\Delta z} = -k\left[C_{A_1}^2 - \frac{(C_T - C_{A_1})}{K_C} \right] \tag{7-68}$$

Effectiveness Factor Calculation

The effectiveness factor (see Problem 7.5 for definition) can be obtained by a material balance at the surface of the catalytic layer. In this case, the surface flux of A acting over an arbitrary area must equal the average reaction rate within the catalyst layer below the arbitrary area. Thus, in terms of the finite difference nodes for an arbitrary area A_c,

$$N_{A_1} A_c = (-r'_A)_{AVG} A_c L \tag{7-69}$$

The effectiveness factor can therefore be expressed as

$$\eta = \frac{(-r'_A)\big|_{AVG}}{(-r'_A)\big|_{SURFACE}} = \frac{N_{A_1}/L}{k\left[C_{A_1}^2 - \frac{(C_T - C_{A_1})}{K_C}\right]} \tag{7-70}$$

Results

The POLYMATH *Simultaneous Nonlinear Equation Solver* can be used to solve Equations (7-63) through (7-68) as applied to the 11 nodes of the problem plus Equation (7-70) for the effectiveness factor. The results are summarized in Table 7–4, and the calculated effectiveness factor is 0.3022.

Table 7–4 Comparison of Implicit Finite Difference and Shooting Method Solutions

z	Implicit Finite Difference Solution		Shooting Method Solution	
	C_A	N_A	C_A	N_A
0	3e–05	4.271e–06	3e–05	4.250e–06
0.02	2.549e–05	3.064e–06	2.526e–05	3.066e–06
0.04	2.165e–05	2.269e–06	2.154e–05	2.239e–06
0.06	1.887e–05	1.661e–06	1.866e–05	1.649e–06
0.08	1.657e–05	1.242e–06	1.644e–05	1.218e–06
0.1	1.493e–05	9.076e–07	1.475e–05	8.954e–07
0.12	1.362e–05	6.632e–07	1.349e–05	6.469e–07
0.14	1.273e–05	4.548e–07	1.258e–05	4.482e–07
0.16	1.209e–05	2.904e–07	1.196e–05	2.823e–07
0.18	1.174e–05	1.361e–07	1.161e–05	1.362e–07
0.2	1.162e–05	0	1.149e–05	–5.150e–10

(b) Shooting Technique Solution

The concentration and flux profiles can be determined during the numerical solution of differential Equation (7-59) with a known initial condition and differential Equation (7-62), which requires determination of an initial condition that satisfies the boundary condition of zero flux when $z = L$, as given in Equation (7-54).

Effectiveness Factor Calculation

The effectiveness factor for the catalyst layer is given by the average rate within the catalytic layer divided by the rate at the surface of the catalytic layer. Thus

$$\eta = \frac{\int_0^L \left(-k_1\left(C_A^2 - \frac{C_B}{K_C}\right)\right)dz}{-k_1\left(C_{As}^2 - \frac{C_{Bs}}{K_C}\right)L} = \frac{\int_0^L \left(C_A^2 - \frac{C_B}{K_C}\right)dz}{\left(C_{As}^2 - \frac{C_{Bs}}{K_C}\right)L} \tag{7-71}$$

which can be differentiated with respect to z, yielding

$$\frac{d\eta}{dz} = \frac{\left(C_A^2 - \frac{C_B}{K_C}\right)}{\left(C_{As}^2 - \frac{C_{Bs}}{K_C}\right)L} \tag{7-72}$$

The initial condition for Equation (7-72) is

$$\eta|_{z=0} = 0 \tag{7-73}$$

and the effectiveness factor is the value of η at $z = L$. Thus the inclusion of differential Equation (7-72) with the boundary condition of Equation (7-73) completes the problem formulation.

Split Boundary Value Solution

The POLYMATH *Simultaneous Ordinary Differential Equation Solver* can be used with a "shooting technique" (see Problem 3.6 for details) to solve the differential equations. With this technique, the initial condition for N_A can be determined.

Initial Condition Estimate for N_A

The shooting technique requires that the initial condition of N_A be determined, which gives a final value of nearly zero. All the techniques require an initial estimate of N_A at the surface of the catalytic layer. Often this proves to be a difficult challenge because an incorrect initial condition may lead to numerical results that give profiles that cannot be optimized easily to improve the initial estimate.

A simple material balance at the catalytic surface can provide an upper bound on this initial value when negligible diffusional resistance is assumed. This balance sets the molar mass transfer rate at the surface equal to the reaction rate in the catalyst layer that is occurring at the surface concentrations of A and B. Thus

$$N_{A_{\text{EST}}}\Big|_{z = 0} = -r_A L = kL\left[C_{As}^2 - \frac{(C_T - C_{As})}{K_C}\right] \tag{7-74}$$

Another alternative that is useful for this problem is to take the calculated value for N_A from the IFD solution in part (a) as an initial estimate for the shooting solution.

Results

The POLYMATH *Simultaneous Differential Equation Solver* can be used with a shooting technique to solve the three ordinary differential equations describing this problem. The results are also summarized in Table 7–4, and the calculated effectiveness factor is 0.3007 versus 0.3022 for the IFD solution.

(c) Comparison of Solution Methods

A comparison of the calculated concentrations and fluxes in Table 7–4 indicates good agreement between the IFD and shooting methods. The IFD solution's accuracy can be improved by more node points within the catalytic layer.

7.8 SIMULTANEOUS MULTICOMPONENT DIFFUSION OF GASES

7.8.1 Concepts Demonstrated

Application of the Stefan-Maxwell equations to describe the multicomponent molecular diffusion of gases.

7.8.2 Numerical Methods Utilized

Numerical integration of a system of simultaneous ordinary differential equations with optimization of two parameters in order to match split boundary conditions.

7.8.3 Problem Statement[*]

Gases A and B are diffusing through stagnant gas C at a temperature of 55 °C and a pressure of 0.2 atmospheres. This process involves molecular diffusion between two points, where the compositions are known, as summarized in Table 7–5. The distance between the points is 10^{-3} m.

(a) Use the Stefan Maxwell equations to calculate the molar fluxes of both gases A and B from point 1 to point 2. Suggestion: An initial approximate solution can be determined by first considering the binary diffusion of only A through component C and then separately considering the binary diffusion of only B through component C.

(b) Plot the mole fractions of the gases as a function of distance from point 1 to point 2.

Table 7–5 Data for Multicomponent Diffusion (from Geankoplis[5] with permission)

Component	Point 1 Concentration kg-mol/m^3	Point 2 Concentration kg-mol/m^3	Diffusivities at 0.2 atm m^2/s
A	2.229×10^{-4}	0	$D_{AC} = 1.075 \times 10^{-4}$
B	0	2.701×10^{-3}	$D_{BC} = 1.245 \times 10^{-4}$
C	7.208×10^{-3}	4.730×10^{-3}	$D_{AB} = 1.47 \times 10^{-4}$

Additional Information and Data

The kinetic theory of gases can be used to derive the Stefan-Maxwell equations

[*] This problem is adapted from Geankoplis[5] with permission.

in the z direction as (see Bird et al.[2] or Geankoplis[5])

$$\frac{dC_i}{dz} = \sum_{j=1}^{n} \frac{(x_i N_j - x_j N_i)}{D_{ij}} \tag{7-75}$$

where C_i represent the concentration of diffusing component i in kg-mol/m^3, x_i is the mole fraction of component i, N_i is the molar flux of component i in kg-mol/m$^2 \cdot$s, n is the number of components, and D_{ij} is the binary molecular diffusivity for components i and j in m^2/s.

Application of Equation (7-75) to a three-component mixture yields the equations

$$\frac{dC_A}{dz} = \frac{(x_A N_B - x_B N_A)}{D_{AB}} + \frac{(x_A N_C - x_C N_A)}{D_{AC}} \tag{7-76}$$

$$\frac{dC_B}{dz} = \frac{(x_B N_A - x_A N_B)}{D_{AB}} + \frac{(x_B N_C - x_C N_B)}{D_{BC}} \tag{7-77}$$

$$\frac{dC_C}{dz} = \frac{(x_C N_A - x_A N_C)}{D_{AC}} + \frac{(x_C N_B - x_B N_C)}{D_{BC}} \tag{7-78}$$

where the appropriate equalities for the binary molecular diffusivities have been substituted for $D_{BA} = D_{AB}$, $D_{CA} = D_{AC}$, and $D_{CB} = D_{BC}$.

Typical boundary conditions for the preceding equations are dictated by the physical or chemical process. For example, if the diffusion is to a catalyst surface where the reaction rate is very fast, then the corresponding concentration of the limiting reactant at the catalyst surface may be assumed to be zero. If the diffusion process leads to a bulk stream, then the concentrations in the bulk stream are usually assumed to be at the bulk stream concentrations and diffusing components not in the bulk stream will have zero concentrations. Often the boundary conditions involving concentrations are split between two locations.

If relationships are known between the fluxes due to reaction stoichiometry or any of the fluxes are zero (stagnant component), then these relationships can be substituted into the preceding equations or expressed separately.

7.8.4   Solution

For the three-component system of this problem, the differential equations of Equations (7-76) to (7-78) directly apply. Since component C is stagnant, then the flux of this component is zero. Thus $N_C = 0$. The problem solution requires that the two fluxes N_A and N_B must be optimized until the boundary conditions of the concentrations of Table 7–5 are satisfied. The initial conditions are the known concentrations at point 1, and the final conditions are the known concentrations at point 2.

(a) & (b) The POLYMATH *Simultaneous Differential Equation Solver* can be used to solve the differential equations as a split boundary value problem with the initial conditions at point 1. In order to converge on the values of fluxes N_A and N_B, error functions for the matching of the boundary conditions at point 2 can be defined by

$$\varepsilon(C_A) = 0 - C_A\big|_{z = 0.001} \tag{7-79}$$

$$\varepsilon(C_B) = 2.701 \times 10^{-3} - C_B\big|_{z = 0.001} \tag{7-80}$$

Note that these error functions should go to zero when convergence is obtained.

Utilizing the preceding error functions and adding definitions for the mole fractions of the three components, one can write an initial POLYMATH equation set as

```
Equations:
d(CA)/d(z)=(xA*NB-xB*NA)/DAB+(xA*NC-xC*NA)/DAC
d(CB)/d(z)=(xB*NA-xA*NB)/DAB+(xB*NC-xC*NB)/DBC
d(CC)/d(z)=(xC*NA-xA*NC)/DAC+(xC*NB-xB*NC)/DBC
NA=2.396e-5
NB=-3.363e-4
DAB=1.47e-4
DBC=1.245e-4
DAC=1.075e-4
CT=0.2/(82.057e-3*328)
errA=CA-0
errB=CB-2.701e-3
NC=0
xB=CB/CT
xA=CA/CT
xC=CC/CT
Initial Conditions:
z(0)=0
CA(0)=0.0002229
CB(0)=0
CC(0)=0.007208
Final Value:
z(f)=0.001
```

Note that the initial estimate for N_A in the preceding POLYMATH equation set is obtained from an application of Fick's law for just simple binary diffusion of A in C while the other diffusional transport is neglected. Thus the initial estimate for N_A is

$$N_A = -D_{AC}\frac{(C_A\big|_2 - C_A\big|_1)}{(z\big|_2 - z\big|_1)} = -1.075 \times 10^{-4}\frac{(0 - 2.229 \times 10^{-4})}{(0.001 - 0)} = 2.396 \times 10^{-5} \tag{7-81}$$

Similarly, for N_B the initial estimate is

$$N_B = -D_{BC} \frac{(C_B|_2 - C_B|_1)}{(z|_2 - z|_1)} = -1.245 \times 10^{-4} \frac{(2.701 \times 10^{-3} - 0)}{(0.001 - 0)} = -3.363 \times 10^{-4} \textbf{ (7-82)}$$

 The POLYMATH problem solution file for the initial solution is found in the *Simultaneous Differential Equation Solver Library* located in directory CHAP7 with file named P7-08AB1.POL.

Optimization of N_A and N_B

A simple way to optimize these two fluxes is first to hold N_B fixed and then to converge upon an improved value of N_A by minimizing the error calculated in Equation (7-79). This iterative shooting method solution can be easily accomplished by trial and error or by the secant method, as discussed in Problem 3.6. Then the improved value of N_A can be held fixed, and an improved value of N_B can be obtained by minimizing the error calculated by Equation (7-80). Note that this simple optimization technique really involves searching along each parameter until a local minimum is obtained in an objective function, and then searching in turn along the other parameter to satisfy another objective function.

Typical progress in the solution for these local searches of this problem is summarized in Table 7–6. Note that the initial optimization holds N_B and searches for the value of N_A. Then the next step involves holding N_A and searching for an improved value of N_B. Convergence is obtained with just two searches for each flux, and the resulting values are found to be $N_A = 2.12 \times 10^{-5}$ kg-mol/ $m^2 \cdot s$ and $N_B = -4.14 \times 10^{-4}$ kg-mol/$m^2 \cdot s$. The resulting mole fraction profiles are nonlinear, as shown in Figure 7–7, and the final result is significantly different from the initial solution calculated from binary diffusion consideration only.

Table 7–6 Iterative Search for Fluxes N_A and N_B

Search	N_A	$\varepsilon(C_A)$	N_B	$\varepsilon(C_B)$
Start	2.396×10^{-5}	-1.692×10^{-5}	-3.363×10^{-4}	-4.170×10^{-4}
1	2.174×10^{-5}	4.224×10^{-8}	-3.363×10^{-4}	-4.196×10^{-4}
2	2.174×10^{-5}	-4.309×10^{-8}	-4.141×10^{-4}	6.325×10^{-8}
3	2.115×10^{-5}	9.811×10^{-10}	-4.141×10^{-4}	-7.510×10^{-7}
4	2.115×10^{-5}	-1.017×10^{-8}	-4.143×10^{-4}	2.827×10^{-7}

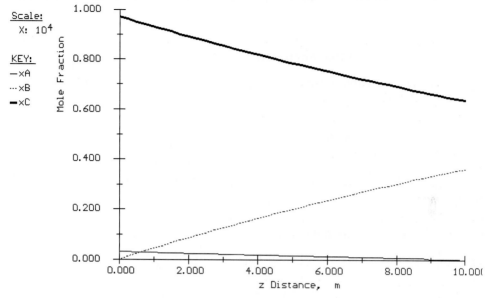

Figure 7–7 Mole Fractions Profiles for Components A, B, and C

7.9 MULTICOMPONENT DIFFUSION OF ACETONE AND METHANOL IN AIR

7.9.1   Concepts Demonstrated

Application of the Stefan-Maxwell equations to describe the multicomponent molecular diffusion of gases.

7.9.2   Numerical Methods Utilized

Numerical integration of a system of simultaneous ordinary differential equations with optimization of two parameters in order to match split boundary conditions.

7.9.3   Problem Statement

Carty and Schrodt[3] have conducted experiments involving the transport from the surface of a mixture of acetone (1) and methanol (2) through a simple diffusion tube into a stream of flowing air (3). Measurements were conducted in a Stefan tube (see Problem 7.1) at 328.5 K, where the pressure was 99.4 kPa. The gas phase composition at the liquid surface was measured to be $x_1 = 0.319$ and $x_2 = 0.528$. The length of the diffusion path was 0.238 m. The binary molecular diffusivities are estimated (see Taylor and Krishna[7], pp. 21–23) to be $D_{12} = 8.48 \times 10^{-6}$ m^2/s, $D_{13} = 13.72 \times 10^{-6}$ m^2/s, and $D_{23} = 19.91 \times 10^{-6}$ m^2/s.

(a)   Calculate the molar fluxes of both acetone and methanol from the liquid surface to the flowing air stream.

(b)   Plot the mole fractions of acetone, methanol, and air from the liquid surface to the flowing air stream.

(c)   Compare the calculated results to the data in the paper by Carty and Schrodt.[3]

(d)   Verify the calculated binary molecular diffusivities used in the problem.

7.10 MULTICOMPONENT DIFFUSION IN A POROUS LAYER COVERING A CATALYST

7.10.1   Concepts Demonstrated

Application of the Stefan-Maxwell equations to describe the multicomponent molecular diffusion of gases and binary diffusion approximations for multicomponent diffusion.

7.10.2   Numerical Methods Utilized

Numerical integration of a system of simultaneous ordinary differential equations with optimization of a single variable in order to match split boundary conditions.

7.10.3   Problem Statement

In a particular reactor for the catalytic oxidation of carbon monoxide to carbon dioxide in an mixture of N_2 and O_2, the very active catalyst is covered by a porous layer through which the reactants and products must diffuse as shown in Figure 7–8. The oxidation reaction at 1 atm and 200 °C is essentially irreversible and given by

$$CO + 1/2\ O_2 \rightarrow CO_2$$

Consider the entrance to the reactor where the composition in the bulk gas stream is 2 mol % CO, 3 mol % O_2 and 95 mol % N_2. The O_2 and CO must diffuse through the porous layer of thickness 0.001 m to the catalyst surface while the product CO_2 must diffuse out through the same layer to the bulk gas stream. The reaction is so rapid that the concentration of the limiting reactant CO is essentially zero at the catalyst surface. The effective molecular diffusivities that take into account the porosity and tortuosity of the porous layer are summarized in Table 7–7. These effective molecular diffusivities can be utilized directly in the diffusional equations.

(a)   Calculate and plot the concentrations of O_2, CO, and CO_2 as a function of the distance into the porous layer to the surface of the catalyst. Use multicomponent diffusion as described by the Stefan-Maxwell equations discussed in Problem 7.8.

(b)   What are the values of the fluxes of the gases of part (a) at the surface of the catalyst considering the z direction of Figure 7–8 to be positive?

(c)   Work part (b) by considering only binary diffusion of CO in N_2 with the effective diffusion coefficient for CO in N_2 from Table 7–7.

(d)   Comment upon the approximate solution of part (c) relative to the more exact solution of parts (a) and (b).

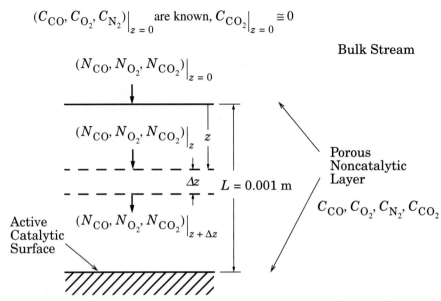

$$(C_{CO}, C_{O_2}, C_{N_2})\big|_{z=0} \quad \text{are known,} \quad C_{CO_2}\big|_{z=0} \cong 0$$

Bulk Stream

$$(N_{CO}, N_{O_2}, N_{CO_2})\big|_{z=0}$$

$$(N_{CO}, N_{O_2}, N_{CO_2})\big|_z \quad z$$

Δz $L = 0.001$ m

Porous
Noncatalytic
Layer

$$C_{CO}, C_{O_2}, C_{N_2}, C_{CO_2}$$

Active
Catalytic
Surface

$$(N_{CO}, N_{O_2}, N_{CO_2})\big|_{z+\Delta z}$$

Figure 7–8 Gas Diffusion through Porous Layer Covering Active Catalyst

Table 7–7 Effective Molecular Diffusivities for Porous Layer at 1 atm and 200 °C

Effective Molecular Diffusivity D_{ij} in m²/s	N_2	O_2	CO	CO_2
N_2		7.20×10^{-6}	6.93×10^{-6}	6.20×10^{-6}
O_2	7.20×10^{-6}		7.61×10^{-6}	6.67×10^{-6}
CO	6.93×10^{-6}	7.61×10^{-6}		6.22×10^{-6}
CO_2	6.20×10^{-6}	6.67×10^{-6}	6.22×10^{-6}	

7.10.4 Solution (Suggestions)

(a) & (b) The reactions will define ratios of fluxes, which can be used in the Stefan-Maxwell equations. A positive flux should be considered to be in the z direction toward the catalyst, as shown in Figure 7–8.

(c) & (d) This problem can be approximated by CO diffusing through stagnant N_2 when the concentrations and diffusional transport of O_2 and CO_2 are neglected.

7.11 SECOND-ORDER REACTION WITH DIFFUSION IN LIQUID FILM

7.11.1   Concepts Demonstrated

Molecular diffusion of two components with a simultaneous irreversible second-order reaction in a finite liquid film.

7.11.2   Numerical Methods Utilized

Solution of simultaneous second-order ordinary differential equations using implicit finite difference techniques, and simultaneous solution of nonlinear algebraic equations and explicit algebraic equations.

7.11.3   Problem Statement

Gas absorption into a liquid film can be enhanced by chemical reaction. Consider the important case in which the reaction in the liquid film is second order and irreversible.

An example where gas A dissolves at the surface of turbulent liquid D is shown in Figure 7–9. Component B, present in the bulk liquid D, undergoes a liquid phase reaction with dissolved A to give liquid product C:

$$A + B \rightarrow C \tag{7-83}$$

where the reaction rate is elementary and the reaction rate expression is given by $r_C = k C_A C_B$. In this system, both components A and B are very dilute solutions in liquid D. The only transport through the liquid film is by diffusion, which is influenced by the chemical reaction. One surface of the liquid film is exposed to gaseous A and thus has a known concentration of A that is designated as C_{As}. Component B has a very low vapor pressure and is not present in the gas phase, and there is no transport of B into the gas phase. The other surface of the liquid film is exposed to the bulk liquid, which is well mixed and has a negligible concentration of dissolved A. Operation is such that there is always excess B within the liquid film and bulk liquid in relation to the amount needed for reaction with A.

Steady-state material balances on a differential volume within the film that include binary diffusion approximations for both A and B in D yield

$$\frac{d^2 C_A}{dx^2} = \frac{k}{D_{AD}} C_A C_B \tag{7-84}$$

and

$$\frac{d^2 C_B}{dx^2} = \frac{k}{D_{BD}} C_A C_B \tag{7-85}$$

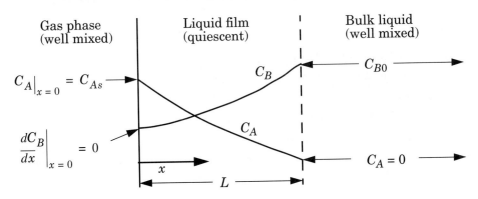

Figure 7–9 Diffusion with Concurrent Reaction of A and B in Liquid Film

where

C_A = concentration of dissolved A in kg-mol/m^3

C_B = concentration of B in kg-mol/m^3

k = reaction rate constant = 1.6×10^{-3} m^3/kg-mol·s

D_{AD} = liquid phase diffusivity of A in D = 2×10^{-10} m^2/s

D_{BD} = liquid phase diffusivity of B in D = 4×10^{-10} m^2/s

L = the total film thickness = 2×10^{-4} m

The gas phase concentration of A results in a liquid phase concentration of C_{As} = 10 kg-mol/m^3 at x = 0. There is no flux of B at x = 0, which means that dC_B/dx = 0 at the film surface. C_B is known to be 10 kg-mol/m^3, and C_A can be assumed to be zero at the interface between the bulk liquid and the liquid film, where L = 2×10^{-4} m.

(a)   Calculate and plot the concentrations of both A and B within the liquid film as a function of x.

(b)   Calculate the and plot the flux of A and B into the film in the x direction as a function of x.

(c)   What is the flux of product C through the film and into the bulk liquid if C has a very low vapor pressure and is not present in the gas phase?

(d)   What would be the maximum flux of A through the film if the reaction rate constant is doubled?

(e)   What would be the maximum flux of A through the film if the reaction rate is negligible?

7.12 SIMULTANEOUS HEAT AND MASS TRANSFER IN CATALYST PARTICLES

7.12.1 Concepts Demonstrated

Simultaneous diffusion with nonisothermal chemical reactions in spherical catalyst particles for first-order reactions.

7.12.2 Numerical Methods Utilized

Solution of simultaneous ordinary differential equations with spit boundary conditions.

7.12.3 Problem Statement

Many porous catalyst systems involve reactions in which both diffusion and effective heat transfer must be included in a complete description of the catalytic reaction. The differential equations that describe the mass transfer in a spherical catalyst particle for an irreversible first-order reaction are given by the material balance on a differential balance within the sphere:

$$\frac{d}{dr}(N_A r^2) = -k''aC_A r^2 \text{ where } N_A = 0 \text{ at } r = 0 \tag{7-86}$$

and by Fick's law for diffusion, which utilizes an effective diffusion coefficient and in which the bulk flow terms are neglected.

$$\frac{dC_A}{dr} = \frac{N_A}{-D_e} \text{ where } N_A = 0 \text{ at } r = 0 \text{ and } C_A = C_{As} \text{ at } r = R_s \tag{7-87}$$

These differential equations and their boundary conditions are discussed in Problem 7.5, where the temperature is assumed constant. Note that the numerical solution also requires the algebraic equation that relates N_A to the combined quantity $N_A r^2$.

$$N_A = \frac{N_A r^2}{r^2} \tag{7-88}$$

The effect of temperature on the reaction rate constant can be described by the Arrhenius expression with a dimensionless activation energy called the Arrhenius number. Thus at any temperature T, the rate constant is given by

$$k''|_T = k''|_{T_s} \exp\left[-\varepsilon\left(\frac{T_s}{T} - 1\right)\right] \tag{7-89}$$

where $k''|_{T_s}$ is the first-order reaction rate constant based on the catalytic area at the particle surface temperature T_s, and ε is the dimensionless Arrhenius number. The value of ε is determined by the temperature dependency of the reaction, given by

$$\varepsilon = \frac{E}{RT_s} \tag{7-90}$$

where E is the activation energy for the reaction and R is the gas constant. The temperature effect on the effective diffusion coefficient is usually neglected.

A similar treatment to the mass balance and application of Fick's law can be made for an energy balance on a differential volume within the pellet and application of Fourier's law for heat conduction. The resulting equations and boundary conditions are

$$\frac{d}{dr}(Q_r r^2) = -k''aC_A r^2 \Delta H_R \text{ where } Q_r = 0 \text{ at } r = 0 \tag{7-91}$$

$$\frac{dT}{dr} = \frac{Q_r}{-k_e} \text{ where } Q_r = 0 \text{ at } r = 0 \text{ and } T = T_s \text{ at } r = R_s \tag{7-92}$$

where ΔH_R is the heat of reaction and k_e is the effective thermal conductivity of the catalyst particle. Note that the numerical solution to these two equations also requires the algebraic equation that relates Q_r to the combined quantity $Q_r r^2$.

$$Q_r = \frac{Q_r r^2}{r^2} \tag{7-93}$$

Thus the combined effects of mass and heat transfer within the spherical catalyst particle can be determined by the simultaneous solution of Equations (7-86) to (7-93), which involves four ODEs and three explicit algebraic equations.

Simplification of Heat Transfer Equations

The solution can be simplified for the heat transfer by considering an energy balance at any radius r, where the heat transfer must correspond to the complete reaction of all of the reactant mass that is transferred at that radius. Thus the energy flux must equal the heat of reaction multiplied by the mass flux in the opposite direction at any radius. Thus

$$Q_r \big|_r = (-\Delta H_R)(-N_A)\big|_r \tag{7-94}$$

where the negative sign with the heat of reaction is necessary because of the thermodynamic sign convention and the negative sign with the flux of A is because the heat flux and mass flux must be in opposite directions.

Substitution of Equations (7-87) and (7-92) into Equation (7-94) yields

$$-k_e \frac{dT}{dr} = -\Delta H_R D_e \frac{dC_A}{dr} \tag{7-95}$$

This equation can be integrated easily with the known boundary conditions to give

$$T = T_s + \frac{\Delta H_R D_e}{k_e}(C_A - C_{As}) \tag{7-96}$$

This algebraic equation can be used to replace Equations (7-91) to (7-93) in calculation of the temperature profile within the catalyst particle.

Nonisothermal Effectiveness Factor

The equivalent effectiveness factor when the catalyst particle is not isothermal can be calculated from

$$\eta = \frac{\int_0^{R_s} k''|_T a C_A (4\pi r^2) dr}{(k''|_{T_s} a C_{As})\left(\frac{4}{3}\pi R_s^3\right)} = \frac{3}{k''|_{T_s} a C_{As} R_s^3} \int_0^{R_s} k''|_{T_s} e^{\left[-\varepsilon\left(\frac{T_s}{T} - 1\right)\right]} a C_A r^2 dr \tag{7-97}$$

which simplifies to

$$\eta = \frac{3}{C_{As} R_s^3} \int_0^{R_s} e^{\left[-\varepsilon\left(\frac{T_s}{T} - 1\right)\right]} C_A r^2 dr \tag{7-98}$$

The concentration C_A is considered to be for a liquid or a gas that is not affected by temperature changes. As in Problem 7.5, the effectiveness expression of Equation (7-98) can be differentiated with respect to r, yielding

$$\frac{d\eta}{dr} = \frac{3 e^{\left[-\varepsilon\left(\frac{T_s}{T} - 1\right)\right]} C_A r^2}{C_{As} R_s^3} \tag{7-99}$$

with the initial condition that $\eta = 0$ at $r = 0$. This differential equation can be solved simultaneously with the other equations for this problem in order to yield the effectiveness factor of η when the integration reaches the final value of r at R_s.

Common Dimensionless Variables

The literature in nonisothermal effectiveness factors utilizes several dimensionless variables. The Thiele modulus, which comes from the form of a combination of Equations (7-86) and (7-87), is given by

$$\phi = R_s \sqrt{\frac{k''|_{T_s} a C_{As}^{n-1}}{D_e}} \tag{7-100}$$

which for a first-order reaction with $n = 1$ simplifies to

$$\phi = R_s \sqrt{\frac{k''|_{T_s} a}{D_e}} \tag{7-101}$$

The heat transfer is characterized by a dimensionless parameter, which is typically designated by β and defined by

$$\beta = \frac{C_{As}(-\Delta H_R) D_e}{k_e T_s} \tag{7-102}$$

A typical numerical solution to the nonisothermal effectiveness factor problem is shown in Figure 7–10, which is a function of ε, ϕ, and β and indicates several interesting features for first-order reactions. For example, some conditions for exothermic reactions where β is positive lead to effectiveness factors that can be greater than unity. This is due to the temperature rise internal to the particle, which accelerates the reaction in spite of the reduction of the reactant concentration due to diffusion into the particle. Another feature is that there are multiple values for the effectiveness factors at certain values of the Thiele modulus.

Consider a first-order reaction of reactant A in a spherical porous catalyst particle for which C_{As} = 0.01 kg-mol/m3, T_s = 400 K, D_e = 10^{-6} m²/s, R_s = 0.01 m, and ΔH_R = -8×10^7 kJ/kg-mol.
(a) Calculate and plot the concentration C_A and the temperature T as a function of radius r for the case where ε = 30, ϕ = 1, and β = 0.2.
(b) Calculate the nonisothermal effectiveness factor for part (a), and compare your result with that found in Figure 7–10.
(c) Select another case for ε = 30 and different values of ϕ and β, where the effectiveness factor η is expected to be less than unity. Verify the point on Figure 7–10 by calculating η.
(d) Figure 7–10 suggests multiple steady states for positive β's when $\phi < 1$. Select fixed values of ϕ and β for ε = 30 from Figure 7–10 where this should occur, and calculate the upper and lower effectiveness factors.
(e) Explain the results of part (d).

7.12.4   Solution (Suggestions)

This problem is similar to Problem 7.5, but it is complicated by need for the energy balance. Figure 7–10 gives some indication of the gradients within the

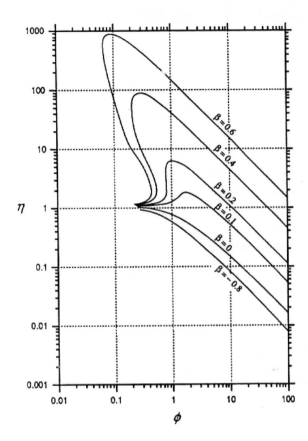

Figure 7–10 Nonisothermal Effectiveness Factor for $\varepsilon = 30$ (Reprinted from Fogler[4], p. 620, with permission)

particle. Increasing values of β lead to large temperature gradients, while higher values of ϕ indicate greater reaction rates. As the reaction rate constant is increased, the reaction becomes more localized nearer the particle surface so that the interior concentrations are nearly zero and the interior temperature profile is nearly constant. The numerical results should indicate these trends.

Multiple steady states can be demonstrated numerically by utilizing different starting points for the solution of the differential equations. This is most conveniently accomplished by utilizing a simple trial and error procedure to converge upon the desired solution that matches the boundary condition if shooting techniques are used. For solutions involving the implicit finite difference approach, the initial estimates of the variables at the various node points usually determine which steady state is reached.

7.13 UNSTEADY-STATE MASS TRANSFER IN A SLAB

7.13.1 Concepts Demonstrated

Unsteady-state mass transfer in a one-dimensional slab having only one face exposed and an initial concentration profile.

7.13.2 Numerical Methods Utilized

Application of the numerical method of lines to solve a partial differential equation, and solution of simultaneous ordinary differential equations and explicit algebraic equations.

7.13.3 Problem Statement[*]

A slab of material with a thickness of 0.004 m has one surface suddenly exposed to a solution containing component A with $C_{A0} = 6 \times 10^{-3}$ kg-mol/m^3 while the other surface is supported by an insulated solid allowing no mass transport. There is an initial linear concentration profile of component A within the slab from $C_A = 1 \times 10^{-3}$ kg-mol/m^3 at the solution side to $C_A = 2 \times 10^{-3}$ kg-mol/m^3 at the solid side. The diffusivity $D_{AB} = 1 \times 10^{-9}$ m^2/s. The distribution coefficient relating between the concentration in the solution adjacent to the slab C_{ALi} and the concentration in the solid slab at the surface C_{Ai} is defined by

$$K = \frac{C_{ALi}}{C_{Ai}} \tag{7-103}$$

where $K = 1.5$. The convective mass transfer coefficient at the slab surface can be considered as infinite.

The unsteady-state diffusion of component A within the slab is described by the partial differential equation

$$\frac{\partial C_A}{\partial t} = D_{AB} \frac{\partial^2 C_A}{\partial x^2} \tag{7-104}$$

The initial condition of the concentration profile for C_A is known to be linear at $t = 0$. Since the differential equation is second order in C_A, two boundary conditions are needed. Utilization of the distribution coefficient at the slab surface gives

$$C_{Ai}\Big|_{x=0} = \frac{C_{A0}}{K} \tag{7-105}$$

[*] Adapted from Geankoplis[6], pp. 471–473, with permission.

and the no diffusional flux condition at the insulated slab boundary gives

$$\frac{\partial C_A}{\partial x}\bigg|_{x = 0.004} = 0 \qquad\qquad \textbf{(7-106)}$$

(a) Calculate the concentrations within the slab after 2500 s. Utilize the numerical method of lines with an interval between nodes of 0.0005 m.

(b) Compare the results obtained with those reported by Geankoplis[6], p. 473, and summarized in Table 7–9.

(c) Plot the concentrations versus time to 20000 s at $x = 0.001, 0.002, 0.003$, and 0.004 m.

(d) Repeat part (a) with an interval between nodes of 0.00025. Compare results with those of part (a).

(e) Repeat parts (a) and (c) for the case where mass transfer is present at the slab surface. The external mass transfer coefficient is $k_c = 1.0 \times 10^{-6}$ m/s.

The Numerical Method of Lines

The method of lines (MOL) is a general technique for the solution of partial differential equations that has been introduced in Problem 3.9. This method utilizes ordinary differential equations for the time derivative and finite differences on the spatial derivatives. The finite difference elements for this problem are shown in Figure 7–11, where the interior of the slab has been divided into $N = 8$ intervals involving $N + 1 = 9$ nodes.

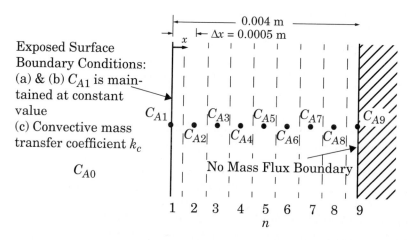

Figure 7–11 Unsteady-State Mass Transfer in a One-Dimensional Slab

Equation (7-104) can be written using a central difference formula for the second derivative and replacing the partial time derivatives with ordinary derivatives

$$\frac{dC_{An}}{dt} = \frac{D_{AB}}{(\Delta x)^2}(C_{An+1} - 2C_{An} + C_{An-1}) \quad \text{for } (2 \leq n \leq 8) \tag{7-107}$$

Boundary Condition for Exposed Surface

The general surface boundary condition is obtained from a mass balance at the interface, which equates the mass transfer to the surface via the mass transfer coefficient to the mass transfer away from the surface due to diffusion within the slab. Thus at any time for mass transfer normal to the slab surface in the x direction,

$$k_c(C_{A0} - KC_{A1}) = -D_{AB}\frac{\partial C_A}{\partial x}\bigg|_{x=0} \tag{7-108}$$

where k_c is the external mass transfer coefficient in m/s and the partition coefficient K is used to have the liquid phase concentration driving force.

The derivative on the right side of Equation (7-108) can be written in finite difference form using the second-order three-point forward difference expression at node 1 [see Equation (A-5) of Appendix A].

$$\frac{\partial C_A}{\partial x}\bigg|_{x=0} = \frac{(-C_{A3} + 4C_{A2} - 3C_{A1})}{2\Delta x} \tag{7-109}$$

Thus substitution of Equation (7-109) into Equation (7-108) yields

$$k_c(C_{A0} - KC_{A1}) = -D_{AB}\frac{(-C_{A3} + 4C_{A2} - 3C_{A1})}{2\Delta x} \tag{7-110}$$

The preceding equation can be directly solved for C_{A1} to give

$$C_{A1} = \frac{2k_c C_{A0}\Delta x - D_{AB}C_{A3} + 4D_{AB}C_{A2}}{3D_{AB} + 2k_c K\Delta x} \tag{7-111}$$

which is the general result. For good mass transfer to the surface where $k_c \to \infty$ in Equation (7-111), the expression for C_{A1} is given by

$$C_{A1} = \frac{C_{A0}}{K} \tag{7-112}$$

Boundary Condition for Insulated Surface (No Mass Flux)

The mass flux is zero at the insulated surface; thus from Fick's law

$$\frac{\partial C_A}{\partial x}\bigg|_{x=0.004} = 0 \tag{7-113}$$

Utilizing the second-order approximation for the three-point forward difference for the preceding derivative [Equation (A-5) of Appendix A] yields

$$\frac{\partial C_{A9}}{\partial x} = \frac{3C_{A9} - 4C_{A8} + C_{A7}}{2\Delta x} = 0 \tag{7-114}$$

which can be solved for C_{A9} to yield

$$C_{A9} = \frac{4C_{A8} - C_{A7}}{3} \tag{7-115}$$

Initial Concentration Profile

The initial profile is known to be linear, so the initial concentrations at the various nodes can be calculated as summarized in Table 7–8.

Table 7–8 Initial Concentration Profile in Slab

x in m	C_A	node n
0	1.0×10^{-3}	1
0.0005	1.125×10^{-3}	2
0.001	1.25×10^{-3}	3
0.0015	1.375×10^{-3}	4
0.002	1.5×10^{-3}	5
0.0025	1.625×10^{-3}	6
0.003	1.75×10^{-3}	7
0.0035	1.825×10^{-3}	8
0.004	2.0×10^{-3}	9

7.13.4 Solution (Partial)

(a), (b), & (c) The problem is solved by the numerical solution of Equations (7-107), (7-112), and (7-115), which results in seven simultaneous ordinary differential equations and two explicit algebraic equations for the nine concentration nodes. This set of equations can be entered into the POLYMATH *Simultaneous Differential Equation Solver*. Note that the equations for nodes 1 and 9 need to use an "if … then … else … " statement in POLYMATH to provide the desired initial values of the concentrations of A. Also, the POLYMATH equation input capability to duplicate equations is convenient for entering the finite difference equations.

The resulting equation set is given by

```
Equations:
d(CA2)/d(t)=DAB*(CA3-2*CA2+CA1)/deltax^2
d(CA3)/d(t)=DAB*(CA4-2*CA3+CA2)/deltax^2
d(CA4)/d(t)=DAB*(CA5-2*CA4+CA3)/deltax^2
d(CA5)/d(t)=DAB*(CA6-2*CA5+CA4)/deltax^2
d(CA6)/d(t)=DAB*(CA7-2*CA6+CA5)/deltax^2
d(CA7)/d(t)=DAB*(CA8-2*CA7+CA6)/deltax^2
d(CA8)/d(t)=DAB*(CA9-2*CA8+CA7)/deltax^2
deltax=0.0005
CA9=if(t==0)then(2.0e-3)else((4*CA8-CA7)/3)
CA0=6.0e-3
K=1.5
DAB=1.0e-9
CA1=if(t==0)then(1.0e-3)else(CA0/K)
Initial Conditions:
t(0)=0
CA2(0)=0.001125
CA3(0)=0.00125
CA4(0)=0.001375
CA5(0)=0.0015
CA6(0)=0.001625
CA7(0)=0.00175
CA8(0)=0.001825
Final Value:
t(f)=2500
```

The concentration variables at t = 2500 are summarized in Table 7–9 where a comparison with the approximate hand calculations by Geankoplis[6] shows reasonable agreement.

Table **7–9** Results for Unsteady-state Mass Transfer in One-Dimensional Slab at t = 2500 s

Distance from Slab Surface in m	n	Geankoplis[6] $\Delta x = 0.001$ m $\;$ C_A in kg-mol/m^3	n	Method of Lines (a) $\Delta x = 0.0005$ m C_A in kg-mol/m^3
0	1	0.004	1	0.004
0.001	2	0.003188	3	0.003169
0.002	3	0.002500	5	0.002509
0.003	4	0.002095	7	0.002108
0.004	5	0.001906	9	0.001977

The calculated concentration profiles for C_A at nodes 3, 5, 7, and 9 to $t = 20000$ s are presented in Figure 7–12, where the dynamics of the interior points show the effects of the initial concentration profile.

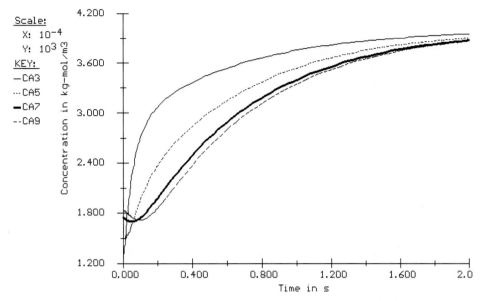

Figure 7–12 Calculated Concentration Profiles for C_A at Selected Node Points

  The POLYMATH problem solution file for parts (a), (b) and (c) is found in the *Simultaneous Differential Equation Solver Library* located in directory CHAP7 with file named P7-13A.POL.

7.14 UNSTEADY-STATE DIFFUSION AND REACTION IN A SEMI-INFINITE SLAB

7.14.1 Concepts Demonstrated

Unsteady-state diffusion with first-order reaction in a one-dimensional semi infinite slab with constant properties.

7.14.2 Numerical Methods Utilized

Application of the numerical method of lines to solve a partial differential equation, and solution of simultaneous ordinary differential equations and explicit algebraic equations.

7.14.3 Problem Statement[*]

Carbon dioxide gas at one atmosphere pressure is being absorbed into an alkaline solution containing a catalyst. The dissolved CO_2 designated as reactant A undergoes diffusion and an irreversible first-order reaction within the solution that is described by the following partial differential equation:

$$\frac{\partial C_A}{\partial t} = D_{AB}\frac{\partial^2 C_A}{\partial x^2} - k'C_A \tag{7-116}$$

where C_A is the concentration of dissolved CO_2 in kg-mol/m^3, t is the time in s, D_{AB} is the diffusivity of CO_2 within the alkaline solution B in m^2/s, x is the distance from the top of the solution in m, and k' is the first-order reaction rate constant in s^{-1}. The partial pressure of CO_2 is $p_{A0} = 1.0132 \times 10^5$ Pa, and the solubility of CO_2 is $S = 2.961 \times 10^{-7}$ kg-mol/Pa.

The initial condition for Equation (7-16) is given by

$$C_A = 0 \text{ at initial time } t = 0 \text{ and for all } x \tag{7-117}$$

and the boundary conditions are

$$C_A = C_{As} \quad \text{for } t > 0 \text{ and } x = 0 \tag{7-118}$$

and

$$C_A = 0 \quad \text{for } t > 0 \text{ and } x = \infty \tag{7-119}$$

Geankoplis[6] provides the following data for this system:

$$C_{As} = p_{A0}S = (1.0132 \times 10^5 \text{ Pa})(2.961 \times 10^{-7} \text{ kg-mol/Pa}) = 0.03 \text{ kg-mol/m}^3$$
$$D_{AB} = 1.5 \times 10^{-9} \text{ m}^2\text{/s and } k' = 35 \text{ s}^{-1}$$

[*] Adapted from Geankoplis[6], pp. 460–461, with permission.

(a) Calculate the concentration of dissolved A within the solution after 0.01 s by utilizing the numerical method of lines with 11 nodes (10 intervals) and an interval between nodes of 1.0×10^{-6} m. Consider the concentration C_A at the deepest node to be zero.

(b) Compute the total amount of A absorbed to a time of 0.01 s and compare your results with the value of 1.458×10^{-7} kg-mol/m^2 as calculated from an analytical solution by Geankoplis[6], p. 461.

(c) Plot the concentration of A versus time to steady state at nodes 2, 3, 4, and 5 and tabulate the concentrations at $t = 0.01$ s.

(d) Repeat parts (a) through (b) with 21 nodes (20 intervals) and an interval between nodes of 5.0×10^{-7} m. Compare the concentrations at equivalent node locations with results of part (c) at $t = 0.01$ s.

(e) Repeat parts (a) and (c) for the case where the total pressure is doubled but the partial pressure of A remains the same. The additional component in the gas phase is not soluble in the solution. The external mass transfer coefficient is $k_G = 1.0 \times 10^{-10}$ kg-mol/s$\cdot$m$^2\cdot$Pa. What is the percent reduction in amount of A absorbed to 0.01s that is due to the external mass transfer resulting from the pressure increase?

7.14.4 Solution (Partial)

(a) The numerical method of lines is introduced in Problem 3.9 and in other problems, including Problem 7.13, for a finite slab that involves only unsteady-state diffusion. A similar treatment here can be utilized for 10 finite difference elements at the slab surface by writing Equation (7-116) as

$$\frac{dC_{An}}{dt} = \frac{D_{AB}}{(\Delta x)^2}(C_{An+1} - 2C_{An} + C_{An-1}) - k'C_{An} \quad \text{for } (2 \leq n \leq 10) \qquad \textbf{(7-120)}$$

where the second-order central difference approximation of the second derivative is used from Equation (A-9) of Appendix A.

The initial and boundary Equations (7-117) and (7-118) can be written as

$$C_{A1} = 0 \text{ when } t = 0 \qquad \textbf{(7-121)}$$

and

$$C_{A1} = p_{CO_2}S_{CO_2} = (1.0132 \times 10^5)(2.961 \times 10^{-7}) = 0.03 \text{ kg-mol/m}^3 \qquad \textbf{(7-122)}$$

for $t > 0$.

The boundary condition of Equation (7-119) under the assumption stated in part (a) becomes

$$C_{A_{11}} = 0 \text{ for } t \geq 0 \tag{7-123}$$

The finite difference equations for the 11 nodes as given in Equations (7-120) to (7-123) can be solved by the POLYMATH *Simultaneous Differential Equation Solver*. Equations (7-121) and (7-122) can be entered using the "if ... then ... else ... " capability, as follows:

```
Equations:
d(CA2)/d(t)=DAB*(CA3-2*CA2+CA1)/deltax^2-kprime*CA2
d(CA3)/d(t)=DAB*(CA4-2*CA3+CA2)/deltax^2-kprime*CA3
d(CA4)/d(t)=DAB*(CA5-2*CA4+CA3)/deltax^2-kprime*CA4
d(CA5)/d(t)=DAB*(CA6-2*CA5+CA4)/deltax^2-kprime*CA5
d(CA6)/d(t)=DAB*(CA7-2*CA6+CA5)/deltax^2-kprime*CA6
d(CA7)/d(t)=DAB*(CA8-2*CA7+CA6)/deltax^2-kprime*CA7
d(CA8)/d(t)=DAB*(CA9-2*CA8+CA7)/deltax^2-kprime*CA8
d(CA9)/d(t)=DAB*(CA10-2*CA9+CA8)/deltax^2-kprime*CA9
d(CA10)/d(t)=DAB*(CA11-2*CA10+CA9)/deltax^2-kprime*CA10
DAB=1.5e-9
CA1=if(t==0)then(0)else(0.03)
deltax=1.e-6
kprime=35
CA11=0
Initial Conditions:
t(0)=0
CA2(0)=0
CA3(0)=0
CA4(0)=0
CA5(0)=0
CA6(0)=0
CA7(0)=0
CA8(0)=0
CA9(0)=0
CA10(0)=0
Final Value:
t(f)=0.01
```

Calculations show that the "if ... then ... else ... " statement for C_{A1} in the preceding equation set is really not necessary and that C_{A1} can just be set equal to 0.03 for all time t.

 The POLYMATH problem solution file for part (a) is found in the *Simultaneous Differential Equation Solver Library* located in directory CHAP7 with file named P7-14A.POL.

(b) An unsteady-state material balance at the surface of the slab yields

$$\frac{dQ}{dt} = -D_{AB}\frac{dC_A}{dx}\bigg|_{x=0} \tag{7-124}$$

where Q is the total amount of A transferred to the solution in kg-mol/m². The initial condition is that $Q = 0$ at $t = 0$, and the integration to any time t yields the value of Q over that time period. Equation (7-124) can be written using a second-order forward finite difference approximation as [see Equation (A-5) of Appendix A]

$$\frac{dQ}{dt} = -D_{AB}\frac{(-3C_{A1}+4C_{A2}-C_{A3})}{2\Delta x} \tag{7-125}$$

Integration of Equations (7-125) can be accomplished along with the numerical method of lines solution for the concentrations yielding the corresponding value of Q.

(e) In the case of external mass transfer, the flux of A to the surface at any time t can be calculated by using the mass transfer coefficient as

$$N_A = k_G\left(P_{A_0} - \frac{C_{As}}{S}\right) \tag{7-126}$$

The same flux of A is also given by applying Fick's law at the surface of the slab at any time t.

$$N_A = -D_{AB}\frac{dC_A}{dx} \tag{7-127}$$

Setting the fluxes equal in Equations (7-126) and (7-127) gives the general expression for the effect of external mass transfer at the slab surface.

$$k_G\left(P_{A_0} - \frac{C_{As}}{S}\right) = -D_{AB}\frac{dC_A}{dx} \tag{7-128}$$

Use of the preceding equation in finite difference notation, which incorporates the second-order forward finite difference formula for the derivative dC_A/dx yields

$$k_G\left(P_{A_0} - \frac{C_{A1}}{S}\right) = -D_{AB}\frac{-3C_{A1}+4C_{A2}-C_{A3}}{2\Delta x} \tag{7-129}$$

and Equation (7-129) can be solved for C_{A1} to give

$$C_{A1} = \frac{2k_G P_{A0}\Delta x - D_{AB}C_{A3}+4D_{AB}C_{A2}}{3D_{AB}+(2k_G\Delta x)/S} \tag{7-130}$$

Thus, the finite difference equations for parts (a) and (c) only need to be modified using Equation (7-130) to account for the external mass transfer.

7.15 Diffusion and Reaction in Falling Laminar Liquid Film of Finite Thickness

7.15.1   Concepts Demonstrated

Unsteady-state mass transfer with gas absorption, liquid-phase diffusion, and first-order reaction in a falling Newtonian fluid of finite thickness.

7.15.2   Numerical Methods Utilized

Application of the numerical method of lines to solve a partial differential equation which can be expressed as a system of simultaneous ordinary differential equations.

7.15.3   Problem Statement

Consider the absorption of CO_2 gas into a falling liquid film of alkaline solution in which there is a first-order irreversible reaction. A similar process without flow is discussed in Problem 7.14. The resulting concentration of the dissolved CO_2 in the film is quite small so that the viscosity of the liquid is not affected, and the mass transport in the liquid by bulk flow is negligible. The steady-state laminar flow of a Newtonian fluid down a vertical wall results in a velocity distribution, which is given by

$$v_z = \frac{\rho g \delta^2}{2\mu}\left[1 - \left(\frac{x}{\delta}\right)^2\right] = v_{z_{max}}\left[1 - \left(\frac{x}{\delta}\right)^2\right] \tag{7-131}$$

as has been discussed in Problem 5.3.

A steady-state material balance on a differential volume within the liquid film yields the partial differential equation given by

$$v_z\frac{\partial C_A}{\partial z} = D_{AB}\frac{\partial^2 C_A}{\partial x^2} - k'C_A \tag{7-132}$$

where v_z is the velocity in m/s, C_A is the concentration of dissolved CO_2 in kg-mol/m^3, D_{AB} is the diffusivity of dissolved CO_2 in the alkaline solution with units of m^2/s, and k' is a first-order reaction rate constant for the neutralization reaction in s^{-1}. The numerical values of all variables are the same as in Problem 7.14 except for k'. The film thickness is $\delta = 3\times10^{-4}$ m, $v_{z_{max}} = 0.6$ m/s, and $v_{z_{avg}} = (2/3)v_{z_{max}} = 0.4$ m/s.

The boundary condition for Equation (7-132) in the z direction is that C_A is zero at the point where the film begins to flow down the wall. Thus

$$C_A\big|_{z=0} = 0 \tag{7-133}$$

The first boundary condition in the x direction is that C_A is known at the surface of the film, and this can be expressed as

$$C_A\big|_{x = 0,\, z > 0} = C_{As} = 0.03 \qquad \text{(7-134)}$$

which implies that the external mass transfer coefficient is very large. The second boundary condition in the x direction is that there is no mass transfer at the wall. Thus

$$\frac{\partial C_A}{\partial x}\bigg|_{x = \delta,\, z \geq 0} = 0 \qquad \text{(7-135)}$$

(a) Calculate the concentration of dissolved A at each node point within the liquid when there is no reaction and at $z = 1$ m. Utilize the numerical method of lines with 11 nodes (10 intervals), as shown in Figure 7–13.

(b) Extend part (a) by calculating the average flux of A absorbed by the film in kg-mol/s to $z = 1$ m.

(c) Plot the concentration of A versus z at nodes 3, 5, 7, and 9 for part (a).

(d) Verify the results of part (a) by calculating the molar rate of A absorbed at the film surface and comparing this with the calculated molar rate of A exiting in the liquid film at $z = 1$ m. Consider the film to be 1 m wide.

(e) Repeat parts (a) and (c) for the case where a weak alkaline solution causes an irreversible first-order reaction of A with a rate constant of $k' = 1\ \mathrm{s^{-1}}$.

(f) What is the percentage increase in absorption of A from the gas phase because of the reaction in part (e) relative to that of part (a) that had no reaction?

7.15.4 Solution (Partial)

(a) The numerical method of lines is introduced in Problem 3.9, and it is applied in Problem 7.13 to unsteady-state diffusion in a finite slab with no reaction. However, this current problem is at steady state, with C_A being a function of depth within the film, designated as x, and the distance from the top of the falling film, designated as z. The finite difference elements for this problem are shown in Figure 7–13, where the interior of the slab has been divided into $N = 10$ intervals involving $N + 1 = 11$ nodes.

The method of lines allows ordinary differential equations to describe the variation of C_A with the z direction and can utilize finite elements to describe the variation in the x direction. This treatment gives a working equation set from

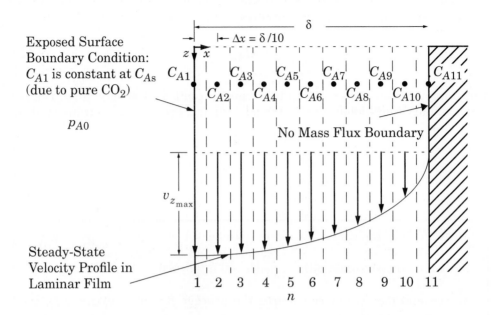

Figure 7–13 Mass Transfer with Reaction within a Falling Laminar Film

Equation (7-132) as

$$\frac{dC_{An}}{dz} = \left(\frac{D_{AB}}{(\Delta x)^2}(C_{An+1} - 2C_{An} + C_{An-1}) - k'C_{An}\right)\Big/ v_{z_n} \quad \text{for } (2 \leq n \leq 10) \quad \textbf{(7-136)}$$

where the second-order central difference approximation of Equation (A-9) is used for the second derivative.

The velocity v_{z_n} in Equation (7-136) varies only with x, and this can be expressed by writing Equation (7-131) as

$$v_{z_n} = v_{z_{max}}\left[1 - \left(\frac{(n-1)\Delta x}{\delta}\right)^2\right] \quad \text{for } (2 \leq n \leq 10) \quad \textbf{(7-137)}$$

Boundary Conditions

The initial condition of Equation (7-133) applies to the C_A in each of the internal finite elements. Thus

$$C_{An} = 0 \text{ at } z = 0 \text{ for } (2 \leq n \leq 10) \quad \textbf{(7-138)}$$

The boundary condition given by Equation (7-134) applies to the first finite element, giving

$$C_{A1} = 0.03 \text{ for } z \geq 0 \quad \textbf{(7-139)}$$

Equation (7-135) involves the derivative of C_A at the wall, which can be obtained using the second-order backward finite difference approximation of Equation (A-7) for the derivative as

$$\frac{\partial C_{A11}}{\partial x} = \frac{3C_{A11} - 4C_{A10} + C_{A9}}{2\Delta x} = 0 \tag{7-140}$$

The preceding equation can be solved for C_{A11} to yield

$$C_{A11} = \frac{4C_{A10} - C_{A9}}{3} \tag{7-141}$$

Sometimes numerical noise may enter into the preceding equation to yield negative values. This can be handled by logic to keep C_{A11} at zero whenever a negative value is calculated.

Numerical Solution

The general finite difference expression for Equation (7-132) with the velocity expression from Equation (7-137) can be combined and written as

$$\frac{dC_{An}}{dz} = \left[\frac{D_{AB}}{(\Delta x)^2}(C_{An+1} - 2C_{An} + C_{An-1}) - k'C_{An}\right] \Bigg/ \left\{v_{z_{max}}\left[1 - \left(\frac{(n-1)\Delta x}{\delta}\right)^2\right]\right\}$$

$$\text{for } (2 \leq n \leq 10) \tag{7-142}$$

The initial conditions are all zero from the boundary condition of Equation (7-138). These nine ordinary differential equations plus Equations (7-139) and (7-141) from the remaining boundary conditions allow the C_A's at the 11 nodes to be calculated as a function of z. Note that more accurate results could be obtained by utilizing more node points.

The POLYMATH *Simultaneous Differential Equation Solver* or any other ODE package can be used to solve this set of equations. The equation set for POLYMATH is given as follows where the capability to duplicate an equation was used during problem entry of the repetitive differential equations so that only the node values needed to be changed:

```
Equations:
d(CA2)/d(z)=(DAB*(CA3-2*CA2+CA1)/deltax^2-kprime*CA2)/(vmax*(1-
   ((2-1)*deltax/delta)^2))
d(CA3)/d(z)=(DAB*(CA4-2*CA3+CA2)/deltax^2-kprime*CA3)/(vmax*(1-
   ((3-1)*deltax/delta)^2))
d(CA4)/d(z)=(DAB*(CA5-2*CA4+CA3)/deltax^2-kprime*CA4)/(vmax*(1-
   ((4-1)*deltax/delta)^2))
d(CA5)/d(z)=(DAB*(CA6-2*CA5+CA4)/deltax^2-kprime*CA5)/(vmax*(1-
   ((5-1)*deltax/delta)^2))
d(CA6)/d(z)=(DAB*(CA7-2*CA6+CA5)/deltax^2-kprime*CA6)/(vmax*(1-
   ((6-1)*deltax/delta)^2))
d(CA7)/d(z)=(DAB*(CA8-2*CA7+CA6)/deltax^2-kprime*CA7)/(vmax*(1-
   ((7-1)*deltax/delta)^2))
d(CA8)/d(z)=(DAB*(CA9-2*CA8+CA7)/deltax^2-kprime*CA8)/(vmax*(1-
   ((8-1)*deltax/delta)^2))
```

```
d(CA9)/d(z)=(DAB*(CA10-2*CA9+CA8)/deltax^2-kprime*CA9)/(vmax*(1-
   ((9-1)*deltax/delta)^2))
d(CA10)/d(z)=(DAB*(CA11-2*CA10+CA9)/deltax^2-kprime*CA10)/
   (vmax*(1-((10-1)*deltax/delta)^2))
DAB=1.5e-9
CA1=0.03
kprime=0
vmax=0.6
delta=3.e-4
CA11=if(4*CA10<CA9)then(0)else((4*CA10-CA9)/3)
deltax=0.1*delta
vavg=(2/3)*vmax
Initial Conditions:
z(0)=0
CA2(0)=0
CA3(0)=0
CA4(0)=0
CA5(0)=0
CA6(0)=0
CA7(0)=0
CA8(0)=0
CA9(0)=0
CA10(0)=0
Final Value:
z(f)=1
```

The POLYMATH problem solution file for part (a) is found in the *Simultaneous Differential Equation Solver Library* located in directory CHAP7 with file named P7-15A.POL.

(b) The average flux of A to a liquid film of height H is given by

$$N_{A_{avg}} = \frac{\int_0^H \left(-D_{AB}\frac{dC_A}{dx}\bigg|_{x=0,\,z} \right) dz}{H} \tag{7-143}$$

where $N_{A_{avg}}$ is the average flux of A transferred to the liquid film in kg-mol/ $m^2 \cdot s$ and H is the film height in m. Equation (7-143) can be differentiated to yield

$$\frac{dN_{A_{avg}}}{dz} = \frac{\left(-D_{AB}\frac{dC_A}{dx}\bigg|_{x=0,\,z} \right)}{H} \tag{7-144}$$

with an initial condition that $N_{A_{avg}} = 0$ at $z = 0$. Integration of this equation to any distance z yields the value of $N_{A_{avg}}$ over the film height.

Equation (7-144) can be written using a second-order forward finite difference approximation as

$$\frac{dN_{A_{avg}}}{dz} = -\frac{D_{AB}}{H}\frac{(-3C_{A1} + 4C_{A2} - C_{A3})}{2\Delta x} \tag{7-145}$$

The integration of the preceding equation for $H = 1$ m simultaneously with the equation set from part (a) to a final value of $z = 1$ m allows the determination of the requested $N_{A_{avg}}$.

(d) An overall steady-state material balance on A within the film when there is no reaction requires that the A that is transferred at the film surface must equal the A that flows out with the film. For a film of height H in m and width W in m, the input is given by

$$M_A = N_{A_{avg}} HW \tag{7-146}$$

where M_A is in kg-mol/s.

The output of A that exits the film at height H can be calculated from

$$M_A = W \int_0^\delta v_z C_A \, dx \tag{7-147}$$

in which v_z varies with z according to Equation (7-131) and C_A is the concentration profile determined from the numerical solution in part (a) at height H.

In order to evaluate Equation (7-147), the numerical values of C_A at the 11 node locations from the solution of part (a) can be used. The integral can be evaluated by fitting the product of $v_z C_A$ versus x with a cubic spline or polynomial and evaluating the integral with the POLYMATH *Curve Fitting and Regression Program*. The comparison of the calculations from Equations (7-146) and (7-147) should be made with $H = W = 1$ m.

REFERENCES

1. Aris, R., *Chem. Eng. Sci.*, *6*, 265 (1957).
2. Bird, R. B., Stewart, W. E., and Lightfoot, E. N. *Transport Phenomena*, New York: Wiley, 1960.
3. Carty, R., and Schrodt, J. T., "Concentration Profiles in Ternary Gaseous Diffusion," *Ind. Eng. Chem. Fundam.*, *14*, 276–278 (1975).
4. Fogler, H. S. *Elements of Chemical Reaction Engineering*, 2nd ed., Englewood Cliffs, NJ: Prentice Hall, 1992.
5. Geankoplis, C. J. *Mass Transport Phenomena*, Minneapolis: C. J. Geankoplis, 1972. (Available from The Ohio State University Bookstore, Columbus, OH)
6. Geankoplis, C. J. *Transport Processes and Unit Operations*, 3rd ed., Englewood Cliffs, NJ: Prentice Hall, 1993.
7. Taylor, R., and Krishna, R. *Multicomponent Mass Transport*, New York: Wiley, 1993.

Chemical Reaction Engineering[*]

8.1 PLUG-FLOW REACTOR WITH VOLUME CHANGE DURING REACTION

8.1.1 Concepts Demonstrated

Calculation of conversion in a gas-phase, isothermal, plug-flow reactor at constant pressure for a reaction with a change in number of moles.

8.1.2 Numerical Methods Utilized

Solution of simultaneous ordinary differential equations.

8.1.3 Problem Statement

The irreversible decomposition of the di-*tert*-butyl peroxide is to be carried out in an isothermal plug-flow reactor in which there is no pressure drop. Symbolically, this reaction can be written $A \rightarrow B + 2C$. The feed consists of di-*tert*-butyl peroxide and inert nitrogen. The reactor volume is 200 dm^3, and the entering volumetric flow rate is maintained constant at 10 dm^3/min. The reaction rate constant k for this first-order reaction is 0.08 min^{-1} which is based on reactant A.

(a) Determine and plot the conversion as a function of reactor volume for a feed stream of pure A at a concentration of 1.0 g-mol/dm^3. Also make a similar conversion plot when the feed is only 5% A with the balance being an inert component.

(b) Repeat (a) for a reaction $3A \rightarrow B$ with the same rate constant that is based on reactant A. All other variables remain the same.

(c) Summarize the results of (a) and (b) and discuss the effect of concentration level and reaction stoichiometry on this first-order reaction.

[*]The notation and equations of this chapter follow those of Fogler, H. S., *Elements of Chemical Reaction Engineering*, 2nd ed., Englewood Cliffs, NJ: Prentice Hall, 1992.

8.1.4   Solution (Partial)

(a) The material balance in moles for the plug-flow reactor is

$$\frac{dX}{dV} = \frac{-r_A}{F_{A0}} \tag{8-1}$$

where the rate law is given by

$$-r_A = kC_A \tag{8-2}$$

Stoichiometric considerations allow the concentration of reactant to be given by

$$C_A = C_{A0}\frac{(1-X)}{(1+\varepsilon X)}\frac{P}{P_0}\left(\frac{T_0}{T}\right) \tag{8-3}$$

where

$$\varepsilon = y_{A0}\delta = y_{A0}(3-1) = 2y_{A0} \tag{8-4}$$

and

$$F_{A0} = C_{A0}v_0 \tag{8-5}$$

Since the reaction is isothermal and there is no pressure drop, the pressure ratio and temperature ratio are unity. Equation (8-3) can be rewritten as

$$C_A = C_{A0}\frac{(1-X)}{(1+\varepsilon X)} \tag{8-6}$$

Equation (8-1) can be simplified by Equations (8-2) and (8-4) through (8-6) to

$$\frac{dX}{dV} = \frac{k(1-X)}{v_0(1+\varepsilon X)} \tag{8-7}$$

where the initial condition is $X = 0$ at $V = 0$. The final value is $V = 200$. The value of ε is calculated from Equation (8-4) to be $\varepsilon = 2$ for the pure reactant case and $\varepsilon = 0.1$ for the 5% reactant case.

The numerical integration of the differential equation given by Equation (8-7) for the pure reactant case can be accomplished by the POLYMATH *Simultaneous Differential Equation Solver* using the equation set

```
Equations:
d(X)/d(V)=k*(1-X)/(v0*(1+epsilon*X))
epsilon=2
k=.08
v0=10
Initial Conditions:
V(0)=0
```

```
X(0)=0
Final Value:
V(f)=200
```

The plot of the conversion versus reactor volume can be obtained easily as shown in Figure 8–1.

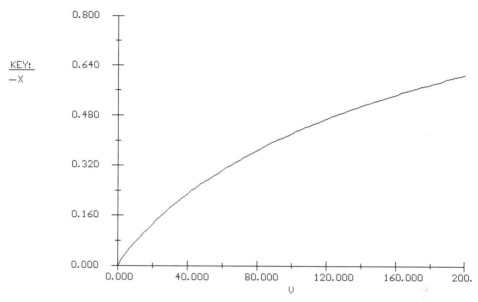

Figure 8–1 Conversion versus Reactor Volume for Pure Reactant

The solution for the dilute reactant case is obtained by changing only $\varepsilon = 0.1$ in the equation set.

 The POLYMATH problem solution files for both cases of part (a) are found in the *Simultaneous Differential Equation Solver Library* located in directory CHAP8 with files named P8-01A1.POL and P8-01A2.POL.

Graphical Comparison of Results

Often it is desirable to plot solutions for different cases for comparisons, such as in this problem solution. This can be accomplished easily by using indices to denote each separate case and solving all cases simultaneously. For part (a) of this problem, a POLYMATH equation set for comparison of both reactant feed cases using index 1 for the pure feed and index 2 for the 5% feed case is given by

```
Equations:
d(X1)/d(V)=k*(1-X1)/(v0*(1+epsilon1*X1))
d(X2)/d(V)=k*(1-X2)/(v0*(1+epsilon2*X2))
epsilon1=2
k=0.08
v0=10
```

```
epsilon2=0.1
Initial Conditions:
V(0)=0
X1(0)=0
X2(0)=0
Final Value:
V(f)=200
```

The graphical result for conversion in both cases is plotted in Figure 8–2. The effect of the concentration level on the conversion is very evident for the pure reactant as compared to the 5% reactant. The increase in moles for the pure reactant greatly increases the volumetric flow rate, thereby reducing the time in the reactor when compared to the 5% reactant. This results in a lower conversion for the pure reactant case.

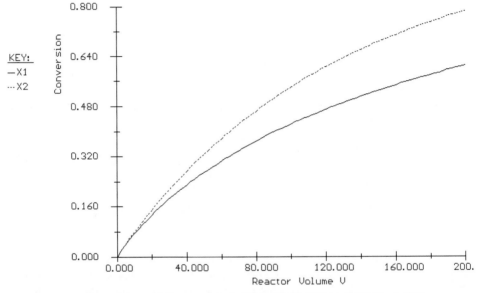

Figure 8–2 Comparison of Conversion for Pure Feed (X1) and 5% Feed (X2)

 The POLYMATH problem solution file for the comparison in part (a) is found in the Simultaneous Differential Equation Solver Library located in directory CHAP8 with file named P8-01A3.POL.

Tabulated Results
There are instances when the actual values are required during problem output. This is illustrated for part (a), where the actual conversions are tabulated by requesting the POLYMATH *Simultaneous Differential Equation Solver* to generate tabular results. A example of the POLYMATH tabular results for 10 increments of *V* is given in Table 8–1 for the two cases of part (a).

Table 8–1 Tabular Output for Conversions

V	X1	X2
0	0	0
20	0.13153	0.14684
40	0.22783	0.27058
60	0.30434	0.37530
80	0.36778	0.46421
100	0.42183	0.53992
120	0.46873	0.60453
140	0.50999	0.65978
160	0.54666	0.70710
180	0.57951	0.74768
200	0.60915	0.78253

(b) For this reaction the same equations can be used as in part (a), where the change in the reaction gives

$$\varepsilon = y_{A0}\delta = y_{A0}\left(\frac{1}{3} - 1\right) = -\frac{2}{3}y_{A0} \tag{8-8}$$

8.2 Variation of Conversion With Reaction Order in a Plug-Flow Reactor

8.2.1   Concepts Demonstrated

The effect of reaction order on conversion in a constant-volume, isothermal, plug-flow reactor.

8.2.2   Numerical Methods Utilized

Solution of ordinary differential equations.

8.2.3   Problem Statement

Consider the effect of reaction order on conversion in a plug-flow reactor whose volume is 1.5 dm^3. The irreversible liquid phase reaction $A \rightarrow B$ is taking place in which the feed concentration is $C_{A0} = 1.0$ g-mol/dm^3 and the volumetric flow rate is $v_0 = 0.9$ dm^3/min. The reaction rate constants for the various reaction orders are fixed at 1.1 for this comparison with units that are consistent with the aforementioned concentration and flow rate.

(a)   Plot the conversion in a plug-flow reactor as a function of reactor volume for a zero-, first-, second-, and third-order reaction to a reactor volume of 1.5.dm^3.
(b)   Repeat part (a) for feed concentrations of C_{A0} at 0.5 and 2.0 g-mol/dm^3.

8.2.4   Solution (Partial)

For the plug-flow reactor, a differential mole balance gives the following differential equation for the conversion of reactant A:

$$\frac{dX}{dV} = \frac{-r_A}{F_{A0}} \qquad (8\text{-}9)$$

where the rate law is

$$-r_A = kC_A^\alpha \qquad (8\text{-}10)$$

and $\alpha = 0, 1, 2,$ or 3 according to the order of the reaction. The usual initial condition for Equation (8-9) is that there is no conversion at the reactor inlet that is expressed as $X = 0$ when $V = 0$.

Stoichiometric considerations at constant temperature and pressure allow the concentration of reactant A to be expressed using conversion as

$$C_A = C_{A0}(1 - X) \qquad (8\text{-}11)$$

Special attention is necessary for the reaction of zero order because the rate proceeds at a constant value until the conversion reaches 1.0 and then the rate becomes zero.

(a) This problem can be conveniently solved with the POLYMATH *Simultaneous Differential Equation Solver.* The special logic required for the zero order reaction can utilize the "if ... then ... else ... " capability within POLYMATH. Also in the solution with POLYMATH, it is convenient to consider all of the reaction orders in a single solution to allow for convenient plotting of the results. This can be accomplished by defining different conversion variables for each reaction order, such as x0 for the conversion from zero order, x1 for conversion for first order, etc. The POLYMATH equation set for part (a) of this problem is given by

```
Equations:
d(x0)/d(V)=if(x0<=1.0)then(k/(v0*CA00))else(0.)
d(x1)/d(V)=k*CA1/(v0*CA10)
d(x2)/d(V)=k*CA2^2/(v0*CA20)
d(x3)/d(V)=k*CA3^3/(v0*CA30)
k=1.1
v0=.9
CA00=1.
CA10=1.
CA20=1.
CA30=1.
CA0=CA00*(1-x0)
CA1=CA10*(1-x1)
CA2=CA20*(1-x2)
CA3=CA30*(1-x3)
Initial Conditions:
V(0)=0
x0(0)=0
x1(0)=0
x2(0)=0
x3(0)=0
Final values:
V(f)=1.5
```

The graphical results for this problem are given in Figure 8–3, where the effect of reaction order on conversion is clearly shown.

 The POLYMATH problem solution file for part (a) is found in the *Simultaneous Differential Equation Solver Library* located in directory CHAP8 with file named P8-02A.POL.

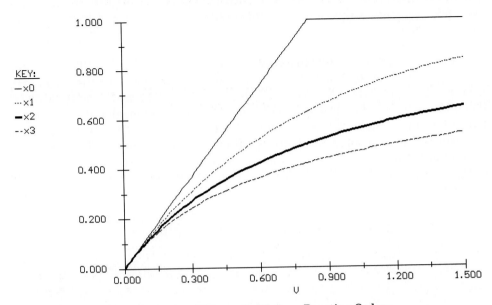

Figure 8–3 Conversion versus Volume for Various Reaction Orders

8.3 GAS PHASE REACTION IN A PACKED BED REACTOR WITH PRESSURE DROP

8.3.1   Concepts Demonstrated

Calculation of conversion and pressure drop in an isothermal, gas phase, packed bed reactor (PBR) with a change in moles during reaction.

8.3.2   Numerical Methods Utilized

Solution of simultaneous ordinary differential equations and explicit algebraic equations.

8.3.3   Problem Statement

The catalytic gas phase reaction $A \overset{k}{\to} B$ is to be carried out in a packed bed reactor under isothermal operation. The reactant is pure A with an inlet concentration of 1 g-mol/dm^3, the entering pressure is 25 atm, and the entering volumetric flow rate is 1 dm^3/min. The first-order reaction rate constant k, which is based on reactant A, is 1 dm^3/kg-cat·min.

(a)   Study the effect of the pressure drop on the conversion for a first-order reaction, and plot the conversion X and relative pressure y as a function of the weight of the packing W for three different values of the pressure drop parameter α. Consider that $(0.05 \text{ kg}^{-1} \leq \alpha \leq 0.2 \text{ kg}^{-1})$. Plot both X and y versus W from zero to $W_{max} = 2$ kg.

(b)   Repeat part (a) for a first-order reaction in which there is a change in the number of moles during the reaction $(\delta > 0)$, $A \to 3B$.

(c)   Repeat part (a) for a first-order reaction in which $(\delta < 0)$, $A \to \frac{1}{3}B$.

(d)   Summarize the results of parts (a), (b), and (c).

8.3.4   Solution (Partial)

(a) The mole balance equation for this reactor is given by

$$\frac{dX}{dW} = \frac{-r'_A}{F_{A0}} \tag{8-12}$$

where the first-order rate law is

$$-r'_A = kC_A \tag{8-13}$$

and the stoichiometry yields

$$C_A = C_{A0}\frac{(1-X)}{(1+\varepsilon X)}\frac{P}{P_0} \tag{8-14}$$

Since there is no change in the number of moles, $\varepsilon = 0$. Defining $y = P/P_0$ and substituting into Equation (8-14) yields

$$C_A = C_{A0}(1-X)y \tag{8-15}$$

where the pressure drop in a packed bed reactor is given by Fogler[4] as

$$\frac{dy}{dW} = \frac{-\alpha}{2}\left(\frac{1+\varepsilon X}{y}\right) \tag{8-16}$$

Suggestions The technique utilized for comparisons in Problem 8.2, in which different conversion variables were used for each case, might be convenient to apply in this problem as well. In this problem, different relative pressures will also need to be used.

(b) Equations (8-12), (8-13), (8-14), (8-15), and (8-16) are again applicable here, but the value of ε changes, as given by $\varepsilon = y_{A0}\delta = y_{A0}(3-1) = 2y_{A0} = 2$.

(c) The solution is the same as in part (a) except that the value of ε is calculated from

$$\varepsilon = y_{A0}\delta = y_{A0}\left(\frac{1}{3}-1\right) = -\frac{2y_{A0}}{3} = -\frac{2}{3} \tag{8-17}$$

8.4 CATALYTIC REACTOR WITH MEMBRANE SEPARATION

8.4.1   Concepts Demonstrated

Calculation of flow rate and concentrations of the reactants and products in an isothermal catalytic reactor with product removal by a membrane where there is pressure drop and mass transfer dependency on local velocity.

8.4.2   Numerical Methods Utilized

Solution of simultaneous ordinary differential equations and explicit algebraic equations.

8.4.3   Problem Statement

The dehydrogenation of a compound (elementary kinetics) is taking place in a selective membrane reactor under isothermal conditions.

$$A \rightleftarrows B + \frac{1}{2}C \qquad\qquad (8\text{-}18)$$

The reaction is reversible, and K_C is the equilibrium constant. The membrane reactor in which the preceding reaction is occurring is shown in Figure 8–4.

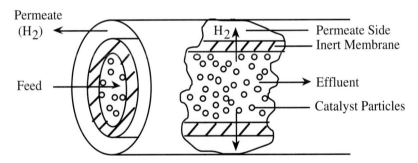

Figure 8–4   Catalytic Reactor with Product Removal by a Membrane (from Fogler,[4] with permission)

The advantage in using a membrane reactor is that by having one of the products selectively pass through the membrane, the reaction is driven toward completion. In this case hydrogen diffuses out through the sides of the membrane, thereby allowing the reaction to proceed further to the right. Isothermal conditions can be assumed.

The rate of the mass transfer across the membrane depends not only on the resistance offered by the membrane, but also on any boundary layers on each side of the membrane. As the flow rate past the membrane surface increases, the boundary layer thickness decreases, as does the resistance. Consequently, the mass transfer coefficient increases. The manner in which the k_C increases with flow rate depends upon the flow geometry. One common correlation relating the mass transfer coefficient with velocity is

$$k_C(@v) = k_C(@v_0)\left(\frac{v}{v_0}\right)^{1/2} \tag{8-19}$$

where $k_C(@v_0)$ is the mass transfer coefficient at volumetric flow rate v_0, and $k_C(@v)$ is the corresponding coefficient at v. However, if transport through the membrane is the limiting step, then the mass transfer coefficient will be independent of velocity.

In this problem the mass transfer dependence on velocity is given by

$$k_C = k_{C0}\left(\frac{v}{v_0}\right)^{1/2} \tag{8-20}$$

and the membrane transport of C is given by $k_{Ca}C_C$ when permeate concentration of C is low.

The following parameter values apply:

$$k = 0.5 \frac{\text{dm}^3}{\text{kg·min}} \qquad v_0 = 50 \text{ dm}^3/\text{min} \qquad K_C = 0.5(\text{kg·mol}/\text{dm}^3)^{1/2} \tag{8-21}$$

$$P_0 = 10 \text{ atm} \qquad F_{A0} = 10 \frac{\text{kg·mol}}{\text{min}} \qquad \alpha = 0.002 \frac{\text{atm}}{\text{kg}} \tag{8-22}$$

$$k_{C0} = 0.1\frac{\text{dm}}{\text{min}} \qquad a = 2 \frac{\text{dm}^2}{\text{kg}} \qquad W = 200 \text{ kg} \tag{8-23}$$

Compare calculated output values of F_B, the molar flow rate of B, for the following:
(a)   Base case
(b)   Base case with no membrane transport
(c)   Base case with no pressure drop
(d)   Base case with no membrane transport and no pressure drop

8.4.4   Solution (Equations)

The applicable equations for this case are as follows:

Mole balances

$$\frac{dF_A}{dW} = r'_A$$

$$\frac{dF_B}{dW} = -r'_A \tag{8-24}$$

$$\frac{dF_C}{dW} = -\frac{1}{2}r'_A - k_C a C_C$$

Rate law

$$r'_A = -k\left(C_A - \frac{C_B C_C^{1/2}}{K_C}\right) \tag{8-25}$$

Stoichiometry

$$C_A = \frac{F_A}{v} \qquad C_B = \frac{F_B}{v} \qquad C_C = \frac{F_C}{v} \tag{8-26}$$

$$v = v_0\left(\frac{F_A + F_B + F_C}{F_{A0}}\right)\frac{P_0}{P} = v_0\left(\frac{F_A + F_B + F_C}{F_{A0}}\right)\frac{1}{y} \tag{8-27}$$

where

$$y = \frac{P}{P_0} \tag{8-28}$$

Pressure drop

$$\frac{dy}{dW} = \frac{-\alpha\left(\dfrac{F_A + F_B + F_C}{F_{A0}}\right)}{2y} \tag{8-29}$$

Mass transfer coefficient

$$k_C = k_{C0}\left(\frac{v}{v_0}\right)^{1/2} \tag{8-30}$$

8.5 SEMIBATCH REACTOR WITH REVERSIBLE LIQUID PHASE REACTION

8.5.1 Concepts Demonstrated

Calculation of conversion in an isothermal liquid phase reaction carried out in a semibatch reactor under both equilibrium and rate-controlling assumptions.

8.5.2 Numerical Methods Utilized

Solution of simultaneous ordinary differential equations and explicit algebraic equations.

8.5.3 Problem Statement

Pure butanol is to be fed into a semibatch reactor containing pure ethyl acetate to produce butyl acetate and ethanol. The reaction

$$CH_3COOC_2H_5 + C_4H_9OH \rightleftarrows CH_3COOC_4H_9 + C_2H_5OH \qquad \text{(8-31)}$$

which can be expressed as

$$A + B \rightleftarrows C + D \qquad \text{(8-32)}$$

is elementary and reversible. The reaction is carried out isothermally at 300 K. At this temperature the equilibrium constant based on concentrations is 1.08 and the reaction rate constant is 9×10^{-5} dm^3/g-mol. Initially there are 200 dm^3 of ethyl acetate in the reactor, and butanol is fed at a rate of 0.05 dm^3/s for a period of 4000 seconds from the start of reactor operation. At the end of the butanol introduction, the reactor is operated as a batch reactor. The initial concentration of ethyl acetate in the reactor is 7.72 g-mol/dm^3, and the feed butanol concentration is 10.93 g-mol/dm^3.

(a) Calculate and plot the concentrations of A, B, C, and D within the reactor for the first 5000 seconds of reactor operation.

(b) Simulate reactor operation in which reaction equilibrium is always attained by increasing the reaction rate constant by a factor of 100 and repeating the calculations and plots requested in part (a). Note that this is a difficult numerical integration.

(c) Compare the conversion of ethyl acetate under the conditions of part (a) with the equilibrium conversion of part (b) during the first 5000 seconds of reactor operation.

(d) If the reactor down time between successive semibatch runs is 2000 seconds, calculate the reactor operation time that will maximize the rate of butyl acetate production.

8.5.4   Solution (Partial)

The mole balance, rate law, and stoichiometry equations applicable to the semi-batch reactor are as follows:

Mole balances

$$\frac{dN_A}{dt} = r_A V \qquad \frac{dN_B}{dt} = r_A V \qquad (8\text{-}33)$$

$$\frac{dN_C}{dt} = -r_A V \qquad \frac{dN_D}{dt} = -r_A V \qquad (8\text{-}34)$$

Rate law

$$-r_A = k\left(C_A C_B - \frac{C_C C_D}{K_e}\right) \qquad (8\text{-}35)$$

Stoichiometry

$$C_A = \frac{N_A}{V} \qquad C_B = \frac{N_B}{V} \qquad (8\text{-}36)$$

$$C_C = \frac{N_C}{V} \qquad C_D = \frac{N_D}{V} \qquad (8\text{-}37)$$

Overall material balance

$$\frac{dV}{dt} = v_0 \qquad (8\text{-}38)$$

Definition of conversion

$$x_A = \frac{N_{A0} - N_A}{N_{A0}} \qquad (8\text{-}39)$$

Definition of production rate of butyl acetate

$$P = \frac{N_C}{(t_p + 2000)} \qquad (8\text{-}40)$$

At equilibrium the net rate is equal to zero or $r_A = 0$. A convenient way to achieve this with the problem described with differential equations is to give the rate constant a large value such, as suggested in part (b).

(a) The equation set for part (a) is as follows for the POLYMATH *Simultaneous Differential Equation Solver*. Note that the POLYMATH "if ... then ... else ... " capability is utilized to introduce the volumetric feed rate of butanol, $v0$, to the reactor during the first 4000 s of operation.

Equations:
d(V)/d(t)=v0
d(NA)/d(t)=rA*V
d(NB)/d(t)=rA*V+v0*CB0
d(NC)/d(t)=-rA*V
d(ND)/d(t)=-rA*V
v0=if(t<=4000)then(.05)else(0.)
CB0=10.93
k=9.e-5
CA=NA/V
CB=NB/V
CC=NC/V
CD=ND/V
Ke=1.08
NA0=200*7.72
P=NC/(t+2000)
xA=(NA0-NA)/NA0
rA=-k*(CA*CB-CC*CD/Ke)
Initial Conditions:
t(0)=0
V(0)=200
NA(0)=1544
NB(0)=0
NC(0)=0
ND(0)=0
Final Value:
t(f)=5000

 The POLYMATH problem solution file for part (a) is found in the *Simultaneous Differential Equation Solver Library* located in directory CHAP8 with file named P8-05A.POL.

(c) A comparison of the rate-based conversion and the equilibrium-based conversion is shown in Figure 8–5. Clearly, the equilibrium conversion is higher than the rate-based conversion at all times, as expected. This graph can be generated by solving both the rate-based case of part (a) and the equilibrium-based case of part (b) in the same POLYMATH equation set. This useful technique for graphical comparisons has been applied in Problem 8.1.

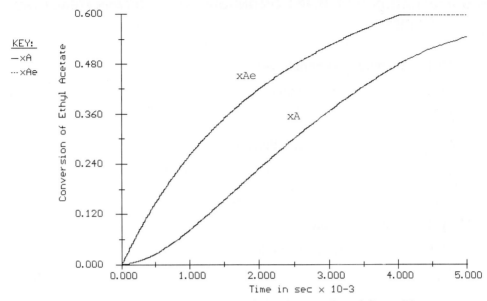

Figure 8–5 Comparison of Rate-Based and Equilibrium-Based Conversions

8.6 OPERATION OF THREE CONTINUOUS STIRRED TANK REACTORS IN SERIES

8.6.1   Concepts Demonstrated

Calculation of the steady-state and dynamic performance of three isothermal liquid-phase continuous stirred tank reactors (CSTRs) in series.

8.6.2   Numerical Methods Utilized

Solution of simultaneous ordinary differential equations and explicit algebraic equations.

8.6.3   Problem Statement

The elementary and irreversible liquid phase reaction

$$A + B \rightarrow C \tag{8-41}$$

is to be carried out in a series of three CSTRs as shown in Figure 8–6.   Species A and B are fed in separate streams to the first CSTR, with the volumetric flow of each stream controlled at 6 dm^3/min. The volume of each CSTR is 200 dm^3, and each reactor is initially filled with inert solvent. The initial concentrations of the reactants are $C_{A0} = C_{B0} = 2.0$ g-mol/dm^3, and the reaction rate coefficient is $k = 0.5$ dm^3/g-mol · min.

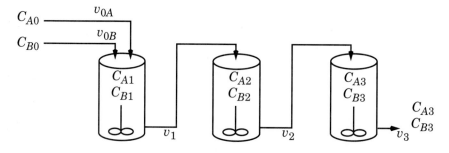

Figure 8–6 Train of Three Continuous Stirred Reactors (CSTRs)

(a) Calculate the steady-state concentrations of all reacting components exiting the third reactor. Suggestion: Set up this problem solution as ordinary differential equations that describe the unsteady-state reactor operation, and integrate these equations to a final time when concentrations do not change (steady state).

(b) Determine the time necessary to reach steady state (i.e., when C_A exiting the third reactor is 99% of the steady-state value).

(c) Plot the concentration of A exiting each tank during start up to the steady-state time determined in part (b).

(d) Consider reactor operation when the feed for species B is split equally between each reactor. Repeat parts (a), (b), and (c).

(e) Compare the results of (a) with that of a plug-flow reactor with the same total volume of 600 dm^3 and the same reactor feed.

8.6.4 Solution

(a)– (c) Material balances can be made on each reactor i. For a liquid phase reaction, the volume change with reaction can be neglected. Thus the unsteady-state balances yield the following ordinary differential equations:

$$\frac{dC_{A1}}{dt} = (v_{0A}C_{A0} - v_1 C_{A1} - kC_{A1}C_{B1}V_1)/V_1 \tag{8-42}$$

$$\frac{dC_{B1}}{dt} = (v_{0B}C_{B0} - v_1 C_{B1} - kC_{A1}C_{B1}V_1)/V_1 \tag{8-43}$$

$$\frac{dC_{A2}}{dt} = (v_1 C_{A1} - v_2 C_{A2} - kC_{A2}C_{B2}V_2)/V_2 \tag{8-44}$$

$$\frac{dC_{B2}}{dt} = (v_1 C_{B1} - v_2 C_{B2} - kC_{A2}C_{B2}V_2)/V_2 \tag{8-45}$$

$$\frac{dC_{A3}}{dt} = (v_2 C_{A2} - v_3 C_{A3} - kC_{A3}C_{B3}V_3)/V_3 \tag{8-46}$$

$$\frac{dC_{B3}}{dt} = (v_2 C_{B2} - v_3 C_{B3} - kC_{A3}C_{B3}V_3)/V_3 \tag{8-47}$$

where

$$C_{A0} = C_{B0} = 2.0; \quad k = 0.5; \quad V_1 = V_2 = V_3 = 200$$
$$v_{0A} = v_{0B} = 6; \quad v_1 = v_2 = v_3 = 12 \tag{8-48}$$

A convenient method for determining steady-state operation is to integrate the differential equations to large time, where the derivatives go to zero. Thus, this corresponds to steady-state operation. An alternate solution would be to solve the simultaneous nonlinear equations that result from setting the time derivatives to zero. Once steady state is determined, then the dynamic response can be determined from the numerical solution.

(d) Material balances for each reactor must be revised to include the fresh feed stream of component B. Thus equations (8-45) and (8-47) become

$$\frac{dC_{B2}}{dt} = (v_{0B}C_{B0} + v_1 C_{B1} - v_2 C_{B2} - kC_{A2}C_{B2}V_2)/V_2$$

$$\frac{dC_{B3}}{dt} = (v_{0B}C_{B0} + v_2 C_{B2} - v_3 C_{B3} - kC_{A3}C_{B3}V_3)/V_3$$

(8-49)

The volumetric flow rates must be altered to $v_{0B} = 2$, $v_1 = 8$, $v_2 = 10$, and $v_3 = 12$.

(e) The material balances for each reactant in a plug-flow reactor yield the following ordinary differential equations, where $v_0 = 12$ dm^3/min. For the combined reactor feed, the initial conditions are given by $C_{A0} = C_{B0} = 1.0$ g-mol/dm^3.

$$\frac{dC_A}{dV} = -\frac{kC_A C_B}{v_0}$$

(8-50)

$$\frac{dC_B}{dV} = -\frac{kC_A C_B}{v_0}$$

(8-51)

The preceding differential equations must be integrated from $V = 0$ to $V = 600$ dm^3.

8.7 DIFFERENTIAL METHOD OF RATE DATA ANALYSIS IN A BATCH REACTOR

8.7.1 Concepts Demonstrated

Determination of reaction order and rate constant by applying the differential method of rate data analysis to batch reactor data.

8.7.2 Numerical Methods Utilized

Fitting a polynomial to experimental data, differentiation of tabular data, and linear and nonlinear regression of algebraic expressions with data.

8.7.3 Problem Statement

Data on the liquid phase bromination of xylene at 17 °C have been reported by Hill.[5] This reaction has been studied by the introduction of iodine (a catalyst) and small quantities of reactant bromine into a batch reactor containing reactant xylene in considerable excess. The concentrations of reactant xylene and the iodine catalyst are approximately constant during the reaction. A material balance on the batch reactor yields

$$\frac{dC_{Br2}}{dt} = -k(C_{Br2})^n \tag{8-52}$$

where C_{Br2} is the concentration of bromine in g-mol/dm^3, k is a pseudo rate constant that depends on the iodine and xylene concentrations, and n is the reaction order.

Equation (8-52) can easily be linearized by taking the ln of each side of the equation, yielding

$$\ln\left(-\frac{dC_{Br2}}{dt}\right) = \ln k + n \ln(C_{Br2}) \tag{8-53}$$

This expression can be used in linear regression to determine k and n.

Nonlinear regression can also be used directly on Equation (8-52) in order to determine k and n.

(a) Find a polynomial that represents the data of C_{Br2} versus t given in Table 8–2.
(b) Prepare a table of $d(C_{Br2})/dt$ versus C_{Br2} at the experimental data points.
(c) Determine k and n from the linearized form of Equation (8-53).
(d) Use nonlinear regression to estimate k and n from the nonlinear form of Equation (8-52).
(e) Compare the results of parts (c) and (d). Discuss any differences.

Table 8–2 Measured Bromine Concentration versus Time

Time, t min	Bromine Concentration g-mol/dm^3	Time, t min	Bromine Concentration g-mol/dm^3
0	0.3335	19.60	0.1429
2.25	0.2965	27.00	0.1160
4.50	0.2660	30.00	0.1053
6.33	0.2450	38.00	0.0830
8.00	0.2255	41.00	0.0767
10.25	0.2050	45.00	0.0705
12.00	0.1910	47.00	0.0678
13.50	0.1794	57.00	0.0553
15.60	0.1632	63.00	0.0482
17.85	0.1500		

8.7.4   Solution (Suggestions)

The regression of experimental data is discussed in Chapter 2.

 The POLYMATH problem data file is found in the *Polynomial, Multiple Linear and Nonlinear Regression Program Library* located in directory CHAP8 with file named P8-07.POL.

8.8 INTEGRAL METHOD OF RATE DATA ANALYSIS IN A BATCH REACTOR

8.8.1 Concepts Demonstrated

Determination of a reaction rate constant by applying the integral method of rate data analysis to batch reactor data.

8.8.2 Numerical Methods Utilized

Linear and nonlinear regression of algebraic expressions with data.

8.8.3 Problem Statement

The gas phase decomposition of dimethyl ether was studied by Hinshelwood and Askey[6] in a constant volume batch reactor at 552 °C.

$$CH_3OCH_3 \rightarrow CH_4 + CO + H_2$$

Experiments were conducted by measuring the total pressure as a function of reaction time. Typical results are given in Table 8–3.

A material balance on the batch reactor for a first-order reaction gives

$$\ln\left(\frac{3P_0 - P}{2P_0}\right) = -kt \qquad \text{(8-54)}$$

where P_0 is the initial pressure, P is the measured pressure, and k is the rate coefficient.

(a) Determine the first-order rate constant and confidence intervals from the data given in Table 8–3 by a direct regression of Equation (8-54).
(b) Solve Equation (8-54) for P and use nonlinear regression to determine the first-order rate constant and the 95% confidence intervals.

Table 8–3 Total Pressure Variation in Decomposition of Dimethyl Ether

Time, t s	Pressure mm Hg	Time, t s	Pressure mm Hg
0	420	182	891
57	584	219	954
85	662	261	1013
114	743	299	1054
145	815		

The POLYMATH problem data file is found in the *Polynomial, Multiple Linear and Nonlinear Regression Program Library* located in directory CHAP8 with file named P8-08.POL.

8.9 Integral Method of Rate Data Analysis—Bimolecular Reaction

8.9.1 Concepts Demonstrated

Determination of reaction order and rate constant by application of the integral method of rate data analysis to batch reactor data.

8.9.2 Numerical Methods Utilized

Linear regression with transformed data.

8.9.3 Problem Statement

The liquid phase reaction between ethylene dibromide and potassium iodide in 99% methanol was investigated by R. T. Dillon[3] at 60 °C.

$$C_2H_4Br_2 + 3KI \rightarrow C_2H_4 + 2KBr + KI_3$$
$$A + 3B \rightarrow C + 2D + E$$

The concentration of dibromide was measured as a function of time in a batch reactor. The data are organized in Table 8–4 in terms of dibromide concentration C_A and calculated conversion X_A as a function of time t. The initial concentrations were $C_{A0} = 0.02864$ kg-mol/m^3 and $C_{B0} = 0.1531$ kg-mol/m^3.

Convenient analysis of these data to determine the reaction order and corresponding rate constant can be based on conversion as discussed by Hill.[5] The material balances on the batch reactor utilizing conversion yield the following:

(a) 0th order

$$C_{A0}X_A = kt \tag{8-55}$$

(b) 1st order with respect to A, 0th order with respect to B

$$\ln[1/(1 - X_A)] = kt \tag{8-56}$$

(c) 1st order with respect to A and B

$$\frac{1}{(\theta_B - 3)C_{A0}} \ln\left(\frac{1 - 3X_A/\theta_B}{1 - X_A}\right) = kt \tag{8-57}$$

where t is the time in s and k is the reaction rate coefficient. The initial concentration ratio of reactants is given by

$$\theta_B = \frac{C_{B0}}{C_{A0}} \tag{8-58}$$

(a) Determine the order of the reaction and the appropriate reaction rate constant from Table 8–4 by fitting Equations (8-55) through (8-57).

(b) Derive Equations (8-55) through (8-57) by making material balances on a batch reactor and applying the reaction stoichiometry.

Table 8–4 Reaction Rate for the Reaction Between Ethylene Dibromide and Potassium Iodide

Time, t s	Conversion, X_A
0	0
29.7	0.2863
40.5	0.3630
47.7	0.4099
55.8	0.4572
62.1	0.4890
72.9	0.5396
83.7	0.5795

 The POLYMATH problem data file is found in the *Polynomial, Multiple Linear and Nonlinear Regression Program Library* located in directory CHAP8 with file named P8-09.POL.

8.10 INITIAL RATE METHOD OF DATA ANALYSIS

8.10.1   Concepts Demonstrated

Determination of the orders of a reaction and the reaction rate constant using initial reaction rate data from a batch reactor.

8.10.2   Numerical Methods Utilized

Nonlinear regression of data to obtain model parameters.

8.10.3   Problem Statement

The initial rate data, given in Table 8–5 as $(-dP_A/dt)$, have been determined for the gas phase reaction $A + 2B \rightarrow C$. It can be assumed that the rate expression is of the form

$$-r_A = k(P_A)^\alpha (P_B)^\beta \tag{8-59}$$

where k is the reaction rate coefficient, α is the reaction order with respect to A, and β is the reaction order with respect to B.

(a)   Use nonlinear regression to find the numerical values for α, β, and k.
(b)   Set the orders of reaction to integer or simple fractional values on the basis of part (a) and then determine the corresponding value of k and the corresponding 95% confidence intervals.

Table 8–5   Initial Reaction Rate versus Initial Pressure Data

Run	P_A torr	P_B torr	$-r_A$ torr/s
1	6	20	0.420
2	8	20	0.647
3	10	20	0.895
4	12	20	1.188
5	16	20	1.811
6	10	10	0.639
7	10	20	0.895
8	10	40	1.265
9	10	60	1.550
10	10	100	2.021

The POLYMATH problem data file is found in the *Polynomial, Multiple Linear and Nonlinear Regression Program Library* located in directory CHAP8 with file named P8-10.POL.

8.11 HALF-LIFE METHOD FOR RATE DATA ANALYSIS

8.11.1 Concepts Demonstrated

Determination of reaction order, rate coefficient, and activation energy by utilizing the half-life method.

8.11.2 Numerical Methods Utilized

Transformation of data and multiple linear regression.

8.11.3 Problem Statement

Hinshelwood and Burk[7] have studied the thermal decomposition of nitrous oxide in a constant volume batch reactor where the reaction is

$$2\,N_2O \rightarrow 2\,N_2 + O_2$$

or

$$2A \rightarrow 2B + C$$

Representative half-life data are given in Table 8–6 for isothermal conditions ($T = 1055$ K) and in Table 8–7 for nonisothermal conditions.

Table 8–6 Half-Life Data at 1055 K

No.	Half-life $t_{1/2}$ s	Temperature T K	Concentration C_{A0} g-mol/dm$^3\times10^3$
1	1048	1055	1.6334
2	919	1055	1.8616
3	704	1055	2.4315
4	537	1055	3.1533
5	474	1055	3.6092
6	409	1055	4.1031
7	382	1055	4.4830
8	340	1055	4.9769

Table 8–7 Half Life Data at Various Temperatures

No	Half-life $t_{1/2}$ s	Temperature T K	Concentration C_{A0} g-mol/dm^3 × 10^3
1	1240	1060	1.2478
2	1352	975	2.5899
3	510	970	3.0250
4	918	1035	4.0495
5	455	1035	4.2599
6	318	1050	5.3060

Fogler[4] gives the integrated design equation for a batch reactor with the rate given by $-r_A = kC_A^\alpha$ and expressed in terms of the half-life as

$$t_{1/2} = \frac{2^{\alpha-1} - 1}{k(\alpha - 1)}\left(\frac{1}{C_{A0}^{\alpha-1}}\right) \tag{8-60}$$

The preceding equation can be linearized to a form suitable for linear regression by taking the natural log of both sides of the equation, giving

$$\ln t_{1/2} = \ln\frac{2^{\alpha-1} - 1}{k(\alpha - 1)} + (1 - \alpha)\ln C_{A0} \tag{8-61}$$

When the temperature of the rate data varies, the Arrhenius expression is used to represent the effect of temperature on the rate constant:

$$k = A \exp[-E/(RT)] \tag{8-62}$$

where typically T is the absolute temperature, R is the gas constant (8.314 kJ/g-mol·K), E is the activation energy (units of kJ/g-mol), and A is the frequency factor with units of the rate constant.

Introduction of the Arrhenius expression into Equation (8-60) yields

$$t_{1/2} = \frac{2^{\alpha-1} - 1}{\{A \exp[-E/(RT)]\}(\alpha - 1)}\left(\frac{1}{C_{A0}^{\alpha-1}}\right) \tag{8-63}$$

which can be used to correlate nonisothermal half-life data.

The corresponding linearized expression for Equation (8-63) is given by

$$\ln t_{1/2} = \ln\left(\frac{2^{\alpha-1} - 1}{\alpha - 1}\right) - \ln A + \frac{E}{RT} + (1 - \alpha)\ln C_{A0} \tag{8-64}$$

(a) Assuming that the rate expression for this reaction is of the form $-r_A = kC_A^\alpha$, apply linear regression to Equation (8-61) to determine the reaction rate constant and the order of the reaction at 1055 K using the data from Table 8–6.

(b) Repeat part (a) by applying nonlinear regression to Equation (8-60) and using the converged parameter values of part (a) as initial estimates for the nonlinear regression.

(c) Utilize linear regression on Equation (8-64) with the combined data from Tables 8–6 and 8–7 to estimate the order of the reaction, the corresponding frequency factor, and the activation energy of the rate coefficient in kJ/g-mol.

(d) Repeat part (c) by applying nonlinear regression to Equation (8-63) and using the converged parameter values of part (c) as initial estimates for the nonlinear regression.

8.11.4   Solution (Suggestions)

The linear and nonlinear regression required for this problem can be accomplished with the POLYMATH *Polynomial, Multiple Linear and Nonlinear Regression Program*. The linear and nonlinear regression of data has been discussed extensively in Chapter 2. The most detailed discussions are found in Problems 2.3 and 2.8.

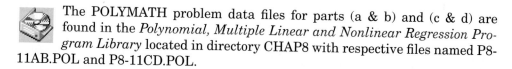 The POLYMATH problem data files for parts (a & b) and (c & d) are found in the *Polynomial, Multiple Linear and Nonlinear Regression Program Library* located in directory CHAP8 with respective files named P8-11AB.POL and P8-11CD.POL.

8.12 METHOD OF EXCESS FOR RATE DATA ANALYSIS IN A BATCH REACTOR

8.12.1   Concepts Demonstrated

Determination of reaction order and kinetic reaction rate constant using the method of excess.

8.12.2   Numerical Methods Utilized

Fitting a polynomial to experimental data, differentiation of tabular data, and linear or nonlinear regression.

8.12.3   Problem Statement

The reaction of acetic acid and cyclohexanol in dioxane solution is catalyzed by sulfuric acid. This reaction was studied in a well-stirred batch reactor at 40 °C. The reaction can be described by $A + B \rightarrow C + D$. For initial equal concentrations of $C_{A0} = C_{B0} = 2$ g-mol/dm^3, the concentration of A changed with time, as shown in Table 8–8. Another experiment was performed in which $C_{A0} = 1$ g-mol/dm^3 and $C_{B0} = 8$ g-mol/dm^3. The results of this run are summarized in Table 8–9.

For a constant volume batch reactor, the material balance on reactant A gives the time derivative that is equal to the reaction rate and rate expression

$$\frac{dC_A}{dt} = -kC_A^\alpha C_B^\beta \tag{8-65}$$

where t is the time in min, k is the reaction rate constant, α is the reaction order with respect to C_A, and β is the reaction order with respect to C_B.

(a)   Use the data in Table 8–8 to calculate the reaction rates given by the derivative dC_A/dt at the given experimental times. From these rates, calculate the reaction rate coefficient and the total reaction order: $\alpha + \beta$.

(b)   Utilize the data in Table 8–9 to calculate the reaction rates at the given experimental times. By assuming that C_B does not change significantly during this experiment, determine the approximate reaction order α with respect to component A.

(c)   Repeat part (b) by calculating the C_B at the various times utilizing the reaction stoichiometry.

(d)   Employ nonlinear regression on the combined reaction rates determined in parts (a) and (b) to determine accurate values for k, α and β. (Adjust α and β to integer values or simple fractions before reporting final results for k.)

Table 8–8 Concentration of Acetic Acid Versus Time for $C_{A0} = 2$ g-mol/dm^3 and $C_{B0} = 2$ g-mol/dm^3

No.	Time t min	Concentration C_A g-mol/dm$^3 \times 10^3$
1	0	2.000
2	120	1.705
3	150	1.647
4	180	1.595
5	210	1.546
6	240	1.501
7	270	1.460
8	300	1.421

Table 8–9 Concentration of Acetic Acid Versus Time for $C_{A0} = 1$ g-mol/dm^3 and $C_{B0} = 8$ g-mol/dm^3

No.	Time t min	Concentration C_A g-mol/dm$^3 \times 10^3$
1	0	1.000
2	30	0.959
3	45	0.939
4	75	0.903
5	120	0.854
6	150	0.824
7	210	0.771
8	255	0.734

The POLYMATH problem data files for parts (a) and (b) are found in the *Polynomial, Multiple Linear and Nonlinear Regression Program Library* located in directory CHAP8 with respective files named P8-12A.POL and P8-12B.POL.

8.13 RATE DATA ANALYSIS FOR A CSTR

8.13.1   Concepts Demonstrated

Determination of the order of reaction and corresponding reaction rate constant and for an irreversible gas-phase reaction carried out in a CSTR.

8.13.2   Numerical Methods Utilized

Regression of a data set to obtain parameters of an algebraic expression.

8.13.3   Problem Statement

A homogeneous irreversible gas phase reaction whose stoichiometry can be represented by $A \rightarrow B + 2C$ is carried out in a 1 dm^3 CSTR at 300 °C and 0.9125 atm. The data for conversion X_A versus the feed flow rate v_0 at reactor conditions where the reactor feed consists of pure reactant A are summarized in Table 8–10.
A material balance on reactant A for this CSTR yields

$$v_0 = \frac{VkC_{A0}^n(1 + X_A)^n}{C_{A0}X_A(1 + 2X_A)^n} = \frac{VkC_{A0}^{n-1}(1 + X_A)^n}{X_A(1 + 2X_A)^n} \tag{8-66}$$

where v_0 is the volumetric flow rate in dm^3/s, V is the reactor volume in dm^3, k is the rate constant, $C_{A0} = 0.1942$ g-mol/dm^3, and n is the order of the reaction. Note that the factor of 2 in the denominator of the preceding equation accounts for the volumetric change during reaction.

Determine the reaction order with respect to A (an integer value) and calculate the corresponding value of the reaction rate coefficient.

Table 8–10 Conversion Data from a CSTR[a]

v_0 dm^3/s	X_A	v_0 dm^3/s	X_A
250	0.45	5	0.8587
100	0.5562	2.5	0.8838
50	0.6434	1	0.9125
25	0.7073	0.5	0.8587
10	0.7874		

[a]These data were generated using the results of McCracken and Dickson[10].

The POLYMATH problem data file is found in the *Polynomial, Multiple Linear and Nonlinear Regression Program Library* located in directory CHAP8 with file named P8-13.POL.

8.14 DIFFERENTIAL RATE DATA ANALYSIS FOR A PLUG-FLOW REACTOR

8.14.1   Concepts Demonstrated

Differential analysis of the reaction order and corresponding reaction rate constant for a liquid phase reaction carried out in a plug-flow reactor.

8.14.2   Numerical Methods Utilized

Fitting and differentiation of tabular data, and multiple linear or nonlinear regression of data.

8.14.3   Problem Statement[*]

Oxygen is transported to living tissues by being bound to hemoglobin in arterial blood and transported to individual cells through tissue capillaries. The initial phases of deoxygenation of hemoglobin at the capillary-cell interface are assumed to be irreversible, where the deoxygenation is represented by

$$HbO_2 \rightarrow Hb + O_2$$

or

$$A \overset{k}{\rightarrow} B + C$$

The kinetics of this deoxygenation have been studied in a 0.158-cm-diameter plug-flow reactor. A solution of HbO_2 with a concentration of 1×10^{-5} g-mol/cm^3 was fed to the reactor at a rate of 19.60 cm^3/s. Oxygen concentrations were measured by electrodes placed at 2.5-cm intervals along the length of the tube. The resulting data are summarized in Table 8–11 for a single run at a constant feed rate of the HbO_2 solution.

The material balance for hemoglobin in this plug-flow reactor is given by the differential equation

$$\frac{dX_A}{dV} = \frac{-r_A}{F_{A0}} \qquad (8\text{-}67)$$

where X_A is the conversion of HbO_2, V is the reactor volume in cm^3, r_A is the reaction rate expression in g-mol/cm^3, and F_{A0} is the feed rate of A to the reactor in g-mol/s.

In the differential method of rate data analysis, the differential equation form of the material balance (reactor design equation) is fitted to the experimental data. In this case, direct use of Equation (8-67) requires the derivative information from the conversion versus reactor volume measurements.

[*] Adapted from Fogler, H. S. *Elements of Chemical Reaction Engineering*, 1st ed., Englewood Cliffs, NJ: Prentice Hall, 1986.

(a)   Fit a polynomial to the data in Table 8–11 and calculate the value for dX_A/dV at the various electrode positions.

(b)   Assuming that the rate expression for this reaction is of the form $-r_A = kC_A^\alpha$, use nonlinear regression directly on Equation (8-57) to determine the reaction order α as an integer or simple fraction. Find the corresponding value of the reaction rate constant k.

(c)   Verify that the reaction order obtained in part (b) is correct by analytically integrating Equation (8-67) and then showing that a linear plot can be obtained with the original data. Some transformations of the original data may be necessary.

Table 8–11 Deoxygenation of Hemoglobin as Function of Distance from Reactor Inlet

Electrode Position z in cm	Conversion of HbO$_2$ X_A
0.0	0.000
2.5	0.0096
5.0	0.0192
7.5	0.0286
10.0	0.0380
12.5	0.0472
15.0	0.0564
17.5	0.0655

The POLYMATH problem data file is found in the *Polynomial, Multiple Linear and Nonlinear Regression Program Library* located in directory CHAP8 with file named P8-14.POL.

8.15 INTEGRAL RATE DATA ANALYSIS FOR A PLUG-FLOW REACTOR

8.15.1 Concepts Demonstrated

Integral analysis of the reaction order and corresponding reaction rate constant for a gas-phase reaction carried out in a plug-flow reactor.

8.15.2 Numerical Methods Utilized

Fitting and differentiation of tabular data, multiple linear or nonlinear regression of data.

8.15.3 Problem Statement

Acetaldehyde undergoes decomposition at elevated temperatures.

$$CH_3CHO \rightarrow CH_4 + CO$$

or

$$A \rightarrow C + D$$

In an experiment, pure acetaldehyde vapor was passed through a reaction tube maintained by a surrounding furnace at 510 °C. The reaction tube had an inside diameter of 2.5 cm and a length of 50 cm, and the pressure was atmospheric. Decomposition rates were obtained for various flow rates are summarized in Table 8–12.

Application of the integral analysis of reactor data involves integration of the reactor design equation (material balance) for a particular reaction rate expression. For a general first-order gas phase reaction with the stoichiometry of the acetaldehyde decomposition, the reaction rate expression is given by

$$r_A = -kC_{A0}\frac{(1 - X_A)}{(1 + X_A)} \tag{8-68}$$

where C_{A0} is the feed concentration entering the reactor.

The integral form of the design equation for a first-order reaction is given by Fogler as

$$V = F_{A0}\int_0^{X_A} \frac{dX_A}{(-r_A)} = \frac{F_{A0}}{kC_{A0}}\int_0^{X_A} \frac{(1 + X_A)}{(1 - X_A)}dX_A \tag{8-69}$$

which can be analytically integrated to

$$V = \frac{F_{A0}}{kC_{A0}}\left\{2\ln\left[\frac{1}{(1 - X_A)}\right] - X_A\right\} \tag{8-70}$$

A convenient rearrangement of Equation (8-70) yields the relationship between the molar feed rate and the conversion from which the rate constant can be determined by a linear regression.

$$F_{A0} = \frac{kC_{A0}V}{\left\{2\,\ln\left[\frac{1}{(1-X_A)}\right] - X_A\right\}} \tag{8-71}$$

A similar treatment for a second-order reaction gives

$$V = \frac{F_{A0}}{kC_{A0}^2}\left[4\,\ln(1-X_A) + X_A + \frac{4X_A}{(1-X_A)}\right] \tag{8-72}$$

which can be rearranged for regression as

$$F_{A0} = \frac{kC_{A0}^2 V}{\left[4\,\ln(1-X_A) + X_A + \frac{4X_A}{(1-X_A)}\right]} \tag{8-73}$$

(a) Utilize linear regression on the data of Table 8–12 to estimate a first-order rate constant from Equation (8-71) and second-order rate constant from Equation (8-73). Please use units of h, cm³, and g-mol.

(b) Select the reaction order and corresponding rate constant that best represents the data. Justify your selection.

Table 8–12 Acetaldehyde Decomposition Data

No.	Flow Rate F_{A0} g-mol/h	Conversion X_A	No.	Flow Rate F_{A0} g-mol/h	Conversion X_A
1	9.09	0.524	5	4.55	0.652
2	2.05	0.775	6	3.27	0.705
3	0.909	0.871	7	2.68	0.738
4	1.73	0.797	8	0.682	0.894

 The POLYMATH problem data file is found in the *Polynomial, Multiple Linear and Nonlinear Regression Program Library* located in directory CHAP8 with file named P8-15.POL.

8.16 Determination of Rate Expressions for a Catalytic Reaction

8.16.1 Concepts Demonstrated

Evaluation of catalytic reaction rate expressions with experimental data.

8.16.2 Numerical Methods Utilized

Nonlinear regression of complex expressions and multiple linear regression, including linearization of complex expressions with transformation of data prior to linear regression.

8.16.3 Problem Statement

Cutlip and Peters[2] investigated the catalytic heterogeneous dehydration of tertiary butyl alcohol (A) to isobutylene (B) and water (W). The reaction was studied at atmospheric pressure and various temperatures. The observed reaction rates at a temperature of 533.1 K for various partial pressures of the reactant and the products are given in Table 8–13.

Consideration of a number of different reaction mechanisms with different rate-controlling steps resulted in 24 expressions for reaction rate as function of the partial pressures. After initial screening, the following four rate expressions needed to be evaluated:

$$\text{Model 1} \qquad r = \frac{kK_A P_A}{1 + K_A P_A + K_W P_W} \qquad\qquad (8\text{-}74)$$

$$\text{Model 2} \qquad r = \frac{kK_A P_A}{1 + K_A P_A + K_W P_W + K_B P_B} \qquad\qquad (8\text{-}75)$$

$$\text{Model 3} \qquad r = \frac{kK_A P_A}{(1 + K_A P_A + K_W P_W)^2} \qquad\qquad (8\text{-}76)$$

$$\text{Model 4} \qquad r = \frac{kK_A P_A}{(1 + K_A P_A + K_W P_W + K_B P_B)^2} \qquad\qquad (8\text{-}77)$$

where
r = reaction rate in g-mol/h·g
k = rate constant in g-mol/h·g
P_A, P_W, and P_B = partial pressures of alcohol, water, and butylene, respectively, in atm
K_A, K_W, and K_B = adsorption constants in atm^{-1}

Table 8–13   Reaction Rates as Function of Partial Pressure at 533.1 K

No.	Average Partial Pressures (atm)			Reaction Rate g-mol/h·g
	Alcohol	Water	Butylene	
1	0.7913	0.0177	0.0172	0.005047
2	0.6349	0.0159	0.0156	0.004409
3	0.4788	0.0149	0.0146	0.003857
4	0.3339	0.0157	0.0163	0.003048
5	0.6362	0.0146	0.1736	0.004464
6	0.4864	0.0128	0.3252	0.003671
7	0. 3302	0.0135	0.4819	0.002716
8	0.651	0.1629	0.0104	0.004271
9	0.474	0.3374	0.0122	0.003244
10	0.3167	0.4982	0.01	0.002348
11	0.3506	0.0121	0.314	0.002841
12	0.3973	0.2705	0.0121	0.002903
13	0.3661	0.186	0.0083	0.002995
14	0.3219	0.0117	0.1819	0.002801
15	0.4737	0.0135	0.1821	0.003622
16	0.4857	0.1687	0.0089	0.003523

Cutlip and Peters[2] also considered an empirical power-law rate expression to correlate their data.

Model 5 $\qquad r = k(P_A)^a (P_W)^w (P_B)^b$ $\qquad\qquad$ **(8-78)**

where a, w, and b are unknown exponents.

(a) Select one of the five models for fitting to the rate data of Table 8–13. Use multiple linear regression on a linearized form of the rate expression to determine the model parameters.

(b) Use direct nonlinear regression on the selected model with the same rate data, where the initial estimates of the model parameters are the parameter values from part (a). Note that this procedure provides reasonable initial estimates for the nonlinear regression from the linear regression results.

(c) Repeat parts (a) and (b) for the additional models, and determine the variances of the various nonlinear regressions. Select the "best" model as the one that has the lowest variance.

(d) Compare your results in part (c) with those of Cutlip and Peters.[2]

8.16.4 Solution (Suggestions and Partial Results)

(a) The POLYMATH *Polynomial, Multiple Linear and Nonlinear Regression Program* has options for both multiple linear regression and nonlinear regression. Linear and nonlinear regressions have been discussed in Problems 2.3, 2.5, 2.6, 2.10, 2.11, and 2.12.

In order to apply multiple linear regression to these rate equation, the various expressions must be linearized and the data appropriately transformed. Consider Model 1. This model can be linearized by taking the reciprocal of both sides of the Equation (8-74):

$$\frac{1}{r} = \frac{1}{k} + \frac{1}{kK_A P_A} + \frac{K_W P_W}{kK_A P_A} \tag{8-79}$$

The preceding equation can be expressed in the general form of a linear regression

$$y = a_0 + a_1 x_1 + a_2 x_2 \tag{8-80}$$

by introducing the transformation functions

$$x_1 = 1/P_A \qquad x_2 = P_W/P_A \tag{8-81}$$

where

$$a_0 = \frac{1}{k} \qquad a_1 = \frac{1}{kK_A} \qquad a_2 = \frac{K_W}{kK_A} \tag{8-82}$$

The multiple linear regression for Model 1 yields the values for a_0, a_1, and a_2 from which simple calculations utilizing Equation (8-82) give $k = 0.01102$ g-mol/h · g, $K_A = 1.0575$ atm^{-1} and $K_W = 0.49706$ atm^{-1}.

Model 3 A linearized form of Model 3 can be obtained by first taking the reciprocal of Equation (8-76) followed by taking the square root, yielding

$$\frac{1}{r^{1/2}} = \frac{1}{(kK_A)^{1/2}P_A^{1/2}} + \frac{K_AP_A}{(kK_A)^{1/2}P_A^{1/2}} + \frac{K_WP_W}{(kK_A)^{1/2}P_A^{1/2}} \qquad (8\text{-}83)$$

The appropriate transformation functions for Equation (8-83) are

$$y = 1/(r^{\wedge}0.5)$$
$$x_1 = 1/(P_A^{\wedge}0.5)$$
$$x_2 = P_A^{\wedge}0.5 \qquad (8\text{-}84)$$
$$x_3 = P_W/(P_A^{\wedge}0.5)$$

where the multiple linear function is

$$y = a_1x_1 + a_2x_2 + a_3x_3 \qquad (8\text{-}85)$$

Note that there is no constant a_0 on the right side of Equation (8-85); thus the POLYMATH option "to regress with $a_0 = 0$" should be selected when solving this problem.

(c) Statistical principles for the selection of the best model are discussed in Chapter 2 and by Constantinides.[1]

 The POLYMATH problem data file for this problem, including the transformations for part (a), is found in the *Polynomial, Multiple Linear and Nonlinear Regression Program Library* located in directory CHAP8 with file named P8-16A1.POL.

8.17 PACKED BED REACTOR DESIGN FOR A GAS PHASE CATALYTIC REACTION

8.17.1 Concepts Demonstrated

Calculation of conversion in an isothermal packed bed catalytic reactor for different catalytic rate expressions with pressure drop.

8.17.2 Numerical Methods Utilized

Solution of simultaneous ordinary differential equations.

8.17.3 Problem Statement

The irreversible gas phase catalytic reaction

$$A + B \rightarrow C + D \tag{8-86}$$

is to be carried out in a packed bed reactor with four different catalysts. For each catalyst the rate expression has a different form, as follows:

$$-r'_{A1} = \frac{kC_A C_B}{1 + K_A C_A} \tag{8-87}$$

$$-r'_{A2} = \frac{kC_A C_B}{1 + K_A C_A + K_C C_C} \tag{8-88}$$

$$-r'_{A3} = \frac{kC_A C_B}{(1 + K_A C_A + K_B C_B)^2} \tag{8-89}$$

$$-r'_{A4} = \frac{kC_A C_B}{(1 + K_A C_A + K_B C_B + K_C C_C)^2} \tag{8-90}$$

The molar feed flow rate of A is $F_{A0} = 1.5$ g-mol/min, and the initial concentrations of the reactants are $C_{A0} = C_{B0} = 1.0$ g-mol/dm^3 at the reactor inlet. There is a total of $W = 2$ kg of each catalyst used in the reactor. The reaction rate constant and the various catalytic rate parameters are given by

$$k = 10 \text{ dm}^6/\text{kg} \cdot \text{min}$$

$$K_A = 1 \text{ dm}^3/\text{g-mol}$$

$$K_B = 2 \text{ dm}^3/\text{g-mol}$$

$$K_C = 20 \text{ dm}^3/\text{g-mol}$$

(a)   Calculate and plot the conversion versus catalyst weight for each of the catalytic rate expressions when the reactor operation is at constant pressure. Summarize the expected outlet conversions.

(b)   Repeat part (a) when the pressure ratio within the reactor is given by

$$\frac{dy}{dW} = \frac{-\alpha}{2y} \tag{8-91}$$

where $y = P/P_0$ and α is a constant ($\alpha = 0.4$).

8.17.4   Solution (Partial with Suggestions)

(a) The equations applicable for part (a) are as follows:

Mole balance

$$\frac{dX}{dW} = \frac{-r'_A}{F_{A0}} \tag{8-92}$$

Stoichiometry

$$C_A = C_B = C_{A0}(1 - X)$$
$$C_C = C_D = C_{A0}X \tag{8-93}$$

It is also convenient to solve the four cases at the same time in order to make a single plot comparing the four conversions. This strategy is discussed in Problem 8.2.

The four cases for part (a) can be integrated with the POLYMATH *Simultaneous Differential Equation Solver* utilizing the following equation set:

```
Equations:
d(XA1)/d(W)=(k/FA0)*(CA0*(1-XA1))^2/(1+KA*CA0*(1-XA1))
d(XA2)/d(W)=(k/FA0)*(CA0*(1-XA2))^2/(1+KA*CA0*(1-XA2)
   +KC*CA0*XA2)
d(XA3)/d(W)=(k/FA0)*(CA0*(1-XA3))^2/(1+KA*CA0*(1-XA3)+KB*CA0*(1-
   XA3))^2
d(XA4)/d(W)=(k/FA0)*(CA0*(1-XA4))^2/(1+KA*CA0*(1-XA4)+KB*CA0*(1-
   XA4)+KC*CA0*XA4)^2
k=10
FA0=1.5
CA0=1
KA=1
KC=20
KB=2
Initial Conditions:
W(0)=0
XA1(0)=0
XA2(0)=0
```

```
XA3(0)=0
XA4(0)=0
Final Value:
W(f)=2
```

A plot of the conversions versus the weight of catalyst W is given in Figure 8–7.

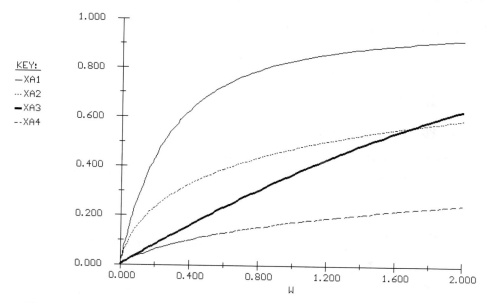

Figure 8–7 Calculated Conversions versus Catalyst Weight for Four Different
Catalysts

 The POLYMATH problem solution file for part (a) is found in the *Simultaneous Differential Equation Solver Library* located in directory CHAP8 with file named P8-17A.POL.

(b) When the pressure varies within the reactor, the stoichiometric equations relating the concentrations to conversion need to be altered to include the change of pressure by

$$C_A = C_B = C_{A0}(1-X)y$$
$$C_C = C_D = C_{A0}Xy \tag{8-94}$$

The differential equation for pressure drop given by Equation (8-91) must be solved simultaneously with the differential equations for conversion.

8.18 CATALYST DECAY IN A PACKED BED REACTOR MODELED BY A SERIES OF CSTRS

8.18.1   Concepts Demonstrated

Determination of the change of reactant and product concentration and catalyst decay with time in a packed bed reactor that is approximated by a series of CSTRs with and without pressure drop.

8.18.2   Numerical Methods Utilized

Solution of simultaneous ordinary differential equations.

8.18.3   Problem Statement

A gas phase catalytic reaction $A \xrightarrow{k} B$ is carried out in a packed bed reactor where the catalyst activity is decaying. The reaction with deactivation follows the rate expression given by

$$-r_A = akC_A \tag{8-95}$$

where a is the catalyst activity that follows either the deactivation kinetics

$$\frac{da}{dt} = -k_{d1}a \tag{8-96}$$

or

$$\frac{da}{dt} = -k_{d2}aC_B \tag{8-97}$$

The packed bed reactor can be approximated by three CSTRs in series, as shown in Figure 8–8.

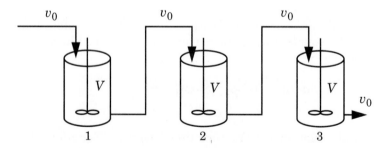

Figure 8–8 Reactor Approximation by a Train of Three CSTRs

The volumetric flow rate to each reactor is v_0. The reactor feed is pure A at concentration C_{A0} to the first reactor. The pressure drop can be neglected. At time zero there is only inert gas in all of the reactors.

The following parameter values apply:

$$k_{d1} = 0.01 \text{ min}^{-1} \qquad k_{d2} = 1.0 \frac{\text{dm}^3}{\text{g-mol} \cdot \text{min}}$$

$$k = 0.9 \frac{\text{dm}^3}{\text{dm}^3 \text{(of catalyst)} \text{min}}$$

$$C_{A0} = 0.01 \frac{\text{g-mol}}{\text{dm}^3} \qquad V = 10 \text{ dm}^3 \qquad v_0 = 5 \frac{\text{dm}^3}{\text{min}}$$

(a)   Plot the concentration of A in each of the three reactors as a function of time to 60 minutes using the activity function given by Equation (8-96). Create a separate plot for the activities in all three reactors.

(b)   Repeat part (a) for the activity function as given in Equation (8-97).

(c)   Compare the outlet concentration of A for parts (a) and (b) at 60 minutes deactivation to that from a plug-flow packed bed reactor with no deactivation (total volume of 30 dm^3) and the three CSTR reactors in series model with no deactivation.

8.18.4   Partial Solution

The respective material balances on components A and B yield the following differential equations for the first CSTR, where the subscript 1 indicates the concentrations in the first reactor:

$$\frac{dC_{A1}}{dt} = \frac{(C_{A0} - C_{A1}) v_0}{V} + r_{A1}$$

$$\frac{dC_{B1}}{dt} = \frac{-C_{B1} v_0}{V} - r_{A1}$$

(8-98)

For the second and third CSTR, where $i = 2$ and 3, the balances yield

$$\frac{dC_{Ai}}{dt} = \frac{(C_{A(i-1)} - C_{A(i)}) v_0}{V} + r_{Ai}$$

$$\frac{dC_{Bi}}{dt} = \frac{(C_{B(i-1)} - C_{B(i)}) v_0}{V} + r_{Ai}$$

(8-99)

These equations, together with Equations (8-95) and (8-96) or (8-97), provide the equations that need to be solved simultaneously in this problem.

(a) For this part, only the material balances involving A are needed in addition to Equations (8-95) and (8-96). This set of equations can be entered into the POLYMATH *Simultaneous Differential Equation Solver* as follows:

Equations:
d(a1)/d(t)=-.01*a1
d(a2)/d(t)=-.01*a2
d(a3)/d(t)=-.01*a3
d(CA2)/d(t)=(CA1-CA2)*5/10-a2*.9*CA2
d(CA3)/d(t)=(CA2-CA3)*5/10-a3*.9*CA3
d(CA1)/d(t)=(.01-CA1)*5/10-a1*.9*CA1
Initial Conditions:
t(0)=0
a1(0)=1
a2(0)=1
a3(0)=1
CA2(0)=0
CA3(0)=0
CA1(0)=0
Final Value:
t(f)=60

The resulting graph of the concentration of A in each of the three reactors is given in Figure 8–9.

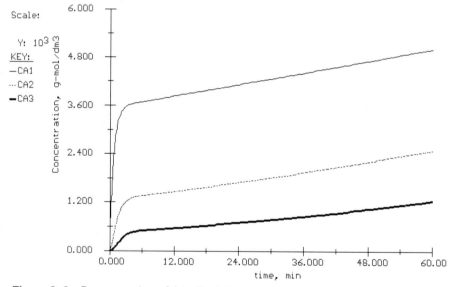

Figure 8–9 Concentration of A in Each Reactor

  The POLYMATH problem solution file for part (a) is found in the *Simultaneous Differential Equation Solver Library* located in directory CHAP8 with file named P8-18A.POL

   (b) The material balances involving both components A and B are needed along with Equations (8-95) and (8-97). Note that the activities in each reactor will be different because of the dependency of the activity relationship of Equation (8-97) on the concentration of B.

8.19   DESIGN FOR CATALYST DEACTIVATION IN A STRAIGHT-THROUGH REACTOR

8.19.1   Concepts Demonstrated

The effect of various types of catalyst deactivation on a catalytic gas phase reaction in a straight-through transport reactor where the catalyst bed is moving through the reactor.

8.19.2   Numerical Methods Utilized

Solution of simultaneous ordinary differential equations.

8.19.3   Problem Statement[*]

The gas phase cracking of a Salina light gas oil can be represented by a lumped parameter kinetic model that incorporates the general reaction

$$\text{Gas Oil (g)} \rightarrow \text{Products (g)} + \text{coke (s)}$$

which can be expressed as the gas phase reaction

$$A \rightarrow B \tag{8-100}$$

A typical application is to carry out this reaction in a straight-through transport reactor. This reactor, which contains a catalyst that decays as a result of coking, is shown in Figure 8–10. The catalyst particles are assumed to move with the mean gas velocity given by $u = 8.0$ m/s in this case. The reaction is to be carried out at 750 °F under constant temperature and pressure. The entering concentration of A is 0.2 kg-mol/m^3. The reactor length is 6 m. For this problem, the volume change with reaction, pressure drop, and temperature variation may be neglected.

The catalytic activity, denoted by a, is usually defined as the ratio of the reaction rate for a catalyst subjected to deactivation for time t to the reaction rate for fresh catalyst. For a moving bed of catalyst traveling through the reactor with a plug-flow velocity u, the time that the catalyst has been in the reactor when it reaches reactor height z is given by

$$t = \frac{z}{u} \tag{8-101}$$

Three types of catalyst deactivation are to be examined in this problem. The first is catalyst coking, which has the form

$$a_1 = \frac{1}{1 + A't^{1/2}} \tag{8-102}$$

[*] This problem adapted from Fogler[4] with permission.

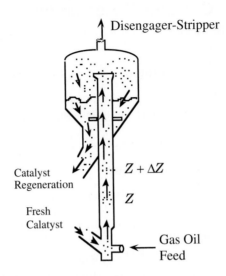

Figure 8–10 Straight-Through Transport Reactor (from Fogler,[4] with permission)

where A' is the coking parameter.

The second type of rate law is deactivation by sintering:

$$\frac{da_2}{dt} = -k_{d2}a_2^2 \tag{8-103}$$

The third type is deactivation by poisoning:

$$\frac{da_3}{dt} = -k_{d3}a_3C_B \tag{8-104}$$

The catalytic reaction rate per unit volume of catalyst bed for this problem is simply the particular activity multiplied by the catalytic reaction rate expression

$$-r_A = \frac{akC_A}{1 + K_AC_A} \tag{8-105}$$

where $k = 30$ s^{-1} and $K_A = 5.0$ m^3/kg-mol.

(a) Plot the conversion of A and the catalyst activity versus the reactor length z for the case of no catalyst deactivation.

(b) Repeat part (a) for the three types of catalyst deactivation, where $A' = 12$ s$^{-1/2}$, $k_{d2} = 17.5$, s^{-1} and $k_{d3} = 140$ dm^3/mol·s.

8.19.4 Solution

The following equations are needed for the base case where A_C is the cross-sectional area of the reactor:

Mole balance (expressed in terms of conversion x_A)

$$v_0 C_{A0} \frac{dx_A}{dz} = -r_A A_C \quad \text{I.C. } x_A = 0 \text{ at } z = 0 \tag{8-106}$$

Rate law (modified by the catalytic activity)

$$-r_A = \frac{akC_A}{1 + K_A C_A} \tag{8-107}$$

Stoichiometry (for gas phase reaction with $\varepsilon = 0$)

$$v = v_0$$

$$u = \frac{v_0}{A_C} \tag{8-108}$$

$$C_A = C_{A0}(1 - x_A)$$

$$C_B = C_{A0} x_A$$

(a) No Deactivation The base case with no deactivation is described by the preceding equations, where the activity is constant at unity. Equation (8-106) can be rearranged with u from Equation Set (8-108) to yield

$$\frac{dx_A}{dz} = \frac{akC_A}{(1 + K_A C_A)u} \quad \text{I.C. } x_A = 0 \text{ at } z = 0 \tag{8-109}$$

The POLYMATH *Simultaneous Differential Equation Solver* can be used to solve the preceding differential equation with the equation set given by

```
Equations:
d(xA)/d(z)=a*30*(1-xA)/((1+5*.2*(1-xA))*8)
a=1
Initial Conditions:
z(0)=0
xA(0)=0
Final Value:
z(f)=6
```

 The POLYMATH problem solution file for part (a) is found in the *Simultaneous Differential Equation Solver Library* located in directory CHAP8 with file named P8-19A.POL.

(b) Deactivation

Coking The introduction of Equation (8-101) into Equation (8-102) yields the activity as a function of the reactor length z:

$$a_1 = \frac{1}{1 + \left(\frac{A'}{u^{1/2}}\right) z^{1/2}} \tag{8-110}$$

where the subscript 1 indicates deactivation by coking.

Sintering The derivative of Equation (8-101)

$$dt = \frac{dz}{u} \tag{8-111}$$

can be used to rewrite Equation (8-103) can be written in terms of the reactor length z as

$$\frac{da_2}{dz} = \frac{-k_{d2}a_2^2}{u} \tag{8-112}$$

where the subscript 2 is used for the deactivation by sintering.

Poisoning Equation (8-104) can also be expressed in terms of z by using

$$\frac{da_3}{dz} = \frac{-k_{d3}a_3 C_B}{u} \tag{8-113}$$

where the subscript 3 is used for the deactivation by poisoning.

POLYMATH Solution It is convenient to enter the equation sets for all four cases of catalytic activity that have been considered in this problem simultaneously by using the subscript notation developed previously.

```
Equations:
d(xA1)/d(z)=a1*30*(1-xA1)/((1+5*.2*(1-xA1))*8)
d(xA)/d(z)=a*30*(1-xA)/((1+5*.2*(1-xA))*8)
d(xA2)/d(z)=a2*30*(1-xA2)/((1+5*.2*(1-xA2))*8)
d(a2)/d(z)=-17.5*a2*a2/8
d(xA3)/d(z)=a3*30*(1-xA3)/((1+5*.2*(1-xA3))*8)
d(a3)/d(z)=-140.*a3*(.2*xA3)/8
a=1
a1=1/(1+(12/(8^.5))*z^(.5))
Initial Conditions:
z(0)=0
xA1(0)=0
xA(0)=0
xA2(0)=0
a2(0)=1
xA3(0)=0
a3(0)=1
Final Value:
z(f)=6
```

The graph of conversion x_A versus reactor height is given in Figure 8–11, and the profiles for the activities are given in Figure 8–12.

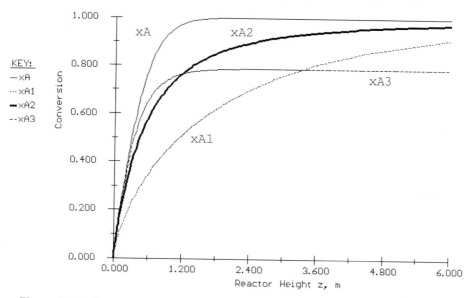

Figure 8–11 Conversion versus Reactor Height

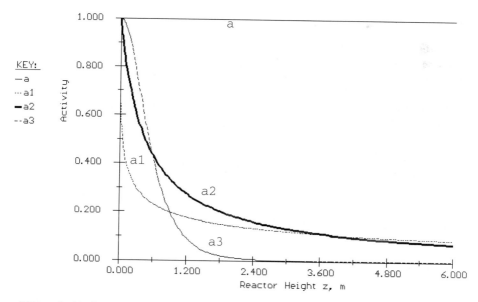

Figure 8–12 Activity Factors versus Reactor Height

 The POLYMATH problem solution file for part (b) is found in the *Simultaneous Differential Equation Solver Library* located in directory CHAP8 with file named P8-19B.POL.

8.20 ENZYMATIC REACTIONS IN A BATCH REACTOR

8.20.1   Concepts Demonstrated

Enzymatic reaction kinetics, batch reactor design equations for enzymatic reactions, pseudo-steady-state hypothesis and the validity of its application.

8.20.2   Numerical Methods Utilized

Solution of simultaneous ordinary differential equations.

8.20.3   Problem Statement

One of the simplest models of enzyme catalysis is the second-order reversible binding of a substrate S (reactant) to an enzyme catalyst E to give an enzyme-substrate complex $E \cdot S$, which then decomposes by a first-order reaction to give product P and regenerates the enzyme catalyst E.   This model of enzyme catalysis is referred to as Michaelis-Menten kinetics when the pseudo-steady-state hypothesis is utilized. (See Fogler[4] for more details.) The elementary steps in the reaction sequence are given by

$$S + E \underset{k_2}{\overset{k_1}{\rightleftarrows}} E \cdot S \overset{k_3}{\rightarrow} E + P \tag{8-114}$$

Material balances can be made on a constant volume batch reactor in the preceding reactions, where the initial enzyme concentration and substrate concentration are C_{E_0} and C_{S_0}, respectively. No product concentration C_P is initially present. Thus

$$\frac{dC_S}{dt} = -k_1 C_S C_E + k_2 C_{E \cdot S} \quad \text{I.C.} \quad C_S = C_{S_0} \text{ at } t = 0 \tag{8-115}$$

$$\frac{dC_{E \cdot S}}{dt} = k_1 C_S C_E - k_2 C_{E \cdot S} - k_3 C_{E \cdot S} \quad \text{I.C.} \quad C_{E \cdot S} = 0 \text{ at } t = 0 \tag{8-116}$$

$$\frac{dC_P}{dt} = k_3 C_{E \cdot S} \quad \text{I.C.} \quad C_P = 0 \text{ at } t = 0 \tag{8-117}$$

$$C_{E_0} = C_E + C_{E \cdot S} \tag{8-118}$$

The preceding set of equations describing the interactions in a batch reactor cannot be solved analytically. The pseudo-steady-state hypothesis (also known as the quasi-steady-state assumption) is normally used by assuming that the enzyme-substrate complex is an active intermediate present at very low concentration and the time-derivative of this concentration is zero:

$$\frac{dC_{E\cdot S}}{dt} = 0 \tag{8-119}$$

The preceding equation is a normal assumption of the pseudo-steady-state hypothesis that leads to the Michaelis-Menton rate expression for a batch reactor.

(a) Show that the pseudo-steady-state hypothesis results in the following batch reactor design expression:

$$\frac{dC_S}{dt} = \frac{-k_3 C_{E_0} C_S}{K + C_S} = -\frac{dC_P}{dt} \quad \text{where} \quad K = \frac{k_2 + k_3}{k_1} \tag{8-120}$$

(b) Analytically integrate the preceding equation for a batch reactor to show that

$$k_3 C_{E_0} t = C_{S_0} - C_S + K \ln\left(\frac{C_{S_0}}{C_S}\right) \tag{8-121}$$

(c) Verify the validity of the pseudo steady-state hypothesis by using numerical techniques to solve the system of differential and algebraic equations [Equations (8-115) to (8-118)] that rigorously describe the steps in the enzymatic reaction. The following parameter values apply: C_{S_0} = 1.0 g-mol/dm^3, E_{S_0} = 10^{-3} g-mol/dm^3, t_{final} = 48.0 h, k_1 = 2.0×10^3 dm^3/g-mol·h, k_2 = 3.0×10^5 h^{-1}, and k_3 = 1.0×10^4 h^{-1}. Please discuss your conclusions.

8.20.4   Solution (Suggestions)

(c) The numerical comparison of the pseudo-steady-state treatment with the more rigorous solution can be easily accomplished by carrying out both numerical solutions of parts (a) and (c) in the same numerical simulation utilizing the POLYMATH *Simultaneous Differential Equation Solver*.

8.21 ISOTHERMAL BATCH REACTOR DESIGN FOR MULTIPLE REACTIONS

8.21.1   Concepts Demonstrated

Calculation of concentration profiles in a constant volume batch reactor at constant temperature in which a number of simultaneous elementary reactions are occurring.

8.21.2   Numerical Methods Utilized

Numerical integration of a system of simultaneous ordinary differential equations subject to known initial conditions.

8.21.3   Problem Statement[*]

A seven-step mechanism can be used to describe the atmospheric reactions involving formaldehyde and nitrogen oxides at 1 atm and $T = 298\ °C$, as presented by Seinfeld[11]. These photochemical reactions lead to the formation of ozone.

Step 1	$NO_2 + hv \rightarrow NO + O$	$k_1 = 0.533\ \text{min}^{-1}$
Step 2	$O + O_2 + M \rightarrow O_3 + M$	$k_2 = 2.21 \times 10^{-5}\ \text{ppm}^{-2}\ \text{min}^{-1}$
Step 3	$O_3 + NO \rightarrow NO_2 + O_2$	$k_3 = 26.7\ \text{ppm}^{-1}\ \text{min}^{-1}$
Step 4a	$HCHO + hv \rightarrow 2HO_2\cdot + CO$	$k_{4a} = 1.6 \times 10^{-3}\ \text{min}^{-1}$
Step 4b	$HCHO + hv \rightarrow H_2 + CO$	$k_{4b} = 2.11 \times 10^{-3}\ \text{min}^{-1}$
Step 5	$HCHO + OH\cdot \rightarrow HO_2\cdot + CO + H_2O$	$k_5 = 1.62 \times 10^4\ \text{ppm}^{-1}\ \text{min}^{-1}$
Step 6	$HO_2\cdot + NO \rightarrow NO_2 + OH\cdot$	$k_6 = 1.22 \times 10^4\ \text{ppm}^{-1}\ \text{min}^{-1}$
Step 7	$OH\cdot + NO_2 \rightarrow HNO_3$	$k_7 = 1.62 \times 10^4\ \text{ppm}^{-1}\ \text{min}^{-1}$

The photolysis reactions of Steps 1, 4a, and 4b are written for the case where the incoming solar radiation is assumed to be constant at the mid-day value of k_{iM}, thus leading to the given rate constants. A more detailed model would be to have the rate constants for these steps vary with the time of day according to

$$k_i = k_{iM}\sin\left[\frac{2\pi(t-6)}{24}\right] \quad \text{for } \sin\left[\frac{2\pi(t-6)}{24}\right] > 0 \tag{8-122}$$

$$k_i = 0 \quad \text{for all other values of } t$$

[*] This problem was adapted from an exercise developed by Joseph J. Helble.

where t is the time in hours on a 24 hour/day basis and the solar radiation begins at 6 a.m. and ends at 6 p.m.

All of the reactions in the mechanism are considered to be elementary as written. For example, the rate of Step 2 is given by

$$r_2 = k_2 C_O C_{O_2} C_M \tag{8-123}$$

where the concentrations are in ppm. C_M is the concentration of third bodies, which is considered to be at a constant value of 10^6 ppm, and C_{O_2}, the concentration of oxygen, is constant at 0.21×10^6 ppm.

The simplest application of chemical reaction engineering to urban smog simulation involves considering the atmosphere below an inversion layer to be a batch reactor with no input or output terms. The resulting material balance for each of the species in the mechanism can be made and expressed as ordinary differential equations. Thus for ozone

$$\frac{dC_{O_3}}{dt} = k_2 C_O C_{O_2} C_M - k_3 C_{O_3} C_{NO} \tag{8-124}$$

Similar material balances can be made for the other reacting components, including the free radicals. Typical initial conditions for all free radicals are that the concentrations are essentially zero.

Consider a simplified model for photochemical smog that is represented by the seven-step mechanism for formaldehyde.

(a) Verify the concentration profiles for O_3, NO_2, NO and HCHO, which are presented in Figure 8–13 for the given initial conditions and for the constant values of photolysis reaction rate constants of Steps 1, 4a, and 4b. Note that a similar figure is given by Seinfeld.[11]

(b) Repeat part (a) for a single 24-hour day starting at midnight with the same initial conditions but with the three photolysis reactions varying according to the general relationship of Equation (8-122).

(c) Stricter NO_x controls have been suggested as a means of reducing ozone. Comment on the wisdom of this approach after making several additional solutions with different initial concentrations of NO.

8.21.4   Solution (Suggestions)

(a) The various material balances on the batch reactor yield the following ordinary differential equations for the reacting species:

$$\frac{dC_{NO_2}}{dt} = -k_1 C_{NO_2} + k_3 C_{O_3} C_{NO} + k_6 C_{HO_2} \cdot C_{NO} - k_7 C_{OH} \cdot C_{NO_2} \qquad \text{(8-125)}$$

$$\frac{dC_{O_3}}{dt} = k_2 C_O C_{O_2} C_M - k_3 C_{O_3} C_{NO} \qquad \text{(8-126)}$$

$$\frac{dC_{NO}}{dt} = k_1 C_{NO_2} - k_3 C_{O_3} C_{NO} - k_6 C_{HO_2} \cdot C_{NO} \qquad \text{(8-127)}$$

$$\frac{dC_{HO_2 \cdot}}{dt} = 2k_{4a} C_{HCHO} + k_5 C_{HCHO} C_{OH} \cdot - k_6 C_{HO_2} \cdot C_{NO} \qquad \text{(8-128)}$$

$$\frac{dC_{OH \cdot}}{dt} = -k_5 C_{HCHO} C_{OH} \cdot + k_6 C_{HO_2} C_{NO} - k_7 C_{OH} \cdot C_{NO_2} \qquad \text{(8-129)}$$

$$\frac{dC_O}{dt} = k_1 C_{NO_2} - k_2 C_O C_{O_2} C_M \qquad \text{(8-130)}$$

$$\frac{dC_{HCHO}}{dt} = -k_{4a} C_{HCHO} - k_{4b} C_{HCHO} - k_5 C_{HCHO} C_{OH} \cdot \qquad \text{(8-131)}$$

The POLYMATH *Simultaneous Differential Equation Solver* can be used to solve the preceding differential equations. Results are presented in Figure 8–13 for the given initial conditions. All initial conditions that are not specified with particular values are zero. The equation set may be stiff so the Problem Option for the "stiff" integration algorithm should be used when necessary.

(b) The variation of light intensity with time of day can be handled with the "if ... then ... else ... " capability available within POLYMATH. Implementation of the variation of k_1 with the time of day can be accomplished by

```
k1=if(sign(sin(2*3.1416/24*(t-6)))<=0)then(0)else
       ((0.533*60)*sin(2*3.1416/24*(t-6)))
```

where the intrinsic sign function returns either +1, 0, or –1, depending upon the value of the function argument being either positive, zero, or negative, respectively. The net result is that the rate constant only has values between 6 a.m. to 6 p.m. and is zero at other times. This same type of statement can be used for the other photolysis reactions that are sunlight dependent.

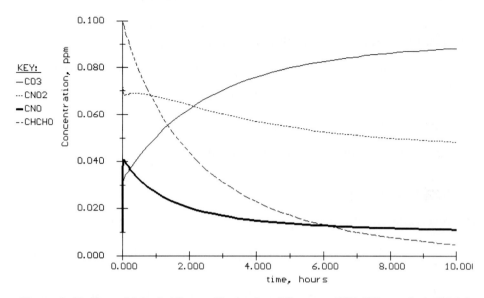

Figure 8–13 Formaldehyde Photooxidation in a Mixture of NO, NO$_2$, and air. Initial Conditions: C_{NO_2} = 0.1 ppm, C_{NO} = 0.01 ppm, C_{HCHO} = 0.1 ppm.

8.22 MATERIAL AND ENERGY BALANCES ON A BATCH REACTOR

8.22.1   Concepts Demonstrated

Calculation of conversion and temperature profile in a heated batch reactor with an endothermic reaction.

8.22.2   Numerical Methods Utilized

Solution of simultaneous ordinary differential equations.

8.22.3   Problem Statement

An irreversible, endothermic, elementary, liquid phase reaction

$$A + B \xrightarrow{k} C \tag{8-132}$$

is to be carried out in an agitated batch reactor that is heated by a steam jacket on the reactor exterior. Initially the concentrations are $C_{A0} = 2.5$ g-mol/dm^3, $C_{B0} = 5$ g-mol/dm^3 and $C_{C0} = 0$. The reactor volume is $V = 1200$ dm^3, and the steam in the surrounding jacket is kept at $T_j = 150$ °C. Additional data are given in Table 8–14.

(a)   Plot the conversion of A, x_A, and the reactor temperature during the first 60 min of reactor operation for an initial reactor temperature of $T_0 = 30$ °C.

(b)   Calculate the minimal initial heating time to assure that $x_A = 0.99$ after 60 min of reactor operation for an initial reactor temperature of $T_0 = 30$ °C.

Table 8–14   Additional Data for Heated Batch Reactor

Activation Energy	$E = 83.6$ kJ/g-mol
Rate Coefficient	$k = 0.001$ dm^3/g-mol·min at 27 °C
Heat of Reaction at 273.13 K	$\Delta H_R = 27.85$ kJ/g-mol
Mean Heat Capacities	$\tilde{C}_{PA} = 14$ J/g-mol·K
	$\tilde{C}_{PB} = 28$ J/g-mol·K
	$\tilde{C}_{PC} = 42$ J/g-mol·K
Heat Transfer Area	$A = 5$ m^2
Heat Transfer Coefficient	$U = 3.76$ kJ/min·m^2·K

8.22.4   Solution (Suggestions)

The equations representing this system are as follows:

Mole balance (Design equation expressed in terms of conversion x_A)

$$\frac{dx_A}{dt} = \frac{-r_A}{C_{A0}} \tag{8-133}$$

Energy balance[4]

$$\frac{dT}{dt} = \frac{UA(T_j - T) + r_A V \Delta H_R}{N_{A0}(C_{PA} + \theta_B C_{PB})} \tag{8-134}$$

Rate law

$$r_A = -k C_{A0}^2 (1 - x_A)(\theta_B - x_A) \tag{8-135}$$

$$k = 0.001 \ \exp\left[\frac{-E}{R}\left(\frac{1}{T} - \frac{1}{300}\right)\right] \tag{8-136}$$

where N_{A0} is the initial molar amount of A that is calculated from $N_{A0} = V C_{A0}$, x_A is the conversion of A, θ_B is the initial molar ratio given by C_{B0}/C_{A0}, and T is the absolute temperature.

These equations, together with the numerical data provided, can be entered into the POLYMATH *Simultaneous Differential Equation Solver*, and a numerical solution can be generated for the time interval indicated. Please be sure to use consistent units in the working equations, particularly for the various temperatures.

8.23 OPERATION OF A COOLED EXOTHERMIC CSTR

8.23.1   Concepts Demonstrated

Material and energy balances on a CSTR with an exothermic reaction and cooling jacket.

8.23.2   Numerical Methods Utilized

Solution of a system of simultaneous nonlinear algebraic equations, and conversion of the system of equations into one equation to examine multiple steady-state solutions.

8.23.3   Problem Statement[*]

An irreversible exothermic reaction $A \xrightarrow{k} B$ is carried out in a perfectly mixed CSTR, as shown in Figure 8–14. The reaction is first order in reactant A and has

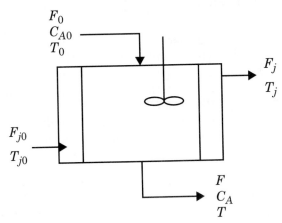

Figure 8–14 Cooled Exothermic CSTR

a heat of reaction given by λ, which is based on reactant A. Negligible heat losses and constant densities can be assumed. A well-mixed cooling jacket surrounds the reactor to remove the heat of reaction. Cooling water is added to the jacket at a rate of F_j and at an inlet temperature of T_{j0}. The volume V of the contents of the reactor and the volume V_j of water in the jacket are both constant. The reaction rate constant changes as function of the temperature according to the equation

$$k = \alpha \exp(-E/RT) \tag{8-137}$$

The feed flow rate F_0 and the cooling water flow rate F_j are constant. The jacket water is assumed to be completely mixed. Heat transferred from the reac-

[*] This problem is adapted from Luyben[8] with permission.

tor to the jacket can be calculated from

$$Q = UA(T - T_j) \tag{8-138}$$

where Q is the heat transfer rate, U is the overall heat transfer coefficient, and A is the heat transfer area. Detailed data for the process from Luyben[8] are shown in the Table 8–15.

Table 8–15 CSTR Parameter Values[a]

F_0	40 ft^3/h	U	150 btu/h $\cdot$ ft^2 $\cdot$ °R
F	40 ft^3/h	A	250 ft^2
C_{A0}	0.55 lb-mol/ft^3	T_{j0}	530 °R
V	48 ft^3	T_0	530 °R
F_j	49.9 ft^3/h	λ	−30,000 btu/lb-mol
C_P	0.75 btu/lb$_m$ $\cdot$ °R	C_j	1.0 btu/lb$_m$ $\cdot$ °R
α	7.08×10^{10} h^{-1}	E	30,000 btu/lb-mol
ρ	50 lb$_m$/ft^3	ρ_j	62.3 lb$_m$/ft^3
R	1.9872 btu/lb-mol $\cdot$ °R	V_j	12 ft^3

[a]Data are from Luyben[8] with permission.

(a) Formulate the material and energy balances that apply to the CSTR and the cooling jacket.

(b) Calculate the steady-state values of C_A, T_j, and T for the operating conditions of Table 8–15.

(c) Identify all possible steady-state operating conditions, as this system may exhibit multiple steady states.

(d) Solve the unsteady-state material and energy balances to identify if any of the possible multiple steady states are unstable.

8.23.4 Solution (Partial)

(a) There are three balance equations that can be written for the reactor and the cooling jacket. These include the material balance on the reactor, the energy balance on the reactor, and the energy balance on the cooling jacket.

Mole balance on CSTR for reactant A

$$F_0 C_{A0} - F C_A - V k C_A = 0 \tag{8-139}$$

Energy balance on the reactor

$$\rho C_p (F_0 T_0 - FT) - \lambda V k C_A - UA(T - T_j) = 0 \tag{8-140}$$

Energy balance on the cooling jacket

$$\rho_j C_j F_j (T_{j0} - T_j) + UA(T - T_j) = 0 \tag{8-141}$$

(b) Introducing the numerical parameter values into Equations (8-139), (8-140), (8-141), and (8-137) and entering the system of equations into the POLYMATH *Nonlinear Algebraic Equation Solver* gives

```
Equations:
f(CA)=40.*(0.55-CA)-48.*k*CA
f(T)=50.*0.75*40.*(530.-T)+30000.*48.*k*CA-150.*250.*(T-Tj)
f(Tj)=62.3*1.0*49.9*(530.-Tj)+150.*250.*(T-Tj)
k=7.08e10*exp(-30000./(1.9872*T))
Initial Conditions:
CA(0)=0.55
T(0)=530
Tj(0)=530
```

A reasonable initial assumption is that there is no reaction; therefore, $C_{Ao} = 0.55$, $T_0 = 530$, and $T_{j0} = 530$. The solution obtained with these initial estimates is summarized in Table 8–16.

Table 8–16   Steady-State Operating Conditions for CSTR

	Solution	
Variable	**Value**	**f( )**
C_A	0.52139	−4.441e−16
T	537.855	1.757e−9
T_j	537.253	−1.841e−9
k	0.0457263	

The POLYMATH data file for part (b) is found in the *Simultaneous Algebraic Equation Solver Library* located in directory CHAP8 with file named P8-23B.POL.

(c) Several different steady states may be possible with exothermic reactions in a CSTR. One possible method to determine these different steady states is to solve the system of nonlinear equations with different initial estimates of the final solution. While this approach is not very sophisticated, it can be of benefit in very complex systems of equations.

Another approach is to convert the system of equations into a single implicit and several explicit or auxiliary equations. (Incidentally, this is a good way to show that a particular system does not have a solution at all.) In this particular case, the material balance of Equation (8-139) can be solved for C_A.

$$C_A = \frac{F_0 C_{A0}}{(F + Vk)} \tag{8-142}$$

Also, the energy balance of Equation (8-141) on the cooling jacket can be solved for T_j.

$$T_j = \frac{\rho_j C_j F_j T_{j0} + UAT}{(\rho_j C_j F_j + UA)} \tag{8-143}$$

Thus the problem has been converted to a single nonlinear equation given by Equation (8-140) and three explicit equations given by Equations (8-137), (8-142), and (8-143). The equation set for POLYMATH solution is

Equations:
```
f(T)=50.*0.75*40.*(530.-T)+30000.*48.*k*CA-150.*250.*(T-Tj)
k=7.08e10*exp(-30000./(1.9872*T))
Tj=(62.3*1.0*49.9*530.+150.*250.*T)/(62.3*1.0*49.9+150.*250.)
CA=40.*0.55/(40.+48.*k)
```
Initial Conditions:
```
T(min)=500, T(max)=700
```

When $f(T)$ is plotted versus T in the range $500 \le T \le 700$, three solutions can be clearly identified, as shown in Figure 8–15. The first one is at low temperature as this is the solution that was initially identified. The second is at an intermediate temperature and the third is at a high temperature. The three resulting solutions are summarized in Table 8–17.

Table 8–17 Multiple Steady-State Solutions for CSTR

	Solution No.		
	1	2	3
T	537.86	590.35	671.28
C_A	0.5214	0.3302	0.03542
T_j	537.25	585.73	660.46

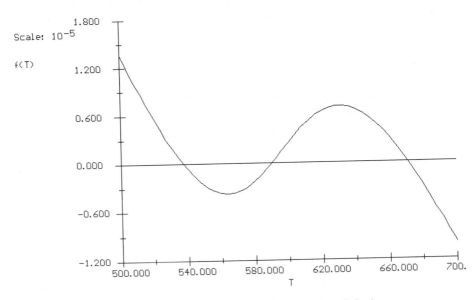

Figure 8–15 Graphical Indication of Multiple Steady-State Solutions

 The POLYMATH data file for part (c) is found in the *Simultaneous Algebraic Equation Solver Library* located in directory CHAP8 with file named P8-23C.POL.

8.24 EXOTHERMIC REVERSIBLE GAS PHASE REACTION IN A PACKED BED REACTOR

8.24.1 Concepts Demonstrated

Design of a gas-phase catalytic reactor with pressure drop for an exothermic reversible gas-phase reaction.

8.24.2 Numerical Methods Utilized

Integration of simultaneous ordinary differential equations with known initial conditions.

8.24.3 Problem Statement

The elementary gas phase reaction $2A \rightleftharpoons C$ is carried out in a packed bed reactor. There is a heat exchanger surrounding the reactor, and there is a pressure drop along the length of the reactor as shown in Figure 8–16.

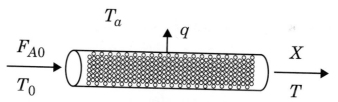

Figure 8–16 Packed Bed Catalytic Reactor

    The various parameters values for this reactor design problem are summarized in Table 8–18.

Table 8–18 Parameter Values for Packed Bed Reactor

C_{PA} = 40.0 J/g-mol·K	R = 8.314 J/g-mol·K
C_{PC} = 80.0 J/g-mol·K	F_{A0} = 5.0 g-mol/min
ΔH_R = - 40,000 J/g-mol	U_a = 0.8 J/kg·min·K
E_A = 41,800 J/g-mol·K	T_a = 500 K
k = 0.5 dm^6/kg·min·mol @ 450 K	α = 0.015 kg^{-1}
K_C = 25,000 dm^3/g-mol @ 450 K	P_0 = 10 atm
C_{A0} = 0.271 g-mol/dm^3	y_{A0} = 1.0 (Pure A feed)
T_0 = 450 K	

(a)   Plot the conversion (X), reduced pressure (y), and temperature ($T \times 10^{-3}$) along the reactor from $W = 0$ kg up to $W = 20$ kg.

(b)   Around 16 kg of catalyst you will observe a "knee" in the conversion profile. Explain why this knee occurs and what parameters affect the knee.

(c)   Plot the concentration profiles for reactant A and product C from $W = 0$ kg up to $W = 20$ kg.

Additional Information

The notation used here and the following equations and relationships for this particular problem are adapted from the textbook by Fogler.[4] The problem is to be worked assuming plug flow with no radial gradients of concentrations and temperature at any location within the catalyst bed. The reactor design will use the conversion of A designated by X and the temperature T which are both functions of location within the catalyst bed specified by the catalyst weight W.

The general reactor design expression for a catalytic reaction in terms of conversion is a mole balance on reactant A given by

$$F_{A0}\frac{dX}{dW} = -r'_A \tag{8-144}$$

The simple catalytic reaction rate expression for this reversible reaction is

$$-r'_A = k\left[C_A^2 - \frac{C_C}{K_C}\right] \tag{8-145}$$

where the rate constant is based on reactant A and follows the Arrhenius expression

$$k = (k@T=450\ \text{K})\exp\frac{E_A}{R}\left[\frac{1}{450} - \frac{1}{T}\right] \tag{8-146}$$

and the equilibrium constant variation with temperature can be determined from van't Hoff's equation with $\Delta\tilde{C}_P = 0$:

$$K_C = (K_C@T=450\ \text{K})\exp\frac{\Delta H_R}{R}\left[\frac{1}{450} - \frac{1}{T}\right] \tag{8-147}$$

The stoichiometry for $2A \rightleftarrows C$ and the stoichiometric table for a gas allow the concentrations to be expressed as a function of conversion and temperature while allowing for volumetric changes due to decrease in moles during the reaction. Therefore,

$$C_A = C_{A0}\left(\frac{1-X}{1+\varepsilon X}\right)\frac{P}{P_0}\frac{T_0}{T} = C_{A0}\left(\frac{1-X}{1-0.5X}\right)y\frac{T_0}{T} \tag{8-148}$$

and

$$y = \frac{P}{P_0}$$

$$C_C = \left(\frac{0.5C_{A0}X}{1 - 0.5X}\right)y\frac{T_0}{T}$$

(8-149)

The pressure drop can be expressed as a differential equation (see Fogler[4] for details):

$$\frac{d\left(\frac{P}{P_0}\right)}{dW} = \frac{-\alpha(1 + \varepsilon X)}{2}\frac{P_0}{P}\frac{T}{T_0}$$

(8-150)

or

$$\frac{dy}{dW} = \frac{-\alpha(1 - 0.5X)}{2y}\frac{T}{T_0}$$

(8-151)

The general energy balance may be written as

$$\frac{dT}{dW} = \frac{U_a(T_a - T) + r_A'\,(\Delta H_R)}{F_{A0}(\sum \theta_i C_{Pi} + X\Delta \tilde{C}_P)}$$

(8-152)

which for only reactant A in the reactor feed simplifies to

$$\frac{dT}{dW} = \frac{U_a(T_a - T) + r_A'\,(\Delta H_R)}{F_{A0}(C_{PA})}$$

(8-153)

This problem is solved with Excel, Maple, MathCAD, MATLAB, Mathematica, and POLYMATH as problem 9 in the Set of Ten Problems discussed in Appendix F.

8.25 TEMPERATURE EFFECTS WITH EXOTHERMIC REACTIONS

8.25.1   Concepts Demonstrated

Reactor design considerations for a reversible exothermic reaction carried out in various ideal reactors.

8.25.2   Numerical Methods Utilized

Solutions of simultaneous nonlinear algebraic equations and simultaneous ordinary differential equations.

8.25.3   Problem Statement

The reversible exothermic and elementary liquid phase reaction

$$A \leftrightarrow R \tag{8-154}$$

is studied in a variety of reactors some of which are adiabatic. The rate expression and rate constants are as follows:

$$r_A = -k_1 C_A + k_2 C_R$$

$$k_1 = 5.2 \times 10^7 e^{-\frac{12000}{RT}} \text{ min}^{-1} \quad \text{and} \quad k_2 = 2.8 \times 10^{18} e^{-\frac{30000}{RT}} \text{ min}^{-1} \tag{8-155}$$

where $R = 1.9872$ cal/g-mol $\cdot$ °K and T is in °K.

    For the base case, the reactor V is 12 dm^3, and the volumetric flow rate $v_0 = 24$ dm^3/min. The feed concentration of A is $C_{A0} = 1$ g-mol/dm^3. Consider the heat capacity of the reacting mixture based on a g-mol of A to be defined as

$$C_P = \sum_{i=1}^{n} \Theta_i \tilde{C}_{pi} \tag{8-156}$$

which is constant. The change in heat capacity during reaction is zero

$$\Delta \tilde{C}_p = 0 \tag{8-157}$$

and the ratio of the heat of reaction to the heat capacity is given by

$$\Delta H_{R,T_R} / C_p = -200 \text{ °C} \tag{8-158}$$

Details of the preceding equations are found in Fogler.[4]

(a) Calculate and plot the conversion and reactor temperature versus volume for an adiabatic plug-flow reactor in which the feed temperature is 60 °C.

(b) Determine what feed temperature would maximize the output conversion in part (a).

(c) If the reactor of part (a) is operated isothermally at 60 °C, what would be the resulting conversion?

(d) If the plug-flow reactor of part (a) were operated at the most optimal temperature profile (where T in the reactor cannot exceed a temperature of 105 °C), what would be the resulting conversion?

(e) The plug-flow reactor of part (a) is to be replaced by a series of three adiabatic CSTRs, with each having one third the volume of the original reactor. What feed temperatures should be used for each CSTR to give the optimal output conversion? What is the overall optimal conversion?

(f) Compare the temperature and conversion profiles down the plug-flow reactor of part (a) to a space time of 0.5 minutes for a feed temperature of 60 °C with different rates of external cooling to an ambient temperature of 20 °C. Consider Ua/C_p to be 0.1, 0.5, and 1.0 g-mol/dm³·min, where C_p is the constant heat capacity of the mixture per g-mol of A. Which heat transfer rate gives the best conversion and why?

(g) For the adiabatic plug-flow reactor of part (a), vary the heat of reaction to one half and to one quarter of the base case value. Plot the temperatures and conversions as a function of reactor space time to 1.0 min. Why would a reduction of the heat of reaction result in higher conversions?

8.26 DIFFUSION WITH MULTIPLE REACTIONS IN POROUS CATALYST PARTICLES

8.26.1 Concepts Demonstrated

Selectivity in multiple reactions occurring with a solid catalyst comprised of two components with different activities and effective diffusivities.

8.26.2 Numerical Methods Utilized

Simultaneous ordinary differential equations with split boundary conditions.

8.26.3 Problem Statement

The catalytic cracking of gas-oil, A, to gasoline, B, is accompanied by the production of light gases, coke, and coke precursors, C. A schematic of this reaction system is indicated in Figure , where both reactions involving reactant A are second order and the reaction involving reactant B is first order.

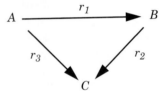

Figure 8–17

Reactions Involved in a Lumped Parameter
Model for Catalytic Cracking

    The industrial catalysts for this catalytic process are typically mixtures of 5.0 weight % zeolite Z in a silica-alumina matrix M. The zeolite catalysis is more catalytically active and is more selective to the desired gasoline product than the matrix, but the zeolite has a much lower effective diffusivities. Table 8–19 presents typical data for these catalytic reaction systems as adapted from Martin et al.[9]

    In a particular application, the catalyst particle can be approximated by a slab of total thickness L (one surface exposed as in Problem 3.6) where the kinetic and transport properties can be considered as the weighted average of the zeolite designated Z and the matrix designated M as given by

$$D_{Ax} = uD_{AZ} + (1-u)D_{AM} \quad \text{and} \quad D_{Bx} = uD_{BZ} + (1-u)D_{BM} \tag{8-159}$$

$$k_{ix} = uk_{iZ} + (1-u)k_{iM} \quad \text{for } i = 1, 2, \text{ and } 3 \tag{8-160}$$

where u is the local mass fraction of zeolite. Note that the total mass fraction of zeolite is always 5% and cannot exceed 20% at any location in the catalyst particle. Thus

$$\frac{1}{L}\int_0^L u \, dx = 0.05 \quad \text{where} \quad u \leq 0.20 \tag{8-161}$$

The zero flux boundary conditions at the particle center are given by

$$\frac{dC_A}{dx} = \frac{dC_B}{dx} = 0 \qquad \text{at} \qquad x = 0 \qquad \textbf{(8-162)}$$

and the boundary conditions for the gas concentrations at the particle surface are given by

$$C_A = 1.6 \times 10^{-5} \, \text{gmol/cm}^3 \quad \text{and} \quad C_B = C_C = 0 \quad \text{at} \quad x = L \qquad \textbf{(8-163)}$$

The selectivity for B is the mass flux of B in the x direction divided by the mass flux of A in the opposite direction. Thus

$$S_B = \frac{-D_{Bx}(dC_B/dx)}{D_{Ax}(dC_A/dx)} \qquad \text{evaluated at} \qquad x = L \qquad \textbf{(8-164)}$$

The appropriate differential equations for the diffusion with reaction are

$$D_{Ax}\frac{d^2C_A}{dx^2} = k_{3x}C_A^2 + k_{1x}C_A^2 \qquad \textbf{(8-165)}$$

$$D_{Bx}\frac{d^2C_B}{dx^2} = k_{2x}C_B - k_{1x}C_A^2 \qquad \textbf{(8-166)}$$

Table 8–19 Parameter Values (Adapted from Martin et al.[9])

$k_{1Z} = 10^8 \, \text{cm}^3\text{/g-mol} \cdot \text{s}$	$k_{1M} = 10^7 \, \text{cm}^3\text{/g-mol} \cdot \text{s}$
$k_{2Z} = 8 \times 10^2 \, \text{s}^{-1}$	$k_{2M} = 10^2 \, \text{s}^{-1}$
$k_{3Z} = 3 \times 10^6 \, \text{cm}^3\text{/g-mol} \cdot \text{s}$	$k_{3M} = 8 \times 10^6 \, \text{cm}^3\text{/g-mol} \cdot \text{s}$
$D_{AZ} = 10^{-6} \, \text{cm}^2\text{/s}$	$D_{AM} = 10^{-3} \, \text{cm}^2\text{/s}$
$D_{BZ} = 10^{-5} \, \text{cm}^2\text{/s}$	$D_{BM} = 10^{-2} \, \text{cm}^2\text{/s}$
$L = 0.002 \, \text{cm}$	

(a) Derive and verify Equations (8-164), (8-165), and (8-166).
(b) Summarize all of the equations needed to determine the selectivity of B for the case where the zeolite catalyst is evenly distributed throughout the catalyst particle.
(c) Repeat part (b) for the case where the zeolite is located in a 20% layer at the surface of the catalyst particle.
(d) Calculate the selectivity for part (b) using the values of Table 8–19.
(e) Calculate the selectivity for part (c) using the values of Table 8–19.
(f) Discuss the benefits of placing the zeolite layer at the surface of the catalyst particle.

REFERENCES

1. Constantinides, A. *Applied Numerical Methods with Personal Computers*, New York: McGraw-Hill, 1987.
2. Cutlip, M. B., and Peters, M. S., "Heterogeneous Catalysis over an Organic Semiconducting Polymer Made from Polyacrylonitrile," *Chem. Eng. Progr. Symp. Ser.*, *64* (89), 1–11 (1968).
3. Dillon, R. T. *J. Am. Chem. Soc.*, *54*, 952 (1932).
4. Fogler, H. S. *Elements of Chemical Reaction Engineering*, 2nd ed., Englewood Cliffs, NJ: Prentice Hall, 1992.
5. Hill, C. G. *An Introduction to Chemical Engineering Kinetics & Reactor Design*, New York: Wiley, 1977.
6. Hinshelwood, C. N., and Askey, P. J. *Proc. Roy. Soc.*, *A115*, 215 (1927).
7. Hinshelwood, C. N., and Burke *Proc. Roy. Soc. (London)*, *A106*, 284 (1924).
8. Luyben, W. L. *Process Modeling Simulation and Control for Chemical Engineers*, 2nd ed., New York: McGraw-Hill, 1990.
9. Martin, G. R., White, C. W., and Dadyburjor, D. B., "Design of Zeolite/Silica-Alumina Catalysts for Triangular Cracking Reactions", *Journal of Catalysis*, *106*, 116–124 (1987).
10. McCracken and Dickson, *Ind. Eng. Chem. Proc. Des. and Dev.*, *6*, 286 (1967).
11. Seinfeld, J. H., *Atmospheric Chemistry and Physics of Air Pollution*, New York: Wiley, 1986.

Useful Constants

Ideal Gas

$R = 8314.34 \text{ m}^3 \cdot \text{Pa/kg-mol} \cdot \text{K}$

$R = 8314.34 \text{ J/kg-mol} \cdot \text{K}$

$R = 8314.34 \text{ kg} \cdot \text{m}^2/\text{s}^2 \cdot \text{kg-mol} \cdot \text{K}$

$R = 0.082057 \text{ m}^3 \cdot \text{atm/kg-mol} \cdot \text{K}$

$R = 1.9872 \text{ btu/lb-mol} \cdot {}^\circ\text{R}$

$R = 1545.3 \text{ ft} \cdot \text{lb}_f/\text{lb-mol} \cdot {}^\circ\text{R}$

$R = 0.7302 \text{ ft}^3 \cdot \text{atm/lb-mol} \cdot {}^\circ\text{R}$

$R = 10.731 \text{ ft}^3 \cdot \text{lb}_f/\text{in}^2 \cdot \text{lb-mol} \cdot {}^\circ\text{R}$

$R = 1.9872 \text{ cal/g-mol} \cdot \text{K}$

$R = 82.057 \text{ cm}^3 \cdot \text{atm/g-mol} \cdot \text{K}$

Stefan-Boltzmann

$\sigma = 5.676 \times 10^{-8} \text{ W/m}^2 \cdot \text{K}^4$

$\sigma = 1.714 \times 10^{-9} \text{ btu/h} \cdot \text{ft}^2 \cdot {}^\circ\text{R}^4$

Useful Conversion Factors

Temperature

$K = {}^\circ\text{C} + 273.15$

${}^\circ\text{R} = {}^\circ\text{F} + 459.67$

${}^\circ\text{F} = 1.8 \times {}^\circ\text{C} + 32$

${}^\circ\text{C} = (5/9) \times ({}^\circ\text{F} - 32)$

$\Delta{}^\circ\text{F} = 1.8 \times \Delta{}^\circ\text{C}$

$\Delta{}^\circ\text{C} = (5/9) \times \Delta{}^\circ\text{F}$

Mass

$1 \text{ kg} = 1000 \text{ g} = 2.2046 \text{ lb}_m$

$1 \text{ metric ton} = 1000 \text{ kg}$

$1 \text{ lb}_m = 453.59 \text{ g} = 0.45359 \text{ kg}$

$1 \text{ short ton} = 2000 \text{ lb}_m$

$1 \text{ long ton} = 2240 \text{ lb}_m$

Length

$1 \text{ m} = 3.2808 \text{ ft} = 100 \text{ cm} = 39.37 \text{ in}$

$1 \text{ μm} = 1 \text{ micron} = 10^{-6} \text{ m} = 10^{-3} \text{ mm} = 10^{-4} \text{ cm}$

$1 \text{ Å} = 10^{-10} \text{ m} = 10^{-4} \text{ μm} = 0.1 \text{ nm}$

$1 \text{ in} = 2.540 \text{ cm}$

$1 \text{ mile} = 5280 \text{ ft}$

Area

$1 \text{ m}^2 = 10^4 \text{ cm} = 10^6 \text{ mm}$

$1 \text{ m}^2 = 10.739 \text{ ft}^2$

$1 \text{ in}^2 = 6.4516 \text{ cm}^2 = 6.4516 \times 10^{-4} \text{ m}^2$

Volume

$1 \text{ m}^3 = 1000 \text{ dm}^3 = 10^6 \text{ cm}^3 = 1000 \text{ liter}$

$1 \text{ m}^3 = 35.313 \text{ ft}^3 = 264.17 \text{ gal (U.S.)}$

$1 \text{ liter} = 1 \text{ dm}^3$

$1 \text{ ft}^3 = 0.028317 \text{ m}^3 = 28.317 \text{ liter}$

$1 \text{ ft}^3 = 7.481 \text{ gal (U.S.)}$

$1 \text{ gal (British)} = 1.2009 \text{ gal (U.S.)}$

$1 \text{ gal (U.S.)} = 3.7854 \text{ liter}$

Force

$1 \text{ N} = 1 \text{ kg} \cdot \text{m/s}^2 = 0.22481 \text{ lb}_f$

$1 \text{ g} \cdot \text{cm/s}^2 = 10^{-5} \text{ N} = 2.2481 \times 10^{-6} \text{ lb}_f$

Pressure

$1 \text{ Pa} = 1 \text{ N/m}^2 = 1 \text{ kg/m} \cdot \text{s}^2$

$1 \text{ bar} = 1 \times 10^5 \text{ Pa}$

$1 \text{ atm} = 1.01325 \times 10^5 \text{ Pa} = 101.325 \text{ kPa}$

$1 \text{ atm} = 760 \text{ mm Hg at } 0 \text{ °C} = 29.921 \text{ in Hg at } 0 \text{ °C}$

$1 \text{ atm} = 33.90 \text{ ft } H_2O \text{ at } 4 \text{ °C}$

$1 \text{ atm} = 14.696 \text{ lb}_f/\text{in}^2$

Molar Volume and Density

1 kg-mol of ideal gas at 0 °C and 1 atm occupies 22.414 m^3

1 g-mol of ideal gas at 0 °C and 1 atm occupies 22.414 liters

1 lb-mol of ideal gas at 32 °F occupies 359.05 ft^3

Air: MW = 28.97

Dry air density at 0 °C and 1 atm = 1.2929 kg/m^3

Dry air density at 32 °F and 1 atm = 0.080711 lb_m/ft^3

Water density at 4.0 °C = 999.972 kg/m^3 = 62.428 lb_m/ft^3

Acceleration of Gravity

g = 9.80665 m/s^2

g = 32.174 ft/s^2

g_c = 32.1740 $ft \cdot lb_m/lb_f \cdot s^2$

Thermodynamics

1 J = 1 N·m= 1 $kg \cdot m^2/s^2$ = 10^7 $g \cdot cm^2/s^2$ = 10^7 erg

1 J = 0.73756 $ft \cdot lb_f$

1 kJ = 1000 J = 0.947813 btu = 0.23900 kcal = 239.00 cal

1 btu = 1.05506 kJ = 1055.06 J = 252.16 cal

1 btu = 778.17 $ft \cdot lb_f$

1 kcal = 1000 cal = 4.1840 kJ

1 cal = 4.1840 J

1 J/s = 1 W = 1.3410 × 10^{-3} hp = 14.340 cal/min

1 hp = 550 $ft \cdot lb_f/s$ = 0.7068 btu/s

1 hp·h = 0.7457 kW·h = 2544.5 btu

1 $kJ/kg \cdot K$ = 0.23884 $btu/lb_m \cdot °F$

1 $btu/lb_m \cdot °F$ = 1 $btu/lb_m \cdot °R$ = 1 $cal/g \cdot °C$ = 1 $cal/g \cdot K$

1 J/kg = 4.2993 × 10^{-4} btu/lb_m = 0.33456 $ft \cdot lb_f/lb_m$

1 kJ/kg-mol = 2.3901 × 10^{-4} kcal/g-mol

Fluid Dynamics

1 kg/m·s = 1 $N \cdot s/m^2$ = 1 Pa·s = 1000 centipoise = 0.67197 $lb_m/ft \cdot s$

1 centipoise = 10^{-3} Pa·s = 10^{-2} g/cm·s = 10^{-2} poise = 2.4191 $lb_m/ft \cdot h$

Heat Transfer

1 $W/m \cdot K$ = 0.57779 $btu/h \cdot ft \cdot °F$ = 2.3901 × 10^{-3} $cal/s \cdot cm \cdot °C$

1 $btu/h \cdot ft \cdot °F$ = 4.1365 × 10^{-3} $cal/s \cdot cm \cdot °C$ = 1.73073 $W/m \cdot K$

1 $W/m^2 \cdot K$ = 0.17611 $btu/h \cdot ft^2 \cdot °F$ = 0.85991 $kcal/h \cdot m^2 \cdot °F$

Mass Transfer

$1 \text{ m}^2/\text{s} = 3.875 \times 10^4 \text{ ft}^2/\text{h} = 10^4 \text{ cm}^2/\text{s}$

$1 \text{ m}^2/\text{h} = 10.764 \text{ ft}^2/\text{h} = 10^4 \text{ cm}^2/\text{h}$

$1 \text{ m/s} = 1.1811 \text{ ft/h} = 0.01 \text{ cm/s}$

$1 \text{ kg-mol/s} \cdot \text{m}^2 = 737.35 \text{ lb-mol/h} \cdot \text{ft}^2 = 0.1 \text{ g-mol/s} \cdot \text{cm}^2$

$1 \text{ kg-mol/s} \cdot \text{m}^2 \cdot \text{atm} = 737.35 \text{ lb-mol/h} \cdot \text{ft}^2 \cdot \text{atm} = 0.1 \text{ g-mol/s} \cdot \text{cm}^2 \cdot \text{atm}$

$1 \text{ kg-mol/s} \cdot \text{m}^2 \cdot \text{kPa} = 7.4712 \times 10^4 \text{ lb-mol/h} \cdot \text{ft}^2 \cdot \text{atm} = 10.1325 \text{ g-mol/s} \cdot \text{cm}^2 \cdot \text{atm}$

Useful Finite Difference Approximations

Table A–1 First-Order Finite Difference Approximations

Difference	First Order Formula	
Forward Difference for First Derivative	$\frac{d}{dx}f(x_i) = \dfrac{f(x_{i+1}) - f(x_i)}{\Delta x}$	(A-1)
Backward Difference for First Derivative	$\frac{d}{dx}f(x_i) = \dfrac{f(x_i) - f(x_{i-1})}{\Delta x}$	(A-2)
Forward Difference for Second Derivative	$\frac{d^2}{dx^2}f(x_i) = \dfrac{f(x_i) - 2f(x_{i+1}) + f(x_{i+2})}{\Delta x^2}$	(A-3)
Backward Difference for Second Derivative	$\frac{d^2}{dx^2}f(x_i) = \dfrac{f(x_i) - 2f(x_{i-1}) + f(x_{i-2})}{\Delta x^2}$	(A-4)

Table A–2 Second-Order Finite Difference Approximations

Difference	Second Order Formula	
Forward Difference for First Derivative	$\frac{d}{dx}f(x_i) = \dfrac{-3f(x_i) + 4f(x_{i+1}) - f(x_{i+2})}{2\Delta x}$	(A-5)
Central Difference for First Derivative	$\frac{d}{dx}f(x_i) = \dfrac{f(x_{i+1}) - f(x_{i-1})}{2\Delta x}$	(A-6)
Backward Difference for First Derivative	$\frac{d}{dx}f(x_i) = \dfrac{3f(x_i) - 4f(x_{i-1}) + f(x_{i-2})}{2\Delta x}$	(A-7)

Table A–2 Second-Order Finite Difference Approximations

Difference	Second Order Formula
Forward Difference for Second Derivative	$$\frac{d^2}{dx^2}f(x_i) = \frac{2f(x_i) - 5f(x_{i+1}) + 4f(x_{i+2}) - f(x_{i+3})}{\Delta x^2} \quad \text{(A-8)}$$
Central Difference for Second Derivative	$$\frac{d^2}{dx^2}f(x_i) = \frac{f(x_{i+1}) - 2f(x_i) + f(x_{i-1})}{\Delta x^2} \quad \text{(A-9)}$$
Backward Difference for Second Derivative	$$\frac{d^2}{dx^2}f(x_i) = \frac{2f(x_i) - 5f(x_{i-1}) + 4f(x_{i-2}) - f(x_{i-3})}{\Delta x^2} \quad \text{(A-10)}$$

Error Function

Table A–3 Error and Complimentary Error Functions

z	erf(z)	erfc(z)
0	0	1
0.1	0.11246278	0.88753722
0.2	0.22270233	0.77729767
0.3	0.32862638	0.67137362
0.4	0.42839185	0.57160815
0.5	0.52049927	0.47950073
0.6	0.60385538	0.39614462
0.7	0.6778004	0.3221996
0.8	0.7421001	0.2578999
0.9	0.79690728	0.20309272
1	0.84269981	0.15730019
1.1	0.88020404	0.11979596
1.2	0.91031291	0.089687086
1.3	0.93400685	0.065993147
1.4	0.95228401	0.047715994
1.5	0.96610402	0.033895983
1.6	0.97634724	0.023652758

Table A–3 Error and Complimentary Error Functions

z	erf(z)	erfc(z)
1.7	0.98378931	0.016210692
1.8	0.98908935	0.010910655
1.9	0.99278927	0.0072107315
2	0.9953211	0.0046788987

Table B–1 Vapor Pressure Data for Sulfur Compounds Present in Petroleum[a]

Pressure mm Hg	Temperature, °C				
	Ethane -thiol	1-Propane thiol	2-Propane- thiol	1-Butane -thiol	2-Butane -thiol
149.41	...	24.275	10.697	51.409	38.962
187.57	0.405	29.563	15.770	57.130	44.549
233.72	5.236	34.891	20.899	62.897	50.185
289.13	10.111	40.254	26.071	68.710	55.866
355.22	15.017	45.663	31.282	74.567	61.597
433.56	19.954	51.113	36.536	80.472	67.370
525.86	24.933	56.605	41.833	86.418	73.195
633.99	29.944	62.139	47.175	92.414	79.063
760.00	35.000	67.719	52.558	98.454	84.981
906.06	40.092	73.341	57.985	104.544	90.945
1074.6	45.221	79.004	63.461	110.682	96.963
1268.0	50.390	84.710	68.979	116.863	103.020
1489.1	55.604	90.464	74.540	123.088	109.133
1740.8	60.838	96.255	80.143	129.362	115.287
2026.0	66.115	102.088	85.795	135.679	121.489

[a]Osborn, A. G., and Douglin, D. R., *J. Chem. Eng. Data*, *11*(4), 502–505 (1966).

Table B–2 Vapor Pressure Data for Sulfur Compounds Present in Petroleum[a]

| | Temperature, °C | | | |
Pressure mm Hg	2-Methyl 1-propane -thiol	2-Methyl 2-propane -thiol	1-Pen- tane -thiol	2-Methyl 1-butane -thiol
71.87	...	...	...	51.339
81.64	...	...	...	54.284
92.52	...	...	...	57.243
104.63	...	...	...	60.219
118.06	...	...	...	63.194
132.95	...	...	...	66.193
149.41	42.207	20.496	76.470	69.207
187.57	47.830	25.785	82.569	75.263
233.72	53.498	31.127	88.721	81.361
289.13	59.211	36.519	94.918	87.510
355.22	64.974	41.959	101.167	93.708
433.56	70.780	47.446	107.457	99.955
525.86	76.641	52.983	113.802	106.253
633.99	82.542	58.573	120.193	112.600
760.00	88.493	64.217	126.638	118.999
906.06	94.493	69.908	133.131	125.446
1074.6	100.539	75.654	139.671	131.944
1268.0	106.640	81.449	146.255	138.492
1489.1	112.785	87.294	152.896	145.089
1740.8	118.972	93.188	159.580	151.733
2026.0	125.212	99.138	166.314	158.428

[a]Osborn, A. G., and Douglin, D. R., *J. Chem. Eng. Data*, *11*(4), 502–505 (1966).

Table B–3 Vapor Pressure Data for Sulfur Compounds Present in Petroleum[a]

| | | Temperature, °C | | | |
Pressure mm Hg	2-Methyl 2-butane thiol	3-Methyl 2-butane thiol	Cyclo pentane thiol	1-Hexane thiol	2-Methyl 2-pentane thiol
71.87	...	42.969	...	80.694	55.855
81.64	...	45.876	...	83.837	58.860
92.52	...	48.791	...	86.991	61.877
104.63	...	51.720	...	90.157	64.907
118.06	...	54.658	...	93.334	67.949
132.95	...	57.613	...	96.530	71.008
149.41	50.888	60.592	80.874	99.733	74.089
187.57	56.725	66.556	87.107	106.168	80.269
233.72	62.625	72.575	93.390	112.658	86.502
289.13	68.578	78.645	99.729	119.198	92.787
355.22	74.579	84.765	106.113	125.789	99.127
433.56	80.638	90.936	112.548	132.429	105.521
525.86	86.749	97.161	119.037	139.121	111.972
633.99	92.914	103.431	125.577	145.866	118.475
760.00	99.132	109.760	132.165	152.659	125.032
906.06	105.401	116.139	138.806	159.507	131.646
1074.6	111.728	122.571	145.501	166.403	138.314
1268.0	118.106	129.051	152.245	173.351	145.037
1489.1	124.537	135.585	159.040	180.349	151.815
1740.8	131.021	142.170	165.887	187.397	158.645
2026.0	137.559	148.805	172.783	194.494	165.531

[a]Osborn, A. G., and Douglin, D. R., *J. Chem. Eng. Data*, *11*(4), 502–505 (1966).

Table B-4 Vapor Pressure Data for Sulfur Compounds Present in Petroleum[a]

Temperature, °C

Pressure mm Hg	2,3-Dimethyl 2-butane thiol	Cyclo hexane thiol	Benzene thiol	1-Heptane thiol
71.87	55.814	83.740	...	101.627
81.64	58.867	87.006	...	104.908
92.52	61.931	90.289	...	108.205
104.63	65.011	93.576	...	111.517
118.06	68.099	96.881	...	114.840
132.95	71.208	100.201	...	118.182
149.41	74.334	103.549	114.543	121.546
187.57	80.613	110.259	121.191	128.269
233.72	86.949	117.023	127.897	135.066
289.13	93.338	123.843	134.649	141.911
355.22	99.783	130.719	141.447	148.807
433.56	106.283	137.654	148.294	155.759
525.86	112.843	144.647	155.194	162.758
633.99	119.458	151.695	162.140	169.812
760.00	126.129	158.803	169.137	176.919
906.06	132.858	165.968	176.188	184.082
1074.6	139.644	173.186	183.278	191.292
1268.0	146.492	180.464	190.426	198.551
1489.1	153.391	187.801	197.623	...
1740.8	160.344	195.196	204.867	...
2026.0	167.355	202.645	212.160	...

[a]Osborn, A. G., and Douglin, D. R., *J. Chem. Eng. Data*, *11*(4), 502–505 (1966).

Table B–5 Vapor Pressure of Propane[a]

No	Temperature °F	Pressure psia
1	−70	7.37
2	−60	9.72
3	−50	12.6
4	−40	16.2
5	−30	20.3
6	−20	25.4
7	−10	31.4
8	0	38.2
9	10	46
10	20	55.5
11	30	66.3
12	40	78
13	50	91.8
14	60	107.1
15	70	124
16	80	142.8
17	90	164
18	100	187
19	110	213
20	120	240

[a]Weast, R. C. (Ed.), *Handbook of Chemistry and Physics*, 56th ed., CRC Press, Cleveland, Ohio (1975).

Table B–6 *t* Distribution for 95% Confidence Intervals

Degrees of Freedom	*t*
1	12.7062
2	4.3027
3	3.1824
4	2.7764
5	2.5706
6	2.4469
7	2.3646
8	2.3060
9	2.2622
10	2.2281
11	2.2010
12	2.1788
13	2.1604
14	2.1448
15	2.1315
16	2.1199
17	2.1098
18	2.1009
19	2.0930
20	2.0860
21	2.0796
22	2.0739
23	2.0687
24	2.0639
25	2.0595

Table B–7 Vapor Pressure of Various Substances[a]

No.	Pressure mm Hg	Ethane	n-Butane	n-Pentane	n-Hexane	n-Heptane	n-Decane
				Temperature °C			
1	10	−142.82	−77.66	−50.08	−25.09	−2.04	57.70
2	50	−127.31	−55.62	−25.41	1.82	26.81	91.23
3	100	−119.24	−44.15	−12.591	15.782	41.769	108.586
4	150	−114.039	−36.75	−4.330	24.781	51.405	119.757
5	200	−110.104	−31.15	1.920	31.586	56.689	128.196
6	250	−106.898	−26.59	7.008	37.126	64.618	135.062
7	300	−104.173	−22.71	11.333	41.834	69.655	140.893
8	400	−99.665	−16.29	18.486	49.617	77.981	150.527
9	500	−95.981	−11.04	24.329	55.975	84.780	158.389
10	600	−92.840	−6.566	29.311	61.392	90.573	165.085
11	700	−90.087	−2.642	33.676	66.139	95.648	170.948
12	760	−88.580	−0.495	36.065	68.736	98.424	174.155
13	800	−87.626	0.865	37.577	70.380	100.181	176.184
14	900	−85.39	4.05	41.11	74.225	104.291	180.93
15	1000	−83.35	6.96	44.36	77.75	108.06	185.28
16	1200	−79.69	12.18	50.16	84.05	114.79	193.02
17	1500	−74.99	18.87	57.59	92.13	123.40	202.88

[a]Thermodynamics Research Center API44 Hydrocarbon Project, *Selected Values of Properties of Hydrocarbon and Related Compounds*, Texas A&M University, College Station, Texas (1978) with permission.

Table **B–8** Heat Capacity of Gaseous Propane[a]

No.	Temperature K	Heat Capacity kJ/kg-mol·K
1	50	34.06
2	100	41.3
3	150	48.79
4	200	56.07
5	273.16	68.74
6	298.15	73.6
7	300	73.93
8	400	94.01
9	500	112.59
10	600	128.7
11	700	142.67
12	800	154.77
13	900	163.35
14	1000	174.6
15	1100	182.67
16	1200	189.74
17	1300	195.85
18	1400	201.21
19	1500	205.89

[a]Thermodynamics Research Center API44 Hydrocarbon Project, *Selected Values of Properties of Hydrocarbon and Related Compounds*, Texas A&M University, College Station, Texas (1978) with permission.

Table B–9 Thermal Conductivity of Gaseous Propane; units: [a]

No.	Temperature °F	Thermal Conductivity $\times 10^6$ cal/s·cm·°C
1	-40	27.69
2	-20	29.75
3	0	32.23
4	20	34.71
5	40	37.19
6	60	39.67
7	80	42.47
8	100	45.46
9	120	48.35
10	200	60.75

[a]Weast, R. C. (Ed.), *Handbook of Chemistry and Physics*, 56th ed., CRC Press, Cleveland, Ohio (1975), p. E-2.

Table B–10 Viscosity of Liquid Propane at Atmospheric Pressure[a]

No.	Temperature °C	Viscosity cp
1	-190	13.8
2	-185	8.75
3	-180	5.94
4	-175	4.25
5	-170	3.17
6	-165	2.45
7	-160	1.95
8	-155	1.6
9	-150	1.34
10	-145	1.14
11	-140	0.981
12	-130	0.76
13	-120	0.612
14	-110	0.508
15	-100	0.432
16	-90	0.373
17	-80	0.326
18	-70	0.287
19	-60	0.255
20	-50	0.227
21	-40	0.204

[a]Thermodynamics Research Center API44 Hydrocarbon Project, *Selected Values of Properties of Hydrocarbon and Related Compounds*, Texas A&M University, College Station, Texas (1978) with permission.

Table B-11 Heat of Vaporization of Propane[a]

No.	Temperature °F	Heat of Vaporization kcal/kg
1	-70	189.5
2	-60	187.0
3	-50	184.5
4	-40	181.5
5	-30	179.0
6	-20	176
7	-10	173.5
8	0	170.5
9	10	168
10	20	165
11	30	162
12	40	159
13	50	156
14	60	153
15	70	149.5
16	80	146.0
17	90	142.5
18	100	138.5
19	110	134
20	120	129

[a]Weast, R. C. (Ed.), *Handbook of Chemistry and Physics*, 56th ed., CRC Press, Cleveland, Ohio (1975) p. E-31.

Table B–12 Heat Capacity of Various Gases[a]

No.	T	Heat Capacity, kJ/kg-mol·K			
	K	Methane	Ethane	Butane	Pentane
1	50	33.26	33.39	38.07	
2	100	33.28	35.65	55.35	
3	150	33.30	38.66	67.32	
4	200	33.51	42.26	76.44	93.55
5	273.16	34.85	49.54	92.30	112.55
6	298.15	35.69	52.47	98.49	120.04
7	300	35.77	52.72	98.95	120.62
8	400	40.63	65.48	124.77	152.55
9	500	46.53	77.99	148.66	182.59
10	600	52.51	89.24	169.28	208.78
11	700	58.20	99.20	187.02	231.38
12	800	63.51	107.99	202.38	250.62
13	900	68.37	115.77	215.73	266.94
14	1000	72.80	122.59	227.36	281.58
15	1100	76.78	128.57	237.48	293.72
16	1200	80.37	133.85	246.27	304.60
17	1300	83.55	138.41	253.93	313.80
18	1400	86.44	142.42	260.58	322.17
19	1500	88.99	145.90	266.40	330.54

[a]Thermodynamics Research Center API44 Hydrocarbon Project, *Selected Values of Properties of Hydrocarbon and Related Compounds*, Texas A&M University, College Station, Texas (1978) with permission.

Table B–13 Thermal Conductivity of Gases[a]

No.	Temp.				Thermal Conductivity $\times 10^6$, cal/s $\cdot$ cm $\cdot$ °C			
	°F	Air	Ammonia	n-Butane	Carbon Dioxide	Ethane	Ethanol	Methane
1	-100					23.97		52.07
2	-40	50.09	43.39		27.90	32.65		61.37
3	-20	52.15	45.87		29.75	35.54		64.55
4	0	54.22	48.35		31.70	38.43		67.86
5	20	56.24	50.83	30.99	33.68	41.33	29.34	71.08
6	40	58.31	53.31	33.06	35.62	44.63	30.99	74.39
7	60	60.34	55.79	35.54	37.61	47.94	32.65	78.11
8	80	62.20	58.68	38.02	39.67	51.24	34.71	81.83
9	100	64.22	61.58	40.91	41.74	54.55	36.78	85.54
10	120	66.04	64.47	43.39	43.81	58.27		89.26
11	200			54.14		74.39		106.62

[a]Weast, R. C. (Ed.), *Handbook of Chemistry and Physics*, 56th Ed., CRC Press, Cleveland, Ohio, p. E-2 (1975).

Table B–14 Viscosity of Various Liquids at Atmospheric Pressure[a]

No.	Temp. °C	Methane	Ethane	*n*-Butane	*n*-Pentane	*n*-Hexane	*n*-Heptane
1	−185	0.225					
2	−180	0.187					
3	−175	0.161	0.982				
4	−170	0.142	0.803				
5	−165	0.127	0.671				
6	−160	0.115	0.572				
7	−155		0.499				
8	−150		0.441				
9	−145		0.395				
10	−140		0.358				
11	−135		0.327				
12	−130		0.300		3.62		
13	−125		0.277		2.88		
14	−120		0.256		2.34		
15	−115		0.237		1.95		
16	−110		0.221		1.66		
17	−105		0.206		1.43		
18	−100		0.194		1.24		
19	−95		0.182		1.09	2.13	
20	−90		0.171	0.63	0.970	1.82	3.76
21	−85		0.162	0.58	0.871	1.58	3.10
22	−80			0.534	0.789	1.38	2.60
23	−75			0.496	0.718	1.22	2.21
24	−70			0.461	0.657	1.09	1.912
25	−65			0.430	0.605	0.975	1.670
26	−60			0.402	0.560	0.885	1.472
27	−55			0.377	0.520	0.807	1.309
28	−50			0.354	0.486	0.739	1.173

Table B–14 Viscosity of Various Liquids at Atmospheric Pressure[a]

No.	Temp. °C	Methane	Ethane	n-Butane	n-Pentane	n-Hexane	n-Heptane
29	−45			0.334	0.454	0.681	1.060
30	−40			0.314	0.427	0.630	0.9624
31	−35			0.297	0.402	0.585	0.8793

[a]Thermodynamics Research Center API44 Hydrocarbon Project, *Selected Values of Properties of Hydrocarbon and Related Compounds*, Texas A&M University, College Station, Texas (1978) with permission.

Table B–15 Heat of Vaporization of Various Compounds[a]

			Heat of Vaporization kcal/kg				
No.	Temp. °F	Difluoro-dichloro-methane	Isobu-tane	Butane	Carbon Disulfide	Carbon Tetra-cloride	Ethyl Ether
1	−40	40.83					
2	−30	40.37					
3	−20	39.89	91.94				
4	−10	39.39	90.56				
5	0	38.87	89.17	94.72	91.94		95.00
6	10	38.32	88.06	93.61	91.39		94.67
7	20	37.74	86.67	92.78	90.67	52.47	94.44
8	30	37.14	85.28	91.94	90.11	52.06	94.11
9	40	36.51	83.89	90.83	89.56	51.78	93.56
10	50	35.84	82.50	89.72	88.89	51.22	93.11
11	60	35.14	81.11	88.61	88.44	50.78	91.89
12	70	34.40	79.72	87.50	87.83	50.04	91.22
13	80	33.62	78.06	86.11	87.17	50.01	90.56
14	90	32.80	76.39	84.44	86.44	49.67	89.72
15	100	31.92	74.72	83.06	85.78	49.28	
16	110	30.99	72.78	81.67	85.11	48.83	
17	120	29.99	70.83	79.72	84.44		
18	130		68.89	78.06			
19	140		66.94	76.39			

[a]Weast, R. C. (Ed.), *Handbook of Chemistry and Physics*, 56th Ed., CRC Press, Cleveland, Ohio, p. E-31 (1975).

Table B–16 Heat Transfer Data External to 3/4-inch OD Tubes[a]

Point	Re	Pr	μ/μ_w	Nu
1	49000	2.3	0.947	277
2	68600	2.28	0.954	348
3	84800	2.27	0.959	421
4	34200	2.32	0.943	223
5	22900	2.36	0.936	177
6	1321	246	0.592	114.8
7	931	247	0.583	95.9
8	518	251	0.579	68.3
9	346	273	0.29	49.1
10	122.9	1518	0.294	56
11	54.0	1590	0.279	39.9
12	84.6	1521	0.267	47
13	1249	107.4	0.724	94.2
14	1021	186	0.612	99.9
15	465	414	0.512	83.1
16	54.8	1302	0.273	35.9

[a]Williams, R. B., and Katz, D. L.,
Trans. ASME, 74, 1307–1320 (1952).

Table B–17 Heat Transfer Data for 21° API Oil[a]

Point	Re	Pr	μ/μ_w	Nu
1	368	545	0.109	15.7
2	381	535	0.112	14.1
3	875	345	0.076	16.0
4	645	348	0.074	15.0
5	90.4	385	0.094	10.5
6	90.8	390	0.101	10.7
7	545	151	0.052	14.6
8	1005	167	0.050	16.1
9	978	168	0.050	16.2
10	1523	160	0.051	17.4
11	1560	158	0.051	17.3
12	2100	159	0.053	21.2
13	2110	159	0.056	21.4
14	67.2	520	0.161	12.5
15	269	507	0.121	13.5
16	655	436	0.114	15.3
17	22.6	3350	3.43	25.0
18	23.8	3160	5.28	21.0
19	20.9	3620	6.02	18.0
20	26.6	2780	6.30	21.2
21	24.7	3100	6.62	18.0
22	29.6	2510	6.62	22.6
23	26.5	2860	8.10	22.5
24	29.2	2470	8.80	21.7
25	27.8	2770	9.75	23.0

[a]Sieder, E.N., and Tate, G. E.,
Ind. and Eng. Chem. 28, 1429
(1936).

Table B–18 Heat Transfer Data of 24.5° API Oil[a]

Point	Re	Pr	μ/μ_w	Nu
1	61.5	567	5.05	19.2
2	64.7	540	4.80	22.6
3	128	532	4.95	26.8
4	121	560	5.25	21.8
5	256	625	4.90	24.2
6	256	522	4.90	27.3
7	335	523	4.90	29.2
8	335	522	4.80	31.4
9	531	513	4.55	40.2
10	531	509	4.80	42.2
11	763	497	4.85	45.3
12	958	493	4.90	42.3
13	969	492	4.90	43.6
14	1255	372	3.70	38.3
15	1260	371	3.75	41.0
16	1090	370	3.75	38.2
17	1085	367	3.60	32.3

[a]Sieder, E.N., and Tate, G. E.,
Ind. and Eng. Chem. 28, 1429
(1936).

Table B–19 Heat transfer Data of 16° API Oil[a]

Point	Re	Pr	μ/μ_w	Nu
1	678	523	0.0052	14.2
2	678	520	0.0057	14.6
3	992	512	0.0044	16.8
4	1030	507	0.0042	16.8
5	476	1050	0.0080	15.3
6	356	1050	0.0077	14.6
7	238	1040	0.0069	12.3
8	173	1050	0.0063	11.9
9	49.0	1490	0.0077	9.3
10	40.6	1920	0.0114	9.3
11	68.1	2200	0.0131	11.9
12	93.0	2450	0.0140	13.1
13	97.4	2340	0.0143	12.4
14	135	2350	0.0152	15.3
15	242	2220	0.0166	16.8
16	58.0	5100	0.0387	15.1
17	111	5170	0.0438	20.0
18	49.0	5150	0.0390	13.8
19	49.8	5150	0.0380	15.1
20	26.6	4900	0.0334	12.6
21	11.7	5800	0.0366	10.9
22	6.83	13700	0.0600	11.3
23	6.10	14300	0.0666	11.7
24	3.78	16300	0.0672	11.2
25	3.63	16700	0.0694	11.2

[a]Sieder, E.N., and Tate, G. E., *Ind. and Eng. Chem. 28*, 1429 (1936).

Table B–20 Heat Transfer Data to a Fluidized Bed[a]

$\dfrac{h_m D_t}{k_g}$	$\dfrac{D_t}{L}$	$\dfrac{D_t}{D_p}$	$\dfrac{1-\varepsilon}{\varepsilon}\dfrac{\rho_s C_s}{\rho_g C_g}$	$\dfrac{D_t G}{\mu_g}$
469	0.636	309	833	256
913	0.636	309	868	555
1120	0.641	309	800	786
234	0.285	309	800	255
487	0.285	309	800	555
709	0.283	309	767	850
581	0.518	683	795	254
650	0.521	683	795	300
885	0.524	683	795	440
672	0.455	1012	867	338
986	0.451	1012	867	565
1310	0.455	1012	867	811
1190	0.944	1130	1608	343
1890	0.974	1130	1608	573
2460	0.985	1130	1608	814
915	0.602	1130	1673	343
1260	0.602	1130	1673	485
1690	0.617	1130	1673	700

[a]Dow, W. M., and Jacob, M., *Chem. Eng. Progr. 47*, 637 (1951).

Table B–21 Activity Coefficients for the System Methanol (1) + 1, 1, 1-trichloroethane (2) at 96.7 kPa[a]

Point	x_1	γ_1	γ_2
1	0.9803	0.9885	7.3622
2	0.9601	0.9983	6.3254
3	0.9013	1.0081	4.2642
4	0.8520	1.0303	3.7621
5	0.6904	1.1826	2.4921
6	0.5852	1.3761	1.9562
7	0.5351	1.4500	1.8410
8	0.4702	1.6217	1.6208
9	0.3721	1.9794	1.3672
10	0.2751	2.5370	1.1506
11	0.1302	3.9803	1.0170
12	0.0603	5.6238	1.0078

[a]Srinivas, Ch., and Venkateshwara Rao, M., *Fluid Phase Equilibria*, *61*, 285 (1991).

Table B–22 Activity Coefficients for the System Ethanol (1) + 1, 1, 1-trichloroethane (2) at 96.7 kPa[a]

Point	x_1	γ_1	γ_2
1	0.1502	2.7815	1.0412
2	0.2802	1.8966	1.1917
3	0.3403	1.6377	1.2700
4	0.5202	1.2138	1.5792
5	0.6198	1.1192	1.8292
6	0.7499	1.0462	2.1097
7	0.8201	1.0121	2.2695
8	0.8498	1.0079	2.1820
9	0.9402	0.9891	1.9823

[a]Laksham, V., Venkateshwara Rao, M., and Prasad, D. H. L., *Fluid Phase Equilibria, 69*, 271 (1991).

Table B–23 Activity Coefficients for the System Propanol (1) +1, 1, 1-trichloroethane (2) at 96.7 kPa[a]

Point	x_1	γ_1	γ_2
1	0.9754	1.0036	5.2313
2	0.9504	1.0084	4.9516
3	0.9352	1.0153	5.0295
4	0.9004	1.0283	4.6721
5	0.8502	1.0502	3.9590
6	0.7001	1.1865	2.4264
7	0.6282	1.2918	2.0725
8	0.5452	1.4699	1.7810
9	0.4451	1.7619	1.5490
10	0.3603	2.1246	1.3695
11	0.1969	3.2094	1.1698
12	0.1302	4.6223	1.0745
13	0.0522	6.5437	0.9998

[a]Kiran Kumar, R., and Venkateshwara Rao, M., *Fluid Phase Equilibria*, 70, 19 (1991).

Table B–24 Reaction Rate Data as Function of Temperature for Hydrogenation of Ethylene[a]

Run	$k \times 10^5$ g-mol/s·atm·cm^3	T °C	Run	$k \times 10^5$ g-mol/s·atm·cm^3	T °C
1	2.70	77	18	1.37	64.0
2	2.87	77	19	0.70	54.5
3	1.48	63.5	20	0.146	39.2
4	0.71	53.3	21	0.159	38.3
5	0.66	53.3	22	0.260	49.4
6	2.44	77.6	23	0.284	40.2
7	2.40	77.6	24	0.323	40.2
8	1.26	77.6	25	0.283	40.2
9	0.72	52.9	26	0.284	40.2
10	0.70	52.9	27	0.277	39.7
11	2.40	77.6	28	0.318	40.2
12	1.42	62.7	29	0.323	40.2
13	0.69	53.7	30	0.326	40.2
14	0.68	53.7	31	0.312	39.9
15	3.03	79.5	32	0.314	39.9
16	3.06	79.5	33	0.307	39.8
17	1.31	64.0			

[a]Smith, J. M., *Chemical Engineering Kinetics*, New York: McGraw-Hill, pp. 41–42, 1970 with permission.

Table C–1   Water (1)-Methyl Alcohol (2) at 59.4 °C[a]

x_1	P (mm Hg)
0.0	145.40
0.2217	317.00
0.2740	342.40
0.3324	368.70
0.3980	393.60
0.5550	450.60
0.6920	497.20
0.7850	530.40
0.8590	557.00
1.0000	609.30

[a]*International Critical Tables*, 1st ed., Vol III, New York: McGraw-Hill, 1928 with permission.

Table C–2 Ethyl Ether (1)-Chloroform (2) at 17 °C[a]

x_1	P (mm Hg)
0.0	143
0.1	143
0.2	143
0.3	151
0.4	166
0.5	193
0.6	230
0.7	273
0.8	317
0.9	360
1.0	397

[a]*lInternational Critical Tables*, 1st ed., Vol III, New York: McGraw-Hill, 1928 with permission.

Table C–3 Toluene (1)-Acetic Acid (2) at 70 °C[a]

x_1	P(mm Hg)
0.000	136.0
0.125	175.3
0.231	195.6
0.3121	204.9
0.4019	213.5
0.4860	218.9
0.5349	221.3
0.5912	223.4
0.6620	225.0
0.7597	225.1
0.8289	222.7
0.9058	216.6
0.9565	210.7
1.0000	202.0

[a]*International Critical Tables*, 1st ed., Vol III, New York: McGraw-Hill, 1928 with permission.

Table C–4 Chloroform (1)-Acetone (2) at 35.17 °C[a]

x_1	P(mm Hg)
0.09	344.50
0.0588	332.40
0.1232	319.70
0.1853	307.30
0.2657	291.60
0.2970	285.70
0.3664	272.20
0.4232	263.20
0.4939	255.40
0.5143	252.80
0.5872	248.40
0.6635	249.20
0.7997	261.90
0.9175	280.10
1.0000	293.10

[a]*International Critical Tables*, 1st ed., Vol III, New York: McGraw-Hill, 1928 with permission.

Table D–1 Transport Properties of Water in SI Units[a]

T	T	ρ	C_p	$\mu \times 10^3$	k	Pr
°C	K	kg/m^3	kJ/kg·K	kg/m·s	W/m·K	
0	273.2	999.6	4.229	1.786	0.5694	13.3
15.6	288.8	998.0	4.187	1.131	0.5884	8.07
26.7	299.9	996.4	4.183	0.860	0.6109	5.89
37.8	311.0	994.7	4.183	0.682	0.6283	4.51
65.6	338.8	981.9	4.187	0.432	0.6629	2.72
93.3	366.5	962.7	4.229	0.3066	0.6802	1.91
121.1	394.3	943.5	4.271	0.2381	0.6836	1.49
148.9	422.1	917.9	4.312	0.1935	0.6836	1.22
204.4	477.6	858.6	4.522	0.1384	0.6611	0.950
260.0	533.2	784.9	4.982	0.1042	0.6040	0.859
315.6	588.8	679.2	6.322	0.0862	0.5071	1.07

[a]Geankoplis, C. J., *Transport Processes and Unit Operations*, 3rd ed., Englewood Cliffs, NJ: Prentice Hall, 1993 with permission.

Table D-2 Transport Properties of Water in English Units[a]

T	ρ	C_p	$\mu \times 10^3$	k	Pr
°F	lb_m/ft^3	$btu/lbm \cdot °F$	$lb_m/ft \cdot s$	$btu/h \cdot ft \cdot °F$	
32	62.4	1.01	1.20	0.329	13.3
60	62.3	1.00	0.760	0.340	8.07
80	62.2	0.999	0.578	0.353	5.89
100	62.1	0.999	0.458	0.363	4.51
150	61.3	1.00	0.290	0.383	2.72
200	60.1	1.01	0.206	0.393	1.91
250	58.9	1.02	0.160	0.395	1.49
300	57.3	1.03	0.130	0.395	1.22
400	53.6	1.08	0.0930	0.382	0.950
500	49.0	1.19	0.0700	0.349	0.859
600	42.4	1.51	0.0579	0.293	1.07

[a]Geankoplis, C. J., *Transport Processes and Unit Operations*, 3rd ed., Englewood Cliffs, NJ: Prentice Hall, 1993 with permission.

Table D–3 Density and Viscosity of Various Liquids in English Units[a]

T	ρ	$\mu \times 10^5$
°F	lb_m/ft^3	$lb_m/ft \cdot sm$
Aniline		
60	64.0	305
80	63.5	240
100	63.0	180
150	61.6	100
200	60.2	62
250	58.9	42
300	57.5	30
Ammonia		
-60	43.9	20.6
-30	42.7	18.2
0	41.3	16.9
30	40.0	16.2
60	38.5	15.0
80	37.5	14.2
100	36.4	13.5
120	35.3	12.6
n-Butyl Alcohol		
60	50.5	225
80	50.0	180
100	49.6	130
150	48.5	68
Benzene		
60	55.2	44.5
80	54.6	38
100	53.6	33
150	51.8	24.5
200	49.9	19.4

Table D–3 Density and Viscosity of Various Liquids in English Units[a]

T	ρ	$\mu \times 10^5$
°F	lb_m/ft^3	$lb_m/ft \cdot sm$
Hydraulic Fluid (MIL-M-5606)		
0	55.0	5550
30	54.0	2220
60	53.0	1110
80	52.5	695
100	52.0	556
150	51.0	278
200	50.0	250
Glycerin		
30	79.7	7200
60	79.1	1400
80	78.7	600
100	78.2	100
Kerosene		
30	48.8	8.0
60	48.1	6.0
80	47.6	4.9
100	47.2	4.2
150	46.1	3.2
Mercury		
40	848	1.11
60	847	1.05
80	845	1.00
100	843	0.960
150	839	0.893
200	835	0.850
250	831	0.806
300	827	0.766
400	819	0.700

Table D–3 Density and Viscosity of Various Liquids in English Units[a]

T	ρ	$\mu \times 10^5$
°F	lb_m/ft^3	$lb_m/ft \cdot sm$
500	811	0.650
600	804	0.606
800	789	0.550

[a]Welty, J. R., Wicks, C. E., and Wilson, R.E., *Fundamentals of Momentum, Heat and Mass Transfer,* 3rd. ed., New York: John Wiley, 1984 with permission.

Table D–4 Condenser and Heat-Exchanger Tube Sizes[a]

OD	BWG	Wall Thickness		Inside Diameter	
in.	Number	in.	mm	ID, in.	mm
5/8	12	0.109	2.77	0.407	10.33
	14	0.083	2.11	0.459	11.66
	16	0.065	1.65	0.495	12.57
	18	0.049	1.25	0.527	13.39
3/4	12	0.109	2.77	0.532	13.51
	14	0.083	2.11	0.584	14.83
	16	0.065	1.65	0.620	15.75
	18	0.049	1.25	0.652	16.56
7/8	12	0.109	2.77	0.657	16.69
	14	0.083	2.11	0.709	18.01
	16	0.065	1.65	0.745	18.92
	18	0.049	1.25	0.777	19.74
1	10	0.134	3.40	0.732	18.59
	12	0.109	2.77	0.782	19.86
	14	0.083	2.11	0.834	21.18
	16	0.065	1.65	0.870	22.10
1 1/4	10	0.134	3.40	0.982	24.94
	12	0.109	2.77	1.032	26.21
	14	0.083	2.11	1.084	27.53
	16	0.065	1.65	1.120	28.45
1 1/2	10	0.134	3.40	1.232	31.29
	12	0.109	2.77	1.282	32.56
	14	0.083	2.11	1.334	33.88
2	10	0.134	3.40	1.732	43.99
	12	0.109	2.77	1.782	45.26

[a]See, for example, Perry, R. H., Green, D. W., and Maloney, J. O., Eds., *Chemical Engineers Handbook*, 7th ed., Section 10, New York: McGraw-Hill, 1997.

Table D–5 Standard Steel Pipe Sizes[a]

Nominal Size in	Outside Diameter in	Schedule Number	Wall Thickness in	Inside Diameter in
1/8	0.405	40	0.068	0.269
		80	0.095	0.215
1/4	0.540	40	0.088	0.364
		80	0.119	0.302
3/8	0.675	40	0.091	0.493
		80	0.126	0.423
1/2	0.840	40	0.109	0.622
		80	0.147	0.546
3/4	1.050	40	0.113	0.824
		80	0.154	0.742
1	1.315	40	0.133	1.049
		80	0.179	0.957
1-1/4	1.660	40	0.191	1.380
		80	0.145	1.278
1-1/2	1.900	40	0.200	1.610
		80	0.179	1.500
2	2.375	40	0.154	2.067
		80	0.218	1.939
2-1/2	2.875	40	0.203	2.469
		80	0.276	2.323
3	3.500	40	0.216	3.068
		80	0.300	2.900
3-1/2	3.500	40	0.318	3.548
		80	0.237	3.364
4	4.500	40	0.237	4.026
		80	0.337	3.826

Table D–5 Standard Steel Pipe Sizes[a]

Nominal Size	Outside Diameter	Schedule Number	Wall Thickness	Inside Diameter
in	in		in	in
5	5.563	40	0.258	5.047
		80	0.375	4.813
6	6.625	40	0.280	6.065
		80	0.432	5.761
8	8.625	40	0.322	7.981
		80	0.500	7.625

[a]See, for example, Perry, R. H., Green, D. W., and Maloney, J. O., Eds., *Chemical Engineers Handbook*, 7th ed., Section 10, New York: McGraw-Hill, 1997.

Table D–6 Effective Surface Roughness[a]

Surface	$\varepsilon(ft)$	$\varepsilon(mm)$
Concrete	0.001-0.01	0.3-3.0
Cast Iron	0.00085	0.25
Galvanized Iron	0.0005	0.15
Commercial Steel	0.00015	0.046
Drawn Tubing	0.000005	0.0015

[a]See, for example, G. G. Brown et. al., *Unit Operations*, New York: John Wiley & Sons, 1950, p. 140.

Table E–1 Physical and Transport Properties of Air[a]

T	C_p	$\mu \times 10^5$	k	ρ	$(g\beta\rho^2)/\mu^2$
°F	btu/lb$_m$·°F	lb$_m$/ft·s	btu/h·ft·°F	lb$_m$/ft^3	1/°F·ft^3
0	0.240	1.09	0.0132	0.0862	4.39×10^6
30	0.240	1.15	0.0139	0.0810	3.28
60	0.240	1.21	0.0146	0.0764	2.48
80	0.240	1.24	0.0152	0.0735	2.09
100	0.240	1.28	0.0156	0.071	1.76
150	0.241	1.36	0.0167	0.0651	1.22
200	0.241	1.45	0.0179	0.0602	0.84

[a]Welty, J.R., Wicks, C.E., and Wilson, R.E., *Fundamentals of Momentum, Heat and Mass Transfer*, 3rd ed., New York: Wiley, 1984 with permission.

Table E–2 Physical and Transport Properties of Carbon Dioxide and Sulfur Dioxide[a]

	T	ρ	C_p	$\mu \times 10^5$	k
	°F	lb_m/ft^3	$btu/lb_m \cdot °F$	$lb_m/ft \cdot s$	$btu/h \cdot ft \cdot °F$
Carbon Dioxide (gas)	0	0.132	0.193	0.865	0.00760
	30	0.124	0.198	0.915	0.00830
	60	0.117	0.202	0.965	0.00910
	80	0.112	0.204	1.00	0.00960
	100	0.108	0.207	1.03	0.0102
	150	0.100	0.213	1.12	0.0115
	200	0.092	0.219	1.20	0.0130
	250	0.0850	0.225	1.32	0.0148
	300	0.0800	0.230	1.36	0.0160
Sulfur Dioxide (gas)	0	0.195	0.142	0.700	0.00460
	100	0.161	0.149	0.890	0.00560
	200	0.136	0.157	1.05	0.00670
	300	0.118	0.164	1.20	0.00790

[a]Welty, J.R., Wicks, C.E., and Wilson, R.E., *Fundamentals of Momentum, Heat and Mass Transfer*, 3rd ed., New York: Wiley, 1984 with permission.

Table E–3 Physical and Transport Properties of Kerosene, Benzene and Ammonia[a]

	T	ρ	C_p	$\mu \times 10^5$	k
	°F	lb_m/ft^3	$btu/lb_m \cdot °F$	$lb_m/ft \cdot s$	$btu/h \cdot ft \cdot °F$
Kerosene (liquid)	30	48.8	0.456	800	0.0809
	60	48.1	0.474	600	0.0805
	80	47.6	0.491	490	0.0800
	100	47.2	0.505	420	0.0797
	150	46.1	0.540	320	0.0788
Benzene (liquid)	60	55.2	0.395	44.5	0.0856
	80	54.6	0.410	38	0.0836
	100	53.6	0.420	33	0.0814
	150	51.8	0.450	24.5	0.0762
	200	49.9	0.480	19.4	0.0711
Ammonia (liquid)	0	41.3	1.08	16.9	0.315
	30	40.0	1.11	16.2	0.312
	60	38.5	1.14	15.0	0.304
	80	37.5	1.16	14.2	0.296
	100	36.4	1.19	13.5	0.287
	120	35.3	1.22	12.6	0.275

[a]Welty, J.R., Wicks, C.E., and Wilson, R.E., *Fundamentals of Momentum, Heat and Mass Transfer*, 3rd ed., New York: Wiley, 1984 with permission.

The CD-ROM

Important resources on the CD-ROM for use with this book that are available on the accompanying CD-ROM include:

1. Polymath software for installation to a personal computer operating under DOS, Windows 3.1, Windows 95, and Windows NT.
2. Acrobat[*] PDF software for installation to a personal computer operating under Windows 3.1, Windows 95, and Windows NT. Acrobat software 3.0 and above is necessary to view and print the enclosed documents in PDF format using most any computer, operating system, and printer.
3. Polymath manual in Adobe Acrobat PDF format.
4. Polymath solutions or data for selected problems as referenced in the book.
5. Polymath data sets for selected tables for use in problem solving.
6. A set of ten representative problems in Chemical Engineering with detailed writeups and problem code for execution by various mathematical software packages. Documentation is in Adobe Acrobat PDF format. Packages include Excel, Maple, Mathcad, MATLAB, Mathematica, and POLYMATH.
7. Description of the POLYMATH software in PDF format.
8. Information on updating of book material in PDF format.
9. Information on updating of POLYMATH software in PDF format.
10. Information on updating of Acrobat software in PDF format.

Using the CD-ROM

The CD that accompanied this book is organized just like any typical hard disk. CD hardware is typically given a drive letter when it is installed into a computer. **All further discussion of the CD will assume that the CD is called drive D**. The software on the CD is provided in subdirectories of drive D that are identified in the following descriptions. It is recommended that the software be installed in the order as discussed since both POLYMATH and the Adobe Acrobat Reader software are essential to the use of the CD-ROM.

---

[*] Acrobat is a trademark of Adobe Systems, Inc. (http://www.adobe.com)

Installation of POLYMATH Software to a Personal Computer

The POLYMATH software must be installed to a hard disk drive. The details of this installation are provided on the file called INSTALL.TXT. This file should be printed and referenced during the installation process. This can be accomplished with utility software such as Notepad or with word processing packages such as WordPerfect and Microsoft Word.

 The POLYMATH software package and associated files are found in D:\POLYMATH. The INSTALL.TXT file gives the essentials of installation while the README.TXT file gives more information on POLYMATH and the License Agreement. The installation is initiated by executing the file INSTALL.EXE from this subdirectory.

Installation of Adobe Acrobat Reader Software to a Personal Computer

The Adobe Acrobat Reader software must also be installed to a hard disk. This software will allow you to read and print additional materials that are on the CD-ROM. This software is also freely available by downloading from www.adobe.com for many different operating systems.

 The Adobe Acrobat Reader software is found in D:\ADOBE. The following files should be executed from the CD-ROM according to your operating system. The installation is self-documented and completely automatic. Use the following:

AR32E301.EXE for Windows 95 and Windows NT

AR16E301.EXE for Windows 3.1.

Accessing the POLYMATH Manual

The POLYMATH Manual is available as a. Adobe Acrobat PDF file found in the directory D:\MANUAL with file name MANUAL.PDF. From Adobe Acrobat, just open this file from the CD-ROM to view or print the manual. You may also search on words or terms of interest using Acrobat.

Learning to Use the POLYMATH Software

The POLYMATH software package is very user-friendly and menu-driven so that it is very easy to use. First-time computer users should make a hard copy of the manual as indicated above. The manual will be a convenient reference guide when using POLYMATH. Then it will be helpful to read through the manual and try many of the QUICK TOUR problems. If you have considerable computer experience, you only need to read the chapters at the back of the manual that discuss the individual programs of interest to you and try some of the QUICK TOUR problems.

Particular attention should be directed at the sections which describe the Library Operations, Library Storage and Library Retrieval as these options are used to provide solutions to many of the problems in this book. These options also allow users to conveniently store their problems for future use. Each POLY-

MATH program will only show the problems in a particular Library which are relevant to that program. For example, a particular Library when accessed from POLYMATH Simultaneous Differential Equation Solver will only indicate those problems that can be solved with that package when there may be many other problems in the same Library for all other POLYMATH programs.

Accessing the POLYMATH Solutions of Selected Problems

Selected problems in this book contain partial or complete solutions using POLYMATH. These solutions are provided on the CD-ROM by using the POLY-MATH Library options. This enables the user to retrieve the problem from a particular Library on the CD-ROM and execute the problem within POLYMATH. It is important to note that the Library stores only the problem and not the solution; however, it is a simple matter to solve the particular problem with POLY-MATH. The solved problem can then be examined with the various plotting and tabular data options, and the problem can easily be modified prior to another solution. Printing of intermediate and final results is quite easy. The interactive nature of POLYMATH greatly enhances the problem-solving experience. The Library option can to used to store problems as desired.

 The solved problems associated with the book are stored on the CD-ROM using the Library in directories D:\CHAP1 ... CHAP8. Identification of the particular POLYMATH program and the Library location is given in each problem.

Accessing the POLYMATH Data Sets for Selected Tables

Selected tables in the chapters and in the appendices have been entered into the POLYMATH *Polynomial, Multiple Linear and Nonlinear Regression Program*. These tables of data can be loaded from the identified Library for use in problem solving so that the user can easily utilize the original data set without data entry.

 The selected tables associated with the book are stored on the CD-ROM using the Library in directory D:\TABLES. Identification of the particular table file is by the table number and title used in the book.

Using the Set of Ten Problems in Chemical Engineering with Other Mathematical Solving Software

While POLYMATH is used to demonstrate numerical problem-solving throughout this book, a number of other mathematical software packages can be utilized to solve these same problems.

The use of mathematical software in chemical engineering was the subject of session at a conference sponsored by the Chemical Engineering Division of the American Society for Engineering Education[*]. This session presented solutions

---

[*] The original materials were distributed at the Chemical Engineering Summer School at Snowbird, Utah on August 13, 1997 in Session 12 entitled "The Use of Mathematical Software in Chemical Engineering."

to ten representative problems in chemical engineering which were solved by different mathematical packages. The papers from this session were enhanced to include detailed solutions papers that utilized Excel[*], Maple[†], Mathcad[‡], MATLAB[•], Mathematica[#], and POLYMATH[¶]. An article based on this problem set and the various solutions is to be published in *Computer Applications in Engineering Education*[**]. The contributing authors were:

Excel - Edward M. Rosen, EMR Technology Group

Maple - Ross Taylor, Clarkson University

MathCAD - John J. Hwalek, University of Maine

MATLAB - Joseph Brule, John Widmann, Tae Han, and Bruce Finlayson, University of Washington

Mathematica - H. Eric Nuttall, University of New Mexico

POLYMATH - Michael B. Cutlip, University of Connecticut and Mordechai Shacham, Ben-Gurion University of the Negev

Many of these ten problems are also problems in this book since the book authors were the organizers and participants in the ASEE session. The authors of these materials have kindly permitted the inclusion of the ten problems paper and the detailed solution manuscripts as PDF files on the CD-ROM. Additionally, the computer files are also available on the CD-ROM so that these problem solutions can be executed by using the appropriate software package.

A summary of the ten problems and the corresponding numbering to the problems in this book is given in Table F–1. Individuals who prefer to use one of these various mathematical software packages will find the detailed writeups and problem files very helpful in utilizing these packages.

 The paper entitled "A Collection of Representative Problems in Chemical Engineering for Solution by Numerical Methods" is found in directory D:\TENPROBS with file named TENPROBS.PDF. The various papers discussing the solutions are located in individual subdirectories with the name of that software package. For example, the Maple paper is found in D:\TENPROBS\MAPLE with the file named MAPLE.PDF. Individual solution files are found in a subdirectory below each package directory such as D:\TENPROBS\MATHCAD with files named PROBLEM1.MCD... PROBLEM10.MCD.

---

[*] Excel is a trademark of Microsoft Corporation (http://www.microsoft.com)
[†] Maple is a trademark of Waterloo Maple, Inc. (http://www.maplesoft.com)
[‡] MathCAD is a trademark of Mathsoft, Inc. (http://www.mathsoft.com)
[•] MATLAB is a trademark of The Math Works, Inc. (http://www.mathworks.com)
[#] Mathematica is a trademark of Wolfram Research, Inc. (http://www.wolfram.com)
[¶] POLYMATH is copyrighted by M. Shacham and M. B. Cutlip (http://www.polymath-software.com)
[**] The Journal *Computer Applications in Engineering Education* is published by John Wiley & Sons, Inc., and the editor is M. F. Iskander. (http://www.journals.wiley.com/1061-3773)

Table F–1 Ten Representative Problem in Chemical Engineering

SUBJECT AREA	PROBLEM TITLE	MATHEMATICAL MODEL	ORIGINAL PROBLEM	THIS BOOK PROBLEM
Introduction to Ch. E.	Molar Volume and Compressibility Factor from Van Der Waals Equation	Single Nonlinear Equation	1	1.1
Introduction to Ch. E.	Steady State Material Balances on a Separation Train	Simultaneous Linear Equations	2	1.5 .
Mathematical Methods	Vapor Pressure Data Representation by Polynomials and Equations	Polynomial Fitting, Linear and Nonlinear Regression	3	1.3 and 2.1
Thermodynamics	Reaction Equilibrium for Multiple Gas Phase Reactions	Simultaneous Nonlinear Equations	4	4.13
Fluid Dynamics	Terminal Velocity of Falling Particles	Single Nonlinear Equation	5	5.6
Heat Transfer	Unsteady State Heat Exchange in a Series of Agitated Tanks	Simultaneous ODE's with known initial conditions.	6	1.15
Mass Transfer	Diffusion with Chemical Reaction in a One Dimensional Slab	Simultaneous ODE's with split boundary conditions.	7	3.6
Separation Processes	Binary Batch Distillation	Simultaneous Differential and Nonlinear Algebraic Equations	8	3.8
Reaction Engineering	Reversible, Exothermic, Gas Phase Reaction in a Catalytic Reactor	Simultaneous ODE's and Algebraic Equations	9	8.24
Process Dynamics and Control	Dynamics of a Heated Tank with PI Temperature Control	Simultaneous Stiff ODE's	10	-

TREE STRUCTURE OF THE CD-ROM

Figure F–1 indicates the tree structure of the CD-ROM.

UPDATING THE BOOK AND OTHER CD-ROM MATERIALS

Corrections and additions to the book content will be avilable via the Internet at www.polymath-software.com/book. The latest description and information on the POLYMATH Numerical Analysis Package is provided at www.polymath-software.com. Details and downloading information on the Adobe Acrobat software is found at www.adobe.com.

 Updating information is also on the CD-ROM in directory D:\UPDATE with files named BOOK.PDF, POLYMATH.PDF, and ADOBE.PDF.

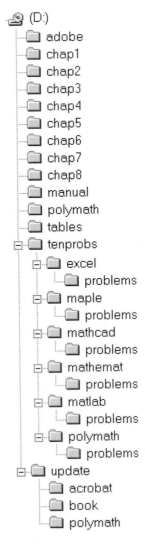

Figure F–1 Directory Tree Structure for CD-ROM

Index

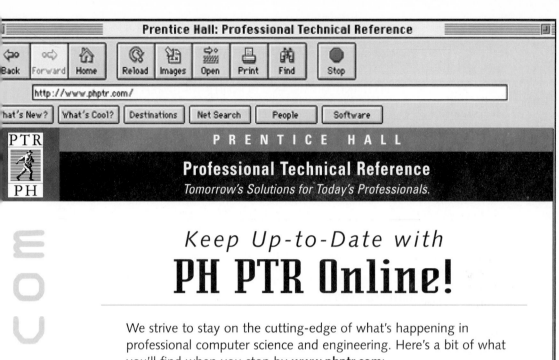

LICENSE AGREEMENT AND LIMITED WARRANTY

READ THE FOLLOWING TERMS AND CONDITIONS CAREFULLY BEFORE OPENING THIS DISK PACKAGE. THIS LEGAL DOCUMENT IS AN AGREEMENT BETWEEN YOU AND PRENTICE-HALL, INC. (THE "COMPANY"). BY OPENING THIS SEALED DISK PACKAGE, YOU ARE AGREEING TO BE BOUND BY THESE TERMS AND CONDITIONS. IF YOU DO NOT AGREE WITH THESE TERMS AND CONDITIONS, DO NOT OPEN THE DISK PACKAGE. PROMPTLY RETURN THE UNOPENED DISK PACKAGE AND ALL ACCOMPANYING ITEMS TO THE PLACE YOU OBTAINED THEM FOR A FULL REFUND OF ANY SUMS YOU HAVE PAID.

1. **GRANT OF LICENSE:** In consideration of your payment of the license fee, which is part of the price you paid for this product, and your agreement to abide by the terms and conditions of this Agreement, the Company grants to you a nonexclusive right to use and display the copy of the enclosed software program (hereinafter the "SOFTWARE") on a single computer (i.e., with a single CPU) at a single location so long as you comply with the terms of this Agreement. The Company reserves all rights not expressly granted to you under this Agreement.

2. **OWNERSHIP OF SOFTWARE:** You own only the magnetic or physical media (the enclosed disks) on which the SOFTWARE is recorded or fixed, but the Company retains all the rights, title, and ownership to the SOFTWARE recorded on the original disk copy(ies) and all subsequent copies of the SOFTWARE, regardless of the form or media on which the original or other copies may exist. This license is not a sale of the original SOFTWARE or any copy to you.

3. **COPY RESTRICTIONS:** This SOFTWARE and the accompanying printed materials and user manual (the "Documentation") are the subject of copyright. You may not copy the Documentation or the SOFTWARE, except that you may make a single copy of the SOFTWARE for backup or archival purposes only. You may be held legally responsible for any copying or copyright infringement which is caused or encouraged by your failure to abide by the terms of this restriction.

4. **USE RESTRICTIONS:** You may not network the SOFTWARE or otherwise use it on more than one computer or computer terminal at the same time. You may physically transfer the SOFTWARE from one computer to another provided that the SOFTWARE is used on only one computer at a time. You may not distribute copies of the SOFTWARE or Documentation to others. You may not reverse engineer, disassemble, decompile, modify, adapt, translate, or create derivative works based on the SOFTWARE or the Documentation without the prior written consent of the Company.

5. **TRANSFER RESTRICTIONS:** The enclosed SOFTWARE is licensed only to you and may not be transferred to any one else without the prior written consent of the Company. Any unauthorized transfer of the SOFTWARE shall result in the immediate termination of this Agreement.

6. **TERMINATION:** This license is effective until terminated. This license will terminate automatically without notice from the Company and become null and void if you fail to comply with any provisions or limitations of this license. Upon termination, you shall destroy the Documentation and all copies of the SOFTWARE. All provisions of this Agreement as to warranties, limitation of liability, remedies or damages, and our ownership rights shall survive termination.

7. **MISCELLANEOUS:** This Agreement shall be construed in accordance with the laws of the United States of America and the State of New York and shall benefit the Company, its affiliates, and assignees.

8. **LIMITED WARRANTY AND DISCLAIMER OF WARRANTY:** The Company warrants that the SOFTWARE, when properly used in accordance with the Documentation, will operate in substantial conformity with the description of the SOFT-

WARE set forth in the Documentation. The Company does not warrant that the SOFT-WARE will meet your requirements or that the operation of the SOFTWARE will be uninterrupted or error-free. The Company warrants that the media on which the SOFT-WARE is delivered shall be free from defects in materials and workmanship under normal use for a period of thirty (30) days from the date of your purchase. Your only remedy and the Company's only obligation under these limited warranties is, at the Company's option, return of the warranted item for a refund of any amounts paid by you or replacement of the item. Any replacement of SOFTWARE or media under the warranties shall not extend the original warranty period. The limited warranty set forth above shall not apply to any SOFTWARE which the Company determines in good faith has been subject to misuse, neglect, improper installation, repair, alteration, or damage by you. EXCEPT FOR THE EXPRESSED WARRANTIES SET FORTH ABOVE, THE COMPANY DISCLAIMS ALL WARRANTIES, EXPRESS OR IMPLIED, INCLUDING WITHOUT LIMITATION, THE IMPLIED WARRANTIES OF MERCHANTABILITY AND FITNESS FOR A PARTICULAR PURPOSE. EXCEPT FOR THE EXPRESS WAR-RANTY SET FORTH ABOVE, THE COMPANY DOES NOT WARRANT, GUARAN-TEE, OR MAKE ANY REPRESENTATION REGARDING THE USE OR THE RESULTS OF THE USE OF THE SOFTWARE IN TERMS OF ITS CORRECTNESS, ACCURACY, RELIABILITY, CURRENTNESS, OR OTHERWISE.

IN NO EVENT, SHALL THE COMPANY OR ITS EMPLOYEES, AGENTS, SUP-PLIERS, OR CONTRACTORS BE LIABLE FOR ANY INCIDENTAL, INDIRECT, SPE-CIAL, OR CONSEQUENTIAL DAMAGES ARISING OUT OF OR IN CONNECTION WITH THE LICENSE GRANTED UNDER THIS AGREEMENT, OR FOR LOSS OF USE, LOSS OF DATA, LOSS OF INCOME OR PROFIT, OR OTHER LOSSES, SUSTAINED AS A RESULT OF INJURY TO ANY PERSON, OR LOSS OF OR DAMAGE TO PROPERTY, OR CLAIMS OF THIRD PARTIES, EVEN IF THE COMPANY OR AN AUTHORIZED REPRESENTATIVE OF THE COMPANY HAS BEEN ADVISED OF THE POSSIBILITY OF SUCH DAMAGES. IN NO EVENT SHALL LIABILITY OF THE COMPANY FOR DAMAGES WITH RESPECT TO THE SOFTWARE EXCEED THE AMOUNTS ACTU-ALLY PAID BY YOU, IF ANY, FOR THE SOFTWARE.

SOME JURISDICTIONS DO NOT ALLOW THE LIMITATION OF IMPLIED WARRANTIES OR LIABILITY FOR INCIDENTAL, INDIRECT, SPECIAL, OR CONSE-QUENTIAL DAMAGES, SO THE ABOVE LIMITATIONS MAY NOT ALWAYS APPLY. THE WARRANTIES IN THIS AGREEMENT GIVE YOU SPECIFIC LEGAL RIGHTS AND YOU MAY ALSO HAVE OTHER RIGHTS WHICH VARY IN ACCORDANCE WITH LOCAL LAW.

ACKNOWLEDGMENT

YOU ACKNOWLEDGE THAT YOU HAVE READ THIS AGREEMENT, UNDER-STAND IT, AND AGREE TO BE BOUND BY ITS TERMS AND CONDITIONS. YOU ALSO AGREE THAT THIS AGREEMENT IS THE COMPLETE AND EXCLUSIVE STATEMENT OF THE AGREEMENT BETWEEN YOU AND THE COMPANY AND SUPERSEDES ALL PROPOSALS OR PRIOR AGREEMENTS, ORAL, OR WRITTEN, AND ANY OTHER COMMUNICATIONS BETWEEN YOU AND THE COMPANY OR ANY REPRESENTATIVE OF THE COMPANY RELATING TO THE SUBJECT MATTER OF THIS AGREEMENT.

Should you have any questions concerning this Agreement or if you wish to contact the Company for any reason, please contact in writing at the address below.

Robin Short
Prentice Hall PTR
One Lake Street
Upper Saddle River, New Jersey 07458

This CD-ROM contains a complete version of the widely utilized POLYMATH Numerical Analysis Package that can be installed on IBM-compatible computers with DOS or Windows operating systems. This interactive software provides solutions to problems that involve:

1. Curve fitting by polynomials and splines
2. Linear and nonlinear regression with statistical analysis
3. Simultaneous linear and nonlinear algebraic equations
4. Simultaneous ordinary differential equations (including stiff systems)
   - Adobe™ Acrobat™ software for viewing, searching, and printing documents
   - POLYMATH user manual in Acrobat PDF format
   - POLYMATH problem solution files that can generated complete solutions to all example problems in the book
   - POLYMATH data files for all book problems and Appendix data tables
   - Solutions to 10 representative book problems using Excel™, Maple™, Mathcad™, Mathematica™, MATLAB™, and ©POLYMATH including detailed writeups by knowledgeable professionals in Acrobat PDF format. Complete problem solution files for each of these packages are provided.

Information on up-dated book materials, the POLYMATH software package, and this CD-ROM can be obtained from:
www.polymath-software.com